Student Solutions Manual

Charles W. Haines
Rochester Institute of Technology

Elementary Differential Equations and Boundary Value Problems

Eighth Edition

William E. Boyce
Richard C. DiPrima
Rensselaer Polytechnic Institute

WILEY

John Wiley & Sons, Inc.

To order books or for customer service call 1-800-CALL-WILEY (225-5945).

ISBN 0-471-43340-3

Printed in the United States of America

10 9 8 7 6 5 4 3 2 1

Printed and bound by Hamilton Printing Company

PREFACE

This supplement has been prepared for use in conjunction with the eighth editions of ELEMENTARY DIFFERNTIAL EQUATIONS AND BOUNDARY VALUE PROBLEMS and ELEMENTARY DIFFERENTIAL EQUATIONS, both by W.E. Boyce and R.C. DiPrima. The supplement contains a sampling of the problems from each section of the text. In most cases the complete details in determining the solutions are given while in the remainder of the problems helpful hints are provided. The problems chosen in each section represent, wherever possible, the variety of applications and types of examples that are covered in the written material of the text, thereby providing the student a complete set of examples from which to learn.

Students should be aware that following these solutions is very different from designing and constructing one's own solution. Using this supplemental resource appropriately for learning differential equations is outlined as follows:

   1. Make an honest attempt to solve the problem without using the guide.
   2. If needed, glance at the beginning of the solution in the guide and then try again to generate the complete solution. Continue using the guide for hints when you reach an impasse.
   3. Compare your final solution with the one provided to see whether yours is more or less efficient than the guide, since there is frequently more than one correct way to solve a problem.
   4. Ask yourself why that particular problem was assigned.

The use of a symbolic computational software package, in many cases, would greatly simplify finding the solution to a given problem, but the details given in this solutions manual are important for the student's understanding of the underlying mathematical principles and applications. In other cases, these software packages are essential for completing the given problem, as the calculations and graphing would be overwhelming using analytical techniques. In these cases, some steps or hints are given and then reference is made to the use of an appropriate software package.

In order to simplify the text, the following abbreviations have been used:

D.E.	differential equation(s)
O.D.E.	ordinary differential equation(s)
P.D.E.	partial differential equation(s)
I.C.	initial condition(s)
I.V.P.	initial value problem(s)
B.C.	boundary condition(s)
B.V.P.	boundary value problem(s)

I wish to express my appreciation to Dr. Josef Torok who has provided invaluable assistance with the preparation of the figures as well as assistance with many of the solutions involving the use of symbolic computational software.

Charles W. Haines
Professor of Mathematics and
    Mechanical Engineering
Rochester Institute of Technology
Rochester, New York
June 2004

CONTENTS

CHAPTER 1

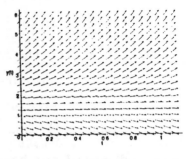

2. For y > 3/2 we see that y' > 0
    and thus y(t) is increasing there.
    For y < 3/2 we have y' < 0 and thus
    y(t) is decreasing there. Hence
    y(t) diverges from 3/2 as t→∞.

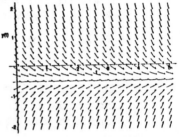

4. Observing the direction field,
    we see that for y>-1/2 we have
    y'<0, so the solution is
    decreasing here. Likewise,
    for y<-1/2 we have y'>0 and
    thus y(t) is increasing here.
    Since the slopes get closer to
    zero as y gets closer to -1/2,
    we conclude that y→-1/2 as t→∞.

7. If all solutions approach 3, then 3 is the equilibrium

    solution and we want $\dfrac{dy}{dt} < 0$ for y > 3 and $\dfrac{dy}{dt} > 0$ for

    y < 3. Thus $\dfrac{dy}{dt} = 3-y$. Of course, there are other possible

    answers, such as $\dfrac{dy}{dt} = 6-2y$.

11. For y = 0 and y = 4 we have y' = 0 and thus y = 0 and
    y = 4 are equilibrium solutions. For y > 4, y' < 0 so if
    y(0) > 4 the solution approaches y = 4 from above. If
    0 < y(0) < 4, then y' > 0 and the solutions "grow" to y = 4
    as t→∞. For y(0) < 0 we see that y' < 0 and the solutions
    diverge from 0.

13. Since y' = y², y = 0 is the equilibrium solution and y' > 0
    for all y. Thus if y(0) > 0, solutions will diverge from
    0 and if y(0) < 0, solutions will aproach y = 0 as t→∞.

16. From Fig. 1.1.6 we see that y = 2 is an equilibrium

    solution, that $\dfrac{dy}{dt} > 0$ for y > 2, and that $\dfrac{dy}{dt} < 0$ for

    y < 2. Only the D.E. in (c) will yield these results.

19. From Fig. 1.1.9 we see that y = 0 and y = 3 are

    equilibrium solutions, that $\dfrac{dy}{dt} < 0$ for y > 3, that

$\dfrac{dy}{dt} > 0$ for $0 < y < 3$, and that $\dfrac{dy}{dt} < 0$ for $y < 0$. Only the D.E. in (h) will yield these results.

21a. Let q(t) be the number of·grams of the chemical in the water at any time. Then $\dfrac{q(t)}{1,000,000}$ represents the concentration of the chemical in the pond at any time and hence $\dfrac{300q(t)}{1,000,000}$ is the rate at which the chemical leaves the pond per hour and 300(.01) represents the amount of the chemical coming into the pond per hour. Thus

$$\dfrac{dq}{dt} = 300(.01) - \dfrac{300q}{1,000,000} = 300(10^{-2} - 10^{-6}q).$$

21b. The equilibrium solution occurs when $q' = 0$, or $q = 10^4$ gm, independent of the amount present at $t = 0$ (all solutions approach the equilibrium solution).

22. Let V be the volume, S the surface area and a the constant of proportionality. Then the D.E. expressing the evaporation is $\dfrac{dV}{dt} = -aS$, $a > 0$. Now $V = \dfrac{4}{3}\pi r^3$ and $S = 4\pi r^2$, so $S = 4\pi\left(\dfrac{3}{4\pi}\right)^{2/3} V^{2/3}$. Thus $\dfrac{dv}{dt} = -kV^{2/3}$, for k>0.

25d.

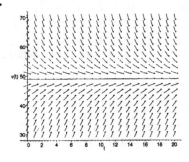

28.

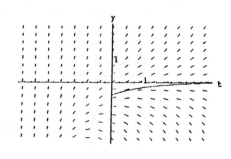

29.

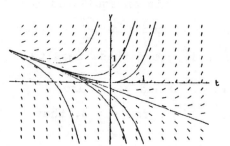

31.

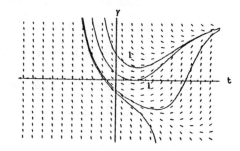

<u>Section 1.2 Page 15</u>

1b. dy/dt = -2y+5 can be rewritten as $\dfrac{dy}{y-5/2}$ = -2dt. Thus

ln$|$y-5/2$|$ = -2t+c_1, or y-5/2 = ce^{-2t}. y(0) = y_0 yields
c = y_0 - 5/2, so y = 5/2 + $(y_0-5/2)e^{-2t}$.
If y_0 > 5/2, the solution starts above the equilibrium
solution and decreases exponentially and approaches 5/2
as t→∞. Conversely, if y_o < 5/2, the solution starts
below 5/2 and grows exponentially and approaches 5/2 from
below as t→∞.

3a. Note that $\dfrac{dy}{dt}$ = -a(y - b/a) and thus $\dfrac{dy}{y-b/a}$ = -a.
Integration, as in the text, yields ln$|$y-b/a$|$ = -at + c,
or y = $\dfrac{b}{a}$ + Ce^{-at}.

3c. If a increases then b/a is smaller. Thus the equilibrium
soluiton is lower and it is reached sooner, since e^{-at}
decays faster for larger a. Similar analysis will yield
the results for ii) and iii).

5a. Rewrite Eq.(ii) as $\dfrac{dy/dt}{y}$ = a and thus ln$|$y$|$ = at+C; or
y_1 = ce^{at}.

5b. If y = y_1(t) + k, then $\dfrac{dy}{dt}$ = $\dfrac{dy_1}{dt}$. Substituting both
these into Eq.(i) we get $\dfrac{dy_1}{dt}$ = a(y_1+k) - b. Since
$\dfrac{dy_1}{dt}$ = ay_1, this leaves ak - b = 0 and thus k = b/a.
Hence y = y_1(t) + b/a is the solution to Eq(i).

5c. Substitution of y_1 = ce^{at} shows this is the same as that
given in Eq.(17).

7b. From Eq.(11) we have p = 900 + $ce^{t/2}$. If p(0) = p_0, then
c = p_0 - 900 and thus p = 900 + $(p_0 - 900)e^{t/2}$. If
p_0 < 900, this decreases, so if we set p = 0 and solve
for T (the time of extinction) we obtain
$e^{T/2}$ = 900/(900-p_0), or T = 2ln[900/(900-p_0)] months.

9a. The solution to this problem is given by Eq.(26), which has
     a limiting velocity of 49 m/sec. Substituting v = 48.02
     (which is 98% of 49) into Eq.(26) yields $e^{-t/5}$ = .02. Solving
     for t we have t = $-5\ln(.02)$ = 19.56 sec.

9b. Use Eq.(29).

11a. If the drag force is proportional to v^2 then F = 98 - kv^2
     is the net force acting on the falling mass (m = 10kg.).
     Thus $10\dfrac{dv}{dt}$ = 98 - kv^2, which has a limiting velocity of
     v^2 = 98/k. Setting $v^2 = 49^2$ gives k = $98/49^2$ and hence
     $$\frac{dv}{dt} = \frac{49^2 - v^2}{10/k} = (49^2 - v^2)/245.$$

11b. From part a we have $\dfrac{dv}{49^2 - v^2} = \dfrac{dt}{245}$, which yields
     $\dfrac{1}{49}\tanh^{-1}\!\left(\dfrac{v}{49}\right) = \dfrac{t}{245} + C_0$. Setting t=0 and v=0, the initial
     point, we then have 0 = 0 + C_0, or C_0 = 0. Thus
     $\tanh^{-1}\!\left(\dfrac{v}{49}\right)$ = t/5, or v(t) = 49tanh(t/5) m/sec.

11c.
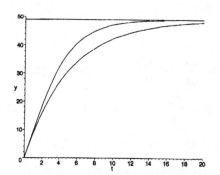

11e. Note that $\int\tanh(x)dx = \ln(\cosh(x) + C$.

12a. $\dfrac{dQ}{dt}$ = -rQ yields $\dfrac{dQ/dt}{Q}$ = -r, or $\ln|Q|$ = -rt + c_1. Thus
     Q = ce^{-rt} and Q(0) = 100 yields c = 100. Hence Q = $100e^{-rt}$.
     Setting t = 1, we have 82.04 = $100e^{-r}$, which yields
     r = .1980/wk or r = .02828/day.

15a. Rewrite the D.E. as $\dfrac{du}{u-T}$ = -kdt and then integrate to find

ln|u-T| = -kt + c. Thus u - T = ±Ce^{-kt}. For t = 0 we have

u$_0$ - T = ±C and thus u(t) = T + (u$_0$ - T)e^{-kt}.

15b. Set u(τ) - T = $\dfrac{u_o - T}{2}$ when t = τ in the solution of part (a).

17a. Rewrite the D.E. as $\dfrac{dQ/dt}{Q-CV}$ = $\dfrac{-1}{CR}$, thus, upon integrating and

simplifying, we get Q = De$^{-t/CR}$ + CV.   Q(0) = 0 ⇒ D = -CV and

thus Q(t) = CV(1 - e$^{-t/CR}$).

17b. $\lim\limits_{t\to\infty}$Q(t) = CV since $\lim\limits_{t\to\infty}$ e$^{-t/CR}$ = 0.

17c. In this case R$\dfrac{dQ}{dt}$ + $\dfrac{Q}{C}$ = 0, Q(t$_1$) = CV. The solution of this

D.E. is Q(t) = Ee$^{-t/CR}$, so Q(t$_1$) = Ee$^{-t_1/CR}$ = CV, or

E = CVe$^{t_1/CR}$. Thus Q(t) = CVe$^{t_1/CR}$e$^{-t/CR}$ = CVe$^{-(t-t_1)/CR}$.

17a. CV = 20, CR = 2.5   17c. CV = 20, CR = 2.5, t$_1$ = 10

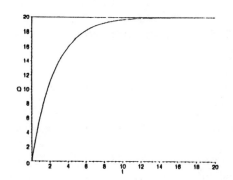   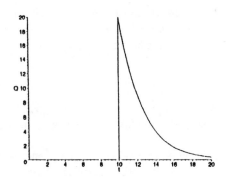

Section 1.3, Page 24

2. The D.E. is second order since there is a second
    derivative of y appearing in the equation. The equation
    is nonlinear due to the y^2 term (as well as due to the y^2
    term multiplying the y″ term).

6. This is a third order D.E. since the highest derivative is
    y‴ and it is linear since y and all its derivatives
    appear to the first power only. The terms t^3 and cos^2 t do
    not affect the linearity of the D.E.

8. For $y_1(t) = e^{-3t}$ we have $y_1(t) = -3e^{-3t}$ and $y_1''(t) = 9e^{-3t}$.
    Substitution of these into the D.E. yields
    $$9e^{-3t} + 2(-3e^{-3t}) - 3(e^{-3t}) = (9-6-3)e^{-3t} = 0.$$

11. Substituting $y_1 = t^{1/2}$ into the D.E. we get
    $$2t^2\left(\frac{-1}{4}t^{-3/2}\right) + 3t\left(\frac{1}{2}t^{-1/2}\right) - t^{1/2} = \left(\frac{-1}{2}\right)t^{1/2} + \left(\frac{3}{2}\right)t^{1/2} - t^{1/2} = 0.$$

14. Recall that if $u(t) = \int_0^t f(s)ds$, then $u'(t) = f(t)$.

16. Differentiating e^{rt} twice and substituting into the D.E.
    yields $r^2e^{rt} - e^{rt} = (r^2-1)e^{rt}$. If $y = e^{rt}$ is to be a
    solution of the D.E. then the last quantity must be zero
    for all t. Thus $r^2-1 = 0$ since e^{rt} is never zero.

19. Differentiating t^r twice and substituting into the D.E.
    yields $t^2[r(r-1)t^{r-2}] + 4t[rt^{r-1}] + 2t^r = [r^2+3r+2]t^r$. If
    $y = t^r$ is to be a solution of the D.E., then the last term
    must be zero for all t and thus $r^2 + 3r + 2 = 0$.

22. The D.E. is second order since there are second partial
    derivatives of $u(x,y)$. The D.E. is nonlinear due to the
    product of $u(x,y)$ times u_x (or u_y).

26. Since $\dfrac{\partial u_1}{\partial t} = -\alpha^2 e^{-\alpha^2 t}\sin x$ and $\dfrac{\partial^2 u_1}{\partial x^2} = -e^{-\alpha^2 t}\sin x$ we have
    $\alpha^2[-e^{-\alpha^2 t}\sin x] = -\alpha^2 e^{-\alpha^2 t}\sin x$, which is true for all t and x.

29a. The free-body diagram is essentially shown in Fig. 1.3.1.
    The gravitational force is shown. The only other force is
    the tension, T, which acts towards the hinge along L.

29b. The component of the gravitational force (mg) along the
    tangent to the circular arc is given by mgsinθ. Since T
    acts perpendicular to the tangent, there is no component
    of the tension in the tangential direction. Newton's
    Second Law states that F = ma. In this problem, since the
    motion is circular, it is appropriate to use polar
    coordinates and thus $a \neq \dfrac{dv}{dt}$ as we have used earlier.
    Since r = L is constant, the linear acceleration tangent

to the circular motion is given by $L\dfrac{d^2\theta}{dt^2}$ and thus Newton's

Second Law gives $-mg\sin\theta = mL\dfrac{d^2\theta}{dt^2}$.

29c. Dividing by mL and rearranging terms gives $\dfrac{d^2\theta}{dt^2} + \dfrac{g}{L}\sin\theta = 0$.

CHAPTER 2

Section 2.1, Page 39

1b. All solutions seem   1a.
to approach a line
in the region where
the negative and
positive slopes
meet each other.

1c. $\mu(t) = \exp(\int 3dt) = e^{3t}$. Thus $e^{3t}(y'+3y) = e^{3t}(t+e^{-2t})$ or

$\dfrac{d}{dt}(ye^{3t}) = te^{3t} + e^t$. Integration of both sides yields

$ye^{3t} = \dfrac{1}{3}te^{3t} - \dfrac{1}{9}e^{3t} + e^t + c$, where integration by parts

is used on the right side, with $u = t$ and $dv = e^{3t}dt$.
Division by e^{3t} gives $y(t) = ce^{-3t} + t/3 - 1/9$, so y
approaches $t/3 - 1/9$ as $t \to \infty$. This is the line
identified in part b.

2c. $\mu(t) = e^{-2t}$.   3c. $\mu(t) = e^t$.

4c. $\mu(t) = \exp(\int \dfrac{dt}{t}) = e^{\ln t} = t$, so $(ty)' = 3t\cos 2t$, and

integration by parts yields the general solution.

6c. The equation must be divided by t so that it is in the
form of Eq.(3): $y' + (2/t)y = (\sin t)/t$. Thus
$\mu(t) = \exp(\int \dfrac{2dt}{t} = t^2$, and $(t^2y)' = t\sin t$. Integration
then yields $t^2y = -t\cos t + \sin t + c$.

7c. $\mu(t) = e^{t^2}$.   8c. $\mu(t) = \exp(\int \dfrac{4tdt}{1+t^2}) = (1+t^2)^2$.

11c. $\mu(t) = e^t$ so $(e^ty)' = 5e^t\sin 2t$. To integrate the right
side you can integrate by parts (twice), use an integral
table, or use a symbolic computational software program
to find $e^ty = e^t(\sin 2t - 2\cos 2t) + c$.

13c. $\mu(t) = e^{-t}$ so that $(e^{-t}y)' = 2te^t$ and thus
$e^{-t}y = 2\int te^t dt + c = 2(te^t - \int e^t dt) + c = 2(te^t - e^t) + c$.
Thus $y(t) = 2(t-1)e^{2t} + ce^t$, so setting $t = 0$ we have
$1 = -2 + c$, or $c = 3$. Hence $y(t) = 2(t-1)e^{2t} + 3e^t$.

15. $\mu(t) = \exp(\int\frac{2dt}{t}) = t^2$ so that $(t^2y)' = t^3 - t^2 + t$.

Integrating and dividing by t^2 gives
$y = t^2/4 - t/3 + 1/2 + c/t^2$. Setting $t = 1$ and $y = 1/2$
we have $c = 1/12$.

18. $\mu(t) = t^2$. Thus $(t^2y)' = t\sin t$ and
$t^2y = -t\cos t + \sin t + c$. Setting $t = \pi/2$ and
$y = 1$ yields $c = \pi^2/4 - 1$.

20. $\mu(t) = \exp\int\frac{t+1}{t}dt = \exp(t + \ln(t)) = te^t$.

21b. $\mu(t) = e^{-t/2}$ so $(e^{-t/2}y)' = 2e^{-t/2}\cos t$. Integrating (see
comments in #11) and dividing by $e^{-t/2}$ yields
$y(t) = -\frac{4}{5}\cos t + \frac{8}{5}\sin t + ce^{t/2}$. Thus $y(0) = -\frac{4}{5} + c = a$,
or $c = a + \frac{4}{5}$ and $y(t) = -\frac{4}{5}\cos t + \frac{8}{5}\sin t + (a + \frac{4}{5})e^{t/2}$.

21c. If $(a + \frac{4}{5}) = 0$, then the solution is oscillatory for all
t, while if $(a + \frac{4}{5}) \neq 0$, the solution is unbounded as
$t \to \infty$. Thus $a_0 = -\frac{4}{5}$.

21a.   25a.

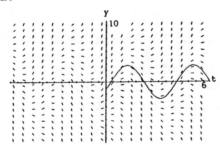   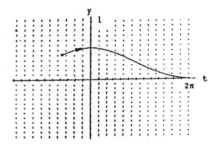

25b. $\mu(t) = \exp\int\frac{2dt}{t} = t^2$, so $(t^2y)' = \sin t$ and
$y(t) = \frac{-\cos t}{t^2} + \frac{c}{t^2}$. Setting $t = -\frac{\pi}{2}$ yields
$\frac{4c}{\pi^2} = a$ or $c = \frac{a\pi^2}{4}$ and hence $y(t) = \frac{a\pi^2/4 - \cos t}{t^2}$, which
is unbounded as $t \to 0$ unless $a\pi^2/4 = 1$ or $a_0 = 4/\pi^2$.

25c. For $a = 4/\pi^2$ $y(t) = \dfrac{1 - \cos t}{t^2}$. To find the limit as

$t \to 0$ L'Hopital's Rule must be used:

$$\lim_{t\to 0}y(t) = \lim_{t\to 0}\frac{\sin t}{2t} = \lim_{t\to 0}\frac{\cos t}{2} = \frac{1}{2}.$$

26b. $\mu(t) = \exp\!\int\dfrac{\cos(t)}{\sin(t)}dt = \exp(\ln(\sin t)) = \sin(t)$ and thus

$(\sin(t)y)' = e^t$. Hence $\sin(t)y = e^t + c$ or $y(t) = \dfrac{e^t + c}{\sin(t)}$.

Setting $t = 1$ and $y = a$ we get $c = a\sin 1 - e$ so
$y(t) = (e^t - e + a\sin 1)/\sin(t)$. If $y(t)$ is to remain
finite as $t \to 0$ the numerator, $e^t - e + a\sin 1$, must
approach 0 as $t \to 0$ and hence $a_0 = (e-1)/\sin 1$.

26c. Using a_0 we have $y(t) = (e^t - 1)/\sin(t)$, which approaches
1 as $t \to 0$, using l'Hospital's Rule.

30. $(e^{-t}y)' = e^{-t} + 3e^{-t}\sin t$ so

$e^{-t}y = -e^{-t} - 3e^{-t}(\dfrac{\sin t + \cos t}{2}) + c$ or

$y(t) = -1 - (\dfrac{3}{2})e^{-t}(\sin t + \cos t) + ce^t.$

If $y(t)$ is to remain bounded as $t \to \infty$, we must have

$c = 0$. Thus $y(0) = -1 - \dfrac{3}{2} + c = y_0$ or $c = y_0 + \dfrac{5}{2} = 0$

and $y_0 = -5/2$.

32. Write the first term of Eq.(47) as $\dfrac{\displaystyle\int_0^t e^{s^2/4}ds}{e^{t^2/4}}$. In applying

L'Hospital's Rule, the derivative of the numerator term
is $e^{t^2/4}$ by the Fundamental Theorem of Calculus. The
derivative of the denominator is $(t/2)e^{t^2/4}$ and thus the
limit is 0 as $t \to \infty$.

33. $\mu(t) = e^{at}$ so the D.E. can be written as
$(e^{at}y)' = be^{at}e^{-\lambda t} = be^{(a-\lambda)t}$. If $a \neq \lambda$, then integration
and solution for y yields $y = [b/(a-\lambda)]e^{-\lambda t} + ce^{-at}$. Then
$\lim_{x\to\infty} y$ is zero since both λ and a are positive numbers.

If $a = \lambda$, then the D.E. becomes $(e^{at}y)' = b$, which yields
$y = (bt+c)/e^{\lambda t}$ as the solution. L'Hopital's Rule gives
$$\lim_{t\to\infty} y = \lim_{t\to\infty} \frac{(bt+c)}{e^{\lambda t}} = \lim_{t\to\infty} \frac{b}{\lambda e^{\lambda t}} = 0.$$

35. There is no unique answer for this situation. One
     possible response is to assume $y(t) = ce^{-2t} + 3 - t$, then
     $y'(t) = -2ce^{-2t} - 1$ and thus $y' + 2y = 5 - 2t$.

39. By Eq.(iii), Prob.38, $y(t) = A(t)\exp(-\int(-2)dt) = A(t)e^{2t}$.
     Differentiating $y(t)$ and substituting into the D.E.
     yields $A'(t) = t^2$ since the terms involving $A(t)$ add to
     zero. Thus $A(t) = t^3/3 + c$, which substituted into $y(t)$
     yields the solution.

42. Since $p(t) = \dfrac{1}{2}$, $y(t) = A(t)\exp(-\int\dfrac{dt}{2}) = A(t)e^{-t/2}$ and
     $A'(t) = (3/2)t^2e^{-t^2/2}$. Integration of A' and substituting
     in $y(t)$ yields then desired solution.

Section 2.2, Page 47

Problems 1 through 20 follow the pattern of the examples
worked in this section. The first eight problems, however,
do not have I.C. so the integration constant, c, cannot be
found.

1.   Write the equation in the form $ydy = x^2dx$. Integrating
     the left side with respect to y and the right side with
     respect to x yields
     $$\frac{y^2}{2} = \frac{x^3}{3} + C, \text{ or } 3y^2 - 2x^3 = c.$$

4.   For $y \neq -3/2$ multiply both sides of the equation by
     $3 + 2y$ to get the separated equation
     $(3+2y)dy = (3x^2-1)dx$. Integration then yields
     $3y + y^2 = x^3 - x + c$.

6.   We need $x \neq 0$ and $|y| < 1$ for this problem to be defined.
     Separating the variables we get $(1-y^2)^{-1/2}dy = x^{-1}dx$.
     Integrating each side yields $\arcsin y = \ln|x|+c$, so
     $y = \sin[\ln|x|+c]$, $x \neq 0$ (note that $|y| < 1$). Also,
     $y = \pm 1$ satisfy the D.E., since both sides are zero.

10a. Separating the variables we get $ydy = (1-2x)dx$, so
     $$\frac{y^2}{2} = x - x^2 + c.$$ Setting $x = 1$ and $y = -2$ we have $2 = c$

and thus $y^2 = 2x - 2x^2 + 4$ or $y = -\sqrt{2x - 2x^2 + 4}$. The negative square root must be used since $y(1) = -2$.

10c. Rewriting $y(x)$ as $-\sqrt{2(2-x)(x+1)}$, we see that y is defined for $-1 \le x \le 2$, However, since y' does not exist for $x = -1$ or $x = 2$, the solution is valid only for the open interval $-1 < x < 2$.

13. Separate variables by factoring the denominator of the right side to get $ydy = \dfrac{2x}{1+x^2}dx$. Integration yields $y^2/2 = \ln(1+x^2)+c$ and use of the I.C. gives $c = 2$. Thus $y = \pm\,[2\ln(1+x^2)+4]^{1/2}$, but we must discard the plus square root because of the I.C. Since $1 + x^2 > 0$, the solution is valid for all x.

15. Separating variables and integrating yields $y + y^2 = x^2 + c$. Setting $y = 0$ when $x = 2$ yields $c = -4$ or $y^2 + y = x^2-4$. To solve for y complete the square on the left side by adding $1/4$ to both sides. This yields $y^2 + y + \dfrac{1}{4} = x^2 - 4 + \dfrac{1}{4}$ or $(y + \dfrac{1}{2})^2 = x^2 - 15/4$. Taking the square root of both sides yields $y + \dfrac{1}{2} = \pm\sqrt{x^2 - 15/4}$, where the positive square root must be taken in order to satisfy the I.C. Thus $y = -\dfrac{1}{2} + \sqrt{x^2 - 15/4}$, which is defined for $x^2 \ge 15/4$ or $x \ge \sqrt{15}/2$. The possibility that $x < -\sqrt{15}/2$ is discarded due to the I.C.

17a. Separating variables gives $(2y-5)dy = (3x^2-e^x)dx$ and integration then gives $y^2 - 5y = x^3 - e^x + c$. Setting $x = 0$ and $y = 1$ we have $1 - 5 = 0 - 1 + c$, or $c = -3$ and thus $y^2 - 5y - (x^3-e^x-3) = 0$. Using the quadratic formula then gives
$$y(x) = \frac{5 \pm \sqrt{25+4(x^3-e^x-3)}}{2} = \frac{5}{2} - \sqrt{\frac{13}{4} + x^3 -e^x}\,,$$ where the negative square root is chosen so that $y(0) = 1$.

17c. The interval of definition for y must be found numerically. Approximate values can be found by plotting

$y_1(x) = \dfrac{13}{4} + x^3$ and $y_2(x) = e^x$ and noting the values of x where the two curves cross.

19a. We start with $\cos 3y\,dy = -\sin 2x\,dx$ and integrate to get $\dfrac{1}{3}\sin 3y = \dfrac{1}{2}\cos 2x + c$. Setting $y = \pi/3$ when $x = \pi/2$ (from the I.C.) we find that $0 = -\dfrac{1}{2} + c$ or $c = \dfrac{1}{2}$, so that $\dfrac{1}{3}\sin 3y = \dfrac{1}{2}\cos 2x + \dfrac{1}{2} = \cos^2 x$ (using the appropriate trigonometric identity). To solve for y we must choose the branch that passes through the point $(\pi/2, \pi/3)$ and thus $3y = \pi - \arcsin(3\cos^2 x)$, or $y = \dfrac{\pi}{3} - \dfrac{1}{3}\arcsin(3\cos^2 x)$.

19c. The solution in part a is defined only for $0 \le 3\cos^2 x \le 1$, or $-\sqrt{1/3} \le \cos x \le \sqrt{1/3}$. Taking the indicated square roots and then finding the inverse cosine of each side yields $.9553 \le x \le 2.1863$, or $\left| x - \pi/2 \right| \le 0.6155$, as the approximate interval.

21. We have $(3y^2 - 6y)\,dy = (1 + 3x^2)\,dx$ so that $y^3 - 3y^2 = x + x^3 - 2$, once the I.C. are used. From the D.E., the integral curve will have a vertical tangent when $3y^2 - 6y = 0$, or $y = 0, 2$. For $y = 0$ we have $x^3 + x - 2 = 0$, which is satisfied for $x = 1$, which is the only zero of the function $w = x^3 + x - 2$. Likewise, for $y = 2$, $x = -1$. Thus the solution is valid on $\left| x \right| < 1$.

23. Separating variables gives $y^{-2}\,dy = (2 + x)\,dx$, so $-y^{-1} = 2x + \dfrac{x^2}{2} + c$. $y(0) = 1$ yields $c = -1$ and thus $y = \dfrac{-1}{\dfrac{x^2}{2} + 2x - 1} = \dfrac{2}{2 - 4x - x^2}$. This gives $\dfrac{dy}{dx} = \dfrac{8 + 4x}{(2 - 4x - x^2)^2}$, so the minimum value is attained at $x = -2$. Note that the solution is defined for $-2 - \sqrt{6} < x < -2 + \sqrt{6}$ (by finding the zeros of the denominator) and has vertical asymptotes at the end points of the interval.

25. Separating variables and integrating yields
$3y + y^2 = \sin 2x + c$. $y(0) = -1$ gives $c = -2$ so that
$y^2 + 3y + (2-\sin 2x) = 0$. The quadratic formula, along
with the I.C., then gives $y = -\dfrac{3}{2} + \sqrt{\sin 2x + 1/4}$, which
is defined for $-.126 < x < 1.697$ (found by solving
$\sin 2x = -.25$ for x and noting $x = 0$ is the initial
point). Thus we have $\dfrac{dy}{dx} = \dfrac{\cos 2x}{(\sin 2x + \frac{1}{4})^{1/2}}$, which yields

$x = \pi/4$ as the only critical point in the above interval.
Using the second derivative test or graphing the solution
indicates the critical point is a maximum.

27a. By sketching the direction field or by using the D.E. we
note that $y' < 0$ for $y > 4$ and y' approaches zero as y
approaches 4. For $0 < y < 4$, $y' > 0$ and again approaches
zero as y approaches 4. Thus $\lim\limits_{t \to \infty} y = 4$ if $y_0 > 0$. For
$y_0 < 0$, $y' < 0$ for all y and hence y becomes negatively
unbounded $(-\infty)$ as t increases. If $y_0 = 0$, then $y' = 0$
for all t, so $y = 0$ for all t.

27b. Separating variables and using a partial fraction
expansion we have $(\dfrac{1}{y} - \dfrac{1}{y-4})dy = \dfrac{4}{3}tdt$. Hence
$\ln\left|\dfrac{y}{y-4}\right| = \dfrac{2}{3}t^2 + c_1$ and thus $\left|\dfrac{y}{y-4}\right| = e^{c_1}e^{2t^2/3} = ce^{2t^2/3}$,
where c is positive. For $y(0) = y_0 = .5$ this becomes
$\dfrac{y}{4-y} = ce^{2t^2/3}$ and thus $c = \dfrac{.5}{3.5} = \dfrac{1}{7}$. Using this value
for c and solving for y yields $y(t) = \dfrac{4}{1 + 7e^{-2t^2/3}}$.
Setting this equal to 3.98 and solving for t yields
$t = 3.29527$.

29. Separating variables yields $\dfrac{cy+d}{ay+b}\,dy = dx$. If $a \neq 0$ and
$ay+b \neq 0$ then $dx = (\dfrac{c}{a} + \dfrac{ad-bc}{a(ay+b)})dy$. Integration then
yields the desired answer.

30c. If $v = y/x$ then $y = vx$ and $\dfrac{dy}{dx} = v + x\dfrac{dv}{dx}$ and thus the

D.E. becomes $v + x\dfrac{dv}{dx} = \dfrac{v-4}{1-v}$. Subtracting v from both

sides yields $x\dfrac{dv}{dx} = \dfrac{v^2-4}{1-v}$.

30d. The last equation in (c) separates into $\dfrac{1-v}{v^2-4}dv = \dfrac{1}{x}dx$. To

integrate the left side use partial fractions to write

$\dfrac{1-v}{v-4} = \dfrac{A}{v-2} + \dfrac{B}{v+2}$, which yields A = -1/4 and B = -3/4.

Integration then gives $-\dfrac{1}{4}\ln|v-2| - \dfrac{3}{4}\ln|v+2| = \ln|x| - k$,

or $\ln|x^4||v-2||v+2|^3 = 4k$ after manipulations using

properties of the ln function. Thus $x^4|v-2||v+2|^3 = C$.

31a. Simplifying the right side of the D.E. gives
$dy/dx = 1 + (y/x) + (y/x)^2$ so the equation is homogeneous.

31b. The substitution y = vx leads to
$v + x\dfrac{dv}{dx} = 1 + v + v^2$ or $\dfrac{dv}{1 + v^2} = \dfrac{dx}{x}$. Solving, we get
arctanv = ln|x| + c. Substituting for v we obtain
arctan(y/x) - ln|x| = c.

33b. Dividing the numerator and denominator of the right side

by x and substituting y = vx we get $v + x\dfrac{dv}{dx} = \dfrac{4v - 3}{2-v}$

which can be rewritten as $x\dfrac{dv}{dx} = \dfrac{v^2 + 2v - 3}{2 - v}$. Note that

v = -3 and v = 1 are solutions of this equation. For
$v \neq 1, -3$ separating variables gives

$\dfrac{2 - v}{(v+3)(v-1)} dv = \dfrac{1}{x}dx$. Applying a partial fraction

decomposition to the left side we obtain

$[\dfrac{1}{4}\dfrac{1}{v-1} - \dfrac{5}{4}\dfrac{1}{v+3}]dv = \dfrac{dx}{x}$, and upon integrating both sides

we find that $\dfrac{1}{4}\ln|v-1| - \dfrac{5}{4}\ln|v+3| = \ln|x| + c$.

Substituting for v and performing some algebraic
manipulations we get the solution in the implicit form
$|y-x| = c|y+3x|^5$. v = 1 and v = -3 yield y = x and
y = -3x, respectively, as solutions also.

35b. As in Prob.33, substituting y = vx into the D.E. we get

$$v + x\frac{dv}{dx} = \frac{1+3v}{1-v}, \text{ or } x\frac{dv}{dx} = \frac{(v+1)^2}{1-v}.$$ Note that v = -1 (or

y = -x) satisfies this D.E. Separating variables yields

$$\frac{1-v}{(v+1)^2}dv = \frac{dx}{x}.$$ Integrating the left side by parts

(let u = 1-v and dw = $\frac{dv}{(v+1)^2}$) we obtain

$$\frac{v-1}{v+1} - \ln|v+1| = \ln|x| + c.$$ Letting v = $\frac{y}{x}$ then yields

$$\frac{y-x}{y+x} - \ln\left|\frac{y+x}{x}\right| = \ln|x| + c, \text{ or } \frac{y-x}{y+x} - \ln|y+x| = c.$$ The

answer in the text can be obtained by integrating the left
side, above, using partial fractions. By differentiating
both answers, it can be verified that indeed both forms
satisfy the D.E.

Section 2.3, Page 59

2. Let S(t) be the amount of salt that is present at any time
    t, then S(0) = 0 is the original amount of salt in the tank,
    2γ is the amount of salt entering per minute, and 2(S/120)
    is the amount of salt leaving per minute (all amounts
    measured in grams). Thus dS/dt = 2γ - 2S/120, S(0) = 0. This
    is a linear equation, which has $e^{t/60}$ as its integrating
    factor. Thus the general solution is S(t) = 120γ + $ce^{-t/60}$.
    S(0) = 0 gives c = -120γ, so S(t) = 120γ(1 - $e^{-t/60}$) and
    hence S(t) → 120γ grams as t → ∞.

3. We must first find the amount of salt that is present after
    10 minutes. For the first 10 minutes (if we let Q(t) be the
    amount of salt in the tank): $\frac{dQ}{dt} = \frac{1}{2}(2) - 2\frac{Q(t)}{100}$, Q(0) = 0.
    This is a linear equation which has the solution
    Q(t) = 50(1 - $e^{-t/50}$), as in Prob. 2, and thus
    Q(10) = 50(1-$e^{-.2}$) = 9.063 lbs. of salt in the tank after
    the first 10 minutes. At this point no more salt is allowed
    to enter, so the new I.V.P. (letting P(t) be the amount of
    salt in the tank after the first 10 minutes) is:
    $\frac{dP}{dt} = (0)(2) - 2\frac{P(t)}{100}$, P(0) = Q(10) = 9.063. The solution
    of this problem is P(t) = 9.063$e^{-.02t}$, which yields
    P(10) = 7.42 lbs.

4.   Salt flows out of the tank at the rate of $\dfrac{Q(t)}{200+t}$ (2) lb/min.

since the volume of water in the tank at any time t is
200 + (1)(t) gallons (due to the fact that water flows into
the tank faster than it flows out). Thus the I.V.P. is

$dQ/dt = (3)(1) - \dfrac{2}{200+t}Q(t)$, Q(0) = 100, which is a linear

equation with $(200+t)^2$ as its integrating factor.

8a. Set $S_0 = 0$ in Eq.(16) (or solve Eq.(15) with S(0) = 0).

8b. Set r = .075, t = 40 and S(t) = $1,000,000 in the answer to
part (a) and then solve for k.

8c. Set k = $2,000, t = 40 and S(t) = $1,000,000 in the answer
to (a) and then solve numerically for r.

9.   If S(t) is the amount of the loan, then dS/dt = .1S - k,
S(0) = $8,000. Solving this for S(t) yields
$S(t) = 8000e^{.1t} - 10k(e^{.1t}-1)$. Setting S = 0 and substitution
of t = 3 gives k = $3,086.64 per year. For 3 years this
totals $9,259.92, so $1,259.92 has been paid in interest.

10. Since we are assuming continuity, either convert the monthly
payment into an annual payment or convert the yearly
interest rate into a monthly interest rate for 240 months.
Then proceed as in Prob. 9.

11a. Using Eq. (15) we have $\dfrac{dS}{dt} = \dfrac{.09}{12}S - 800(1+\dfrac{t}{120})$ or

$\dfrac{dS}{dt} - \dfrac{3}{400}S = -(800+\dfrac{20}{3}t)$, S(0) = 100,000. Using an

integrating factor and integration by parts (or using a D.E.

solver) we get $S(t) = \dfrac{6,080,000}{27} + \dfrac{8000}{9}t + ce^{3t/400}$. Using

the I.C. yields $c = \dfrac{-3,380,000}{27}$. Substituting this value

into S, setting S(t) = 0, and solving numerically for t
yields t ≅ 135.36 months.

14a. We have $\dfrac{dy}{y} = (.1+.2\sin t)dt$, by separating variables, and

thus $y(t) = c\exp(.1t-.2\cos t)$. y(0) = 1 gives $c = e^{.2}$, so
$y(t) = \exp(.2+.1t-.2\cos t)$. Setting y = 2 yields
$\ln 2 = .2 + .1\tau - .2\cos\tau$, which can be solved numerically to
give τ = 2.9632. If $y(0) = y_0$, then as above,

$y(t) = y_0 \exp(.2+.1t-.2\cos t)$. Thus if we set $y = 2y_0$ we get the same numerical equation for τ and hence the doubling time has not changed.

16. If T is the temperature of the coffee at any time t, then
$\dfrac{dT}{dt} = -k(T - 70^\circ)$; $T(0) = 200^\circ$, $T(1) = 190^\circ$. The solution of this linear equation will involve k (the cooling rate) and the integration constant c. Use $T(0) = 200$ to find c and then use $T(1) = 190$ to evaluate k.

18a. Eq.(i) is a linear equation with the integrating factor e^{kt}. Thus $(e^{kt}u)' = k(T_0 + T_1\cos\omega t)e^{kt}$ and hence $e^{kt}u = T_0 e^{kt} + kT_1\int\cos\omega t e^{kt}dt + c$. Evaluating the integral (by parts or by a symbolic software package) and dividing by e^{kt} yields $u(t) = T_0 + kT_1\dfrac{k\cos\omega t + \omega\sin\omega t}{k^2 + \omega^2} + ce^{-kt}$. Note that the last term approaches zero as $t \to \infty$ for any I.C., and that the rest of the solution oscillates about $u(t) = T_0$

18c. Recall that $R\cos[\omega(t-\tau)] = R\cos\omega t\cos\omega\tau + R\sin\omega t\sin\omega\tau$. Comparing this with the oscillatory portion of the above solution we have $R\cos\omega\tau = \dfrac{k^2 T_1}{k^2+\omega^2}$ and $R\sin\omega\tau = \dfrac{k\omega T_1}{k^2+\omega^2}$ since these are the coefficients of $\cos\omega t$ and $\sin\omega t$ respectively. By squaring and adding we find $R^2 = \dfrac{k^2 T_1^2}{k^2+\omega^2}$ and by dividing we find $\tan\omega\tau = \omega/k$.

19a. The required D.E. is $dQ/dt = kr + P - \dfrac{Q(t)}{V}r$, since kr is the rate of water pollutant entering the lake, P is the rate of pollutant entering directly and $Q(t)r/V$ is the rate at which the pollutant leaves the lake. The I.C. is $Q(0) = Vc_0$. Since $c = Q(t)/V$, the I.V.P. may be rewritten $Vc'(t) = kr + P - rc$, $c(0) = c_0$, which has the solution $c(t) = k + \dfrac{P}{r} + (c_0 - k - \dfrac{P}{r})e^{-rt/V}$.

19b. Set $k = 0$, $P = 0$, $t = T$ and $c(T) = .5c_0$ in the solution found in (a).

20a. If we measure x positively upward from the ground, then Eq.(4) of Section 1.1 becomes $m\dfrac{dv}{dt} = -mg$, since there is no air resistance. Thus the I.V.P. for $v(t)$ is $dv/dt = -g$,

$v(0) = 20$, which gives $v(t) = 20 - gt$. Since $\dfrac{dx}{dt} = v(t)$ we get $x(t) = 20t - (g/2)t^2 + c$. Then $x(0) = 30$ gives $c = 30$ and thus $x(t) = 20t - (g/2)t^2 + 30$. At the maximum height $v(t_m) = 0$ and thus $t_m = 20/9.8 = 2.04$ sec., which when substituted in the equation for $x(t)$ yields the maximum height.

21. The I.V.P. in this case is $m\dfrac{dv}{dt} = -\dfrac{1}{30}v - mg$, $v(0) = 20$, where the positive direction is measured upward.

23a. The I.V.P. is $m\dfrac{dv}{dt} = mg - .75v$, $v(0) = 0$ and v is measured positively downward. Since $m = 180/32$, the D.E. becomes $\dfrac{dv}{dt} = 32 - \dfrac{2}{15}v$ and thus $v(t) = 240(1-e^{-2t/15})$ so that $v(10) = 176.7$ ft/sec.

23b. Integration of $v(t)$ as found in (a) yields
$x(t) = 240t + 1800(e^{-2t/15}-1)$, x is measured positively down from the altitude of 5000 feet. Set $t = 10$ to find the distance traveled when the parachute opens.

23c. After the parachute opens the I.V.P. is $m\dfrac{dv}{dt} = mg-12v$,
$v(0) = 176.7$, which has the solution
$v(t) = 161.7e^{-32t/15} + 15$ and where $t = 0$ now represents the time the parachute opens. Letting $t \rightarrow \infty$ yields the limiting velocity of 15 ft/sec.

23d. Integrate $v(t)$ as found in (c) to find
$x(t) = 15t - 75.8e^{-32t/15} + C_2$. $C_2 = 75.8$ since $x(0) = 0$, x now being measured from the point where the parachute opens. Setting $x = 3925.5$ will then yield the length of time the skydiver is in the air after the parachute opens.

26a. As in Prob.21, $m\dfrac{dv}{dt} = -mg - kv$, $v(0) = v_0$.

26b. From part (a) $v(t) = -\dfrac{mg}{k} + [v_0 + \dfrac{mg}{k}]e^{-kt/m}$. As $k \rightarrow 0$ this has the indeterminant form of $-\infty + \infty$. Thus rewrite $v(t)$ as $v(t) = [-mg + (v_0 k + mg)e^{-kt/m}]/k$ which has the indeterminant form of $0/0$, as $k \rightarrow 0$ and hence L'Hospital's Rule may be applied with k as the variable.

27a. The equation of motion is m(dv/dt) = w-R-B which, in this problem, is $\frac{4}{3}\pi a^3\rho(dv/dt) = \frac{4}{3}\pi a^3\rho g - 6\pi\mu av - \frac{4}{3}\pi a^3\rho'g$. The limiting velocity occurs when dv/dt = 0.

27b. Since the droplet is motionless, v = dv/dt = 0, we have the equation of motion $0 = (\frac{4}{3})\pi a^3\rho g - Ee - (\frac{4}{3})\pi a^3\rho'g$, where ρ is the density of the oil and ρ' is the density of air. Solving for e yields the answer.

28. All three parts can be answered from one solution if k represents the resistance and if the method of solution of Example 4 is used. Thus we have
$m\frac{dv}{dt} = mv\frac{dv}{dx} = mg - kv$, v(0) = 0, where we have assumed the velocity is a function of x. The solution of this I.V.P. involves a logarithmic term, and thus the answers to parts (a) and (c) must be found using a numerical procedure.

29a. Use Eq.(30)

29b. Note that 32 ft/sec^2 = 78,545 m/hr^2.

30b. From part a) $\frac{dx}{dt}$ = v = ucosA and hence x(t) = (ucosA)t + d_1. Since x(0) = 0, we have d_1 = 0 and x(t) = (ucosA)t. Likewise $\frac{dy}{dt}$ = -gt + usinA and therefore y(t) = $-gt^2/2$ + (usinA)t + d_2. Since y(0) = h we have d_2 = h and y(t) = $-gt^2/2$ + (usinA)t + h.

30d. Let t_w be the time the ball reaches the wall. Then x(t_w) = L = (ucosA)t_w and thus $t_w = \frac{L}{ucosA}$. For the ball to clear the wall y(t_w) ≥ H and thus (setting $t_w = \frac{L}{ucosA}$, g = 32 and h = 3 in y) we get $\frac{-16L^2}{u^2cos^2A}$ + LtanA + 3 ≥ H.

30e. Setting L = 350 and H = 10 we get $\dfrac{-161.98}{\cos^2 A} + 350\dfrac{\sin A}{\cos A} \geq 7$

or $7\cos^2 A - 350\cos A\sin A + 161.98 \leq 0$. This can be solved numerically or by plotting the left side as a function of A and finding where the zero crossings are.

30f. Setting L = 350, and H = 10 in the answer to part d

yields $\dfrac{-16(350)^2}{u^2\cos^2 A} + 350\tan A = 7$, where we have chosen the

equality sign since we want to just clear the wall.

Solving for u^2 we get $u^2 = \dfrac{1,960,000}{175\sin 2A - 7\cos^2 A}$. Now u will

have a minimum when the denominator has a maximum. Thus $350\cos 2A + 7\sin 2A = 0$, or $\tan 2A = -50$, which yields A = .7954 rad. and u = 106.89 ft./sec.

Section 2.4, Page 75

1. If the equation is written in the form of Eq.(1), then $p(t) = (\ln t)/(t-3)$ and $g(t) = 2t/(t-3)$. These are defined and continuous on the intervals (0,3) and (3,∞), but since the initial point is t = 1, the solution will be continuous on 0 < t < 3.

4. $p(t) = 2t/(2-t)(2+t)$ and $g(t) = 3t^2/(2-t)(2+t)$, which have discontinuities at t = ±2. Since $y_0 = -3$, the solution will be continuous on -∞ < t < -2.

8. Theorem 2.4.2 guarantees a unique solution to the D.E. through any point (t_0, y_0) such that $t_0^2 + y_0^2 < 1$ since $\dfrac{\partial f}{\partial y} = -y/(1-t^2-y^2)^{1/2}$ is defined and continuous only for $1-t^2-y^2 > 0$. Note also that $f = (1-t^2-y^2)^{1/2}$ is defined and continuous in this region as well as on the boundary $t^2+y^2 = 1$. The boundary can't be included in the final region due to the discontinuity of $\dfrac{\partial f}{\partial y}$ there.

11. In this case $f = \dfrac{1+t^2}{y(3-y)}$ and $\dfrac{\partial f}{\partial y} = \dfrac{1+t^2}{y(3-y)^2} - \dfrac{1+t^2}{y^2(3-y)}$, which are both continuous everywhere except for y = 0 and y = 3.

13. The D.E. may be written as ydy = -4tdt so that
$\dfrac{y^2}{2}$ = $-2t^2+c$, or $y^2 = C-4t^2$. The I.C. then yields
$y_0^2 = C$, so that $y^2 = y_0^2 - 4t^2$ or $y = \pm\sqrt{y_0^2-4t^2}$, which is
defined for $4t^2 < y_0^2$ or $|t| < |y_0|/2$. Note that $y_0 \neq 0$
since Theorem 2.4.2 does not hold there.

17. From the direction field (or the given D.E.) it is noted
that for t > 0 and y < 0 that $y' < 0$, so $y \to -\infty$ for
$y_0 < 0$. Likewise, for $0 < y_0 < 3$, $y' > 0$ and $y' \to 0$ as
$y \to 3$, so $y \to 3$ for $0 < y_0 < 3$ and for $y_0 > 3$, $y' < 0$
and again $y' \to 0$ as $y \to 3$, so $y \to 3$ for $y_0 > 3$. For
$y_0 = 3$, $y' = 0$ and $y = 3$ for all t and for $y_0 = 0$, $y' = 0$
and $y = 0$ for all t.

22a. For $y_1 = 1-t$, $y_1' = -1 = \dfrac{-t+[t^2+4(1-t)]^{1/2}}{2}$

$= \dfrac{-t+[(t-2)^2]^{1/2}}{2}$

$= \dfrac{-t+|t-2|}{2} = -1$ if

(t-2) ≥ 0, by the definition of absolute value. Setting
t = 2 in y_1 we get $y_1(2) = -1$, as required.

22b. By Theorem 2.4.2 we are guaranteed a unique solution only
where $f(t,y) = \dfrac{-t+(t^2+4y)^{1/2}}{2}$ and $f_y(t,y) = (t^2+4y)^{-1/2}$ are
continuous. In this case the initial point (2,-1) lies
in the region $t^2 + 4y \leq 0$, in which case $\dfrac{\partial f}{\partial y}$ is not
continuous and hence the theorem is not applicable and
there is no contradiction.

22c. For $y = ct + c^2$ follow the steps of Prob. 22a. If
$y = y_2(t)$ then we must have $ct + c^2 = -t^2/4$ for all t,
which is not possible since c is a constant.

23b. $\phi(t) = t^{-1}$ gives $\phi'(t) = -t^{-2}$ so $\phi' + \phi^2 = 0$. $\phi(t) = ct^{-1}$
gives $\phi'(t) = -ct^{-2}$, so $\phi' + \phi^2 \neq 0$ unless c = 0 or c = 1.

25. $[y_1(t) + y_2(t)]' + p(t)[y_1(t) + y_2(t)] =$
$y_1'(t) + p(t)y_1(t) + y_2'(t) + p(t)y_2(t) = 0 + g(t)$.

27a. For n = 1, we have $y' + [p(t)-q(t)]y = 0$, which is
linear. Thus Eq.(3) gives
$y(t) = c\mu^{-1}(t) = ce^{-\int[p(t)-q(t)]dt}$, since $g(t) = 0$.

27b. Let $v = y^{1-n}$ then $\dfrac{dv}{dt} = (1-n)y^{-n}\dfrac{dy}{dt}$ so $\dfrac{dy}{dt} = \dfrac{1}{1-n}y^n\dfrac{dv}{dt}$, for
$n \neq 1$. Substituting into the D.E. yields
$\dfrac{y^n}{1-n}\dfrac{dv}{dt} + p(t)y = q(t)y^n$ or
$v' + (1-n)p(t)y^{1-n} = (1-n)q(t)$, or
$v' + (1-n)p(t)v = (1-n)q(t)$, which is a linear D.E. for v.

28. n = 3 so $v = y^{-2}$ and $\dfrac{dv}{dt} = -2y^{-3}\dfrac{dy}{dt}$ or $\dfrac{dy}{dt} = -\dfrac{1}{2}\,y^3\dfrac{dv}{dt}$.
Substituting this into the D.E. gives
$-\dfrac{1}{2}y^3\dfrac{dv}{dt} + \dfrac{2}{t}y = \dfrac{1}{t^2}y^3$. Simplifying and using
$y^{-2} = v$ then gives the linear D.E.
$v' - \dfrac{4}{t}v = -\dfrac{2}{t^2}$, where $\mu(t) = \dfrac{1}{t^4}$ and
$v(t) = ct^4 + \dfrac{2}{5t} = \dfrac{2+5ct^5}{5t}$. Thus $y = \pm[5t/(2+5ct^5)]^{1/2}$.

29. n = 2 so $v = y^{-1}$ and $\dfrac{dv}{dt} = -y^2\dfrac{dv}{dt}$. Thus the D.E.
becomes $-y^2\dfrac{dv}{dt} - ry = -ky^2$ or $\dfrac{dv}{dt} + rv = k$. Hence
$\mu(t) = e^{rt}$ and $v = k/r + ce^{-rt}$. Setting v = 1/y then
yields the solution.

32. Since g(t) is continuous on the interval
$0 \leq t \leq 1$ we may solve the I.V.P.
$y_1' + 2y_1 = 1$, $y_1(0) = 0$ on that interval to obtain
$y_1 = 1/2 - (1/2)e^{-2t}$, $0 \leq t \leq 1$. For 1<t, $g(t) = 0$; and
hence we may solve $y_2' + 2y_2 = 0$ to obtain $y_2 = ce^{-2t}$, 1<t.
The solution y of the original I.V.P. must be continuous
at t = 1 (since its derivative must exist) and hence we
need c in y_2 so that y_2 at 1 has the same value as y_1 at
1. Thus
$ce^{-2} = 1/2 - e^{-2}/2$ or $c = (1/2)(e^2-1)$ and we obtain

$$y = \begin{cases} 1/2 - (1/2)e^{-2t} & 0 \le t \le 1 \\ 1/2(e^2-1)e^{-2t} & 1 \le t \end{cases} \qquad \text{and}$$

$$y' = \begin{cases} e^{-2t} & 0 \le t \le 1 \\ (1-e^2)e^{-2t} & 1 < t. \end{cases}$$

Evaluating the two parts of y' at $t_0 = 1$ we see that they are different, and hence y' is not continuous at $t_0 = 1$.

Section 2.5, Page 88

3. From the graph, or by setting $\dfrac{dy}{dt} = y(y-1)(y-2) = 0$, we find that $y = 0,1,2$ are the critical points. The graph of $y(y-1)(y-2)$ is positive for $0 < y < 1$ and $2 < y$ and negative for $1 < y < 2$. Thus $y(t)$ is increasing $(\dfrac{dy}{dt} > 0)$ for $0 < y < 1$ and $2 < y$ and decreasing $(\dfrac{dy}{dt} < 0)$ for $1 < y < 2$. Therefore 0 and 2 are unstable critical points while 1 is an asymptotically stable critical point.

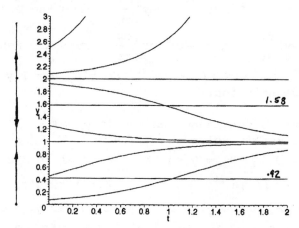

The phase line and several solutions are shown. The inflection points of the solutions (1.58 and .42) are found by determining where $y(y-1)(y-2)$ has its relative max and min points. These lines are are also shown.

5. $\dfrac{dy}{dt}$ is zero only when

$e^{-y}-1=0$, or $y = 0$.

Since $\dfrac{dy}{dt}>0$ for $y<0$

and $\dfrac{dy}{dt}<0$ for $y>0$

we conclude that $y=0$
is an asymptotically
stable critical point.
The graphs of several solutions verifies this.

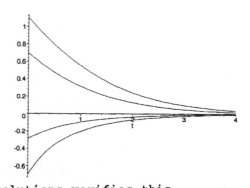

7c. Separate variables to get $\dfrac{dy}{(1-y)^2} = kt$. Integration

yields $\dfrac{1}{1-y} = kt + c$, or $y = 1 - \dfrac{1}{kt + c} = \dfrac{kt + c - 1}{kt + c}$.

Setting $t = 0$ and $y(0) = y_0$ yields $y_0 = \dfrac{c-1}{c}$ or

$c = \dfrac{1}{1-y_0}$. Hence $y(t) = \dfrac{(1-y_0)kt + y_0}{(1-y_0)kt + 1}$. Note that for

$y_0 < 1\ y \rightarrow (1-y_0)k/(1-y_0)k = 1$ as $t \rightarrow \infty$. For $y_0 > 1$
notice that the denominator will have a zero for some
value of t, depending on the values chosen for y_0 and k.
Thus the solution has a discontinuity at that point.

9. Setting $\dfrac{dy}{dt} = 0$ we find $y = 0, \pm 1$ are the critical

points. Since $\dfrac{dy}{dt} > 0$ for $|y| > 1$ while $\dfrac{dy}{dt} < 0$ for

$|y| < 1$ we conclude that $y = -1$ is asymptotically stable,
$y = 0$ is semistable, and $y = 1$ is unstable.

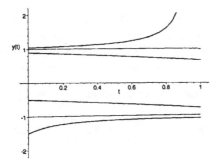

11. $y = b^2/a^2$ and $y = 0$ are the only critical points. For

$0 < y < b^2/a^2$, $\dfrac{dy}{dt} < 0$ and thus $y = 0$ is asymptotically

stable. For $y > b^2/a^2$, $dy/dt > 0$ and thus $y = b^2/a^2$ is
unstable. All solutions that start above $y = b^2/a^2$ will
continue to increase and all solutions that start for y
between 0 and b^2/a^2 will decay to zero. If $y(0) = b^2/a^2$
then $y(t) = b^2/a^2$ for all t, since this is an equilibrium
point.

14. If $f'(y_1) < 0$ then the slope of f is negative at y_1 and
thus $f(y) > 0$ for $y < y_1$ and $f(y) < 0$ for $y > y_1$ since
$f(y_1) = 0$. Hence y_1 is an asysmtotically stable critical
point. A similar argument will yield the result for
$f'(y_1) > 0$.

16a. Setting $\dfrac{dy}{dt} = 0$ we have $ry\ln(K/y) = 0$, so $y = 0$ and $y = K$
are the critical points. The graph for $r = 2$ and $K = 4$ is
shown, and we see that $\dfrac{dy}{dt} > 0$ for $0 < t < K = 4$ and
$\dfrac{dy}{dt} < 0$ for $t > K = 4$. Thus $y = 0$ is unstable and
$y = K$ is asymptotically stable.

16b. By taking the derivative of $y\ln(K/y)$ it can be shown that
the graph of $\dfrac{dy}{dt}$ vs y has a maximum point at $y = K/e$. Thus
$\dfrac{dy}{dt}$ is positive and increasing for $0 < y < K/e$ and hence
$y(t)$ is concave up for that interval. Similarly $\dfrac{dy}{dt}$ is
positive and decreasing for $K/e < y < K$ and thus $y(t)$ is
concave down for that interval.

16c. $\ln(K/y)$ is very large for small values of y and thus
$(ry)\ln(K/y) > ry(1 - y/K)$ for small y. Since $\ln(K/y)$ and
$(1 - y/K)$ are both strictly decreasing functions of y and
since $\ln(K/y) = (1 - y/K)$ only for $y = K$, we may conclude
that $\dfrac{dy}{dt} = (ry)\ln(K/y)$ is never less than $\dfrac{dy}{dt} = ry(1-y/K)$.

17a. If $u = \ln(y/K)$ then $y = Ke^u$ and $\dfrac{dy}{dt} = Ke^u\dfrac{du}{dt}$ so that the
D.E. becomes $du/dt = -ru$.

18a. The D.E. is $dV/dt = k - \alpha\pi r^2$. The volume of a cone of
height L and radius r is given by $V = \pi r^2 L/3$ where $L = hr/a$
from symmetry. Solving for r yields the desired solution.

18b. Equilibrium is given by $k - \alpha\pi r^2 = 0$.

18c. The equilibrium height must be less than h.

20b. Use the results of Problem 14.

20c. Y is defined to be Ey_2, where y_2 was found in part a.

20d. Differentiate Y with respect to E.

21a. Set $\dfrac{dy}{dt} = 0$ and solve for y using the quadratic formula.

21b. Use the results of Prob. 14.

21d. If $h > rK/4$ there are no critical points (see part a) and $\dfrac{dy}{dt} < 0$ for all t.

24a. If $z = x/n$ then $dz/dt = \dfrac{1}{n}\dfrac{dx}{dt} - \dfrac{x}{n^2}\dfrac{dn}{dt}$. Use of Equations (i) and (ii) then gives the D.E. (iii) and the I.C. is $z(0) = 1$ since $n(0) = x(0)$.

24b. Separate variables to get $\dfrac{dz}{z(1-vz)} = -\beta dt$. Using partial fractions this becomes $\dfrac{dz}{z} + \dfrac{vdz}{1-vz} = -\beta dt$. Integration and solving for z yields the answer.

24c. From part b, find $z(20)$ when $\beta = v = 1/8$.

25b. For $a = 0$, $\dfrac{dy}{dt}$ is always negative, so $y = 0$ is semistable. For $a > 0$ we have $\dfrac{dy}{dt} < 0$ for $|y| > \sqrt{a}$ and $\dfrac{dy}{dt} > 0$ for $|y| < -\sqrt{a}$, so $y = \sqrt{a}$ is asymptotically stable and $y = \sqrt{a}$ is unstable.

25c. The graphs are shown for $a = 4$. Note that $\dfrac{d}{dy}(a-y^2) = 0$ for $y = 0$ and thus $y = 0$ is an inflection point for the solution that crosses the x-axis.

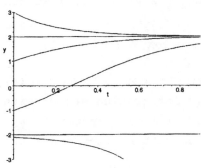

28a. Plot dx/dt vs x and observe that x = p and x = q are
     critical points. Also note that dx/dt > 0 for
     x < min(p,q) and x > max(p,q) while dx/dt < 0 for x
     between min(p,q) and max(p,q). Thus x = min(p,q) is an
     asymptotically stable point while x = max(p,q) is
     unstable. To solve the D.E., separate variables and use
     partial fractions to obtain $\dfrac{1}{q-p}[\dfrac{dx}{q-x} - \dfrac{dx}{p-x}] = \alpha dt$.
     Integration and solving for x yields the solution.

28b. x = p is a semistable critical point and since $\dfrac{dx}{dt} > 0$,
     x(t) is an increasing function. Thus for x(0) = 0, x(t)
     approaches p as t → ∞. To solve the D.E., separate
     variables and integrate.

Section 2.6, Page 99

3. M(x,y) = $3x^2$-2xy+2 and N(x,y) = $6y^2$-x^2+3, so M_y = -2x = N_x
    and thus the D.E. is exact. Integrating M(x,y) with
    respect to x we get ψ(x,y) = x^3 - x^2y + 2x + h(y).
    Taking the partial derivative of this with respect to y
    and setting it equal to N(x,y) yields -x^2+h'(y) =
    $6y^2$-x^2+3, so that h'(y) = $6y^2$ + 3 and h(y) = $2y^3$ + 3y.
    Substitute this h(y) into ψ(x,y) and recall that the
    equation which defines y(x) implicitly is ψ(x,y) = c.
    Thus x^3 - x^2y + 2x + $2y^3$ + 3y = c is the equation that
    implicitly defines the solution.

5. Writing the equation in the form M(x,y)dx + N(x,y)dy = 0
    gives M(x,y) = ax + by and N(x,y) = bx + cy. Thus
    M_y = b = N_x and the equation is exact. Integrating M(x,y)
    with respect to x yields ψ(x,y) = (a/2)x^2 + bxy + h(y).
    Differentiating ψ with respect to y (x constant) and
    setting $ψ_y$(x,y) = N(x,y) we find that h'(y) = cy and thus
    h(y) = (c/2)y^2. Hence the solution is given by
    (a/2)x^2 + bxy + (c/2)y^2 = k.

7. M_y(x,y) = e^xcosy - 2sinx = N_x(x,y) and thus the D.E. is
    exact. Integrating M(x,y) with respect to x gives
    ψ(x,y) = e^xsiny + 2ycosx + h(y). Finding $ψ_y$(x,y) from
    this and setting that equal to N(x,y) yields h'(y) = 0

and thus h(y) is a constant. Hence an implicit solution of the D.E. is $e^x siny + 2ycosx = c$. The solution y = 0 is also valid since it satisfies the D.E. for all x.

9.   If you try to find $\psi(x,y)$ by integrating M(x,y) with respect to x you must integrate by parts. Instead find $\psi(x,y)$ by integrating N(x,y) with respect to y to obtain $\psi(x,y) = e^{xy}cos2x - 3y + g(x)$. Now find g(x) by differentiating $\psi(x,y)$ with respect to x and set that equal to M(x,y), which yields $g'(x) = 2x$ or $g(x) = x^2$.

12. As long as $x^2 + y^2 \neq 0$, we can simplify the equation by multiplying both sides by $(x^2 + y^2)^{3/2}$. This gives the exact equation $xdx + ydy = 0$. The solution to this equation is given implicitly by $x^2 + y^2 = c$. If you apply Theorem 2.6.1 and its construction without the simplification, you get $(x^2 + y^2)^{-1/2} = C$ which can be written as $x^2 + y^2 = c$ under the same assumption required for the simplification.

14. $M_y = 1$ and $N_x = 1$, so the D.E. is exact. Integrating M(x,y) with respect to x yields
$\psi(x,y) = 3x^3 + xy - x + h(y)$. Differentiating this with respect to y and setting $\psi_y(x,y) = N(x,y)$ yields
$h'(y) = -4y$ or $h(y) = -2y^2$. Thus the implicit solution is $3x^3 + xy - x - 2y^2 = c$. Setting x = 1 and y = 0 gives c = 2 so that $2y^2 - xy + (2+x-3x^3) = 0$ is the implicit solution satisfying the given I.C. Use the quadratic formula to find y(x), where the negative square root is used in order to satisfy the I.C. The solution will be valid for $24x^3 + x^2 - 8x - 16 > 0$.

15. We want $M_y(x,y) = 2xy + bx^2$ to be equal to $N_x(x,y) = 3x^2 + 2xy$. Thus we must have b = 3. This gives $\psi(x,y) = \frac{1}{2}x^2y^2 + x^3y + h(y)$ and consequently $h'(y) = 0$. After multiplying through by 2, the solution is given implicitly by $x^2y^2 + 2x^3y = c$.

19. $M_y(x,y) = 3x^2y^2$ and $N_x(x,y) = 1 + y^2$ so the equation is not exact by Theorem 2.6.1. Multiplying by the integrating factor $\mu(x,y) = 1/xy^3$ we get
$x + \frac{(1+y^2)}{y^3}y' = 0$, which is an exact equation since

$M_y = N_x = 0$ (it is also separable). In this case

$\psi = \dfrac{1}{2}x^2 + h(y)$ and $h'(y) = y^{-3} + y^{-1}$ so that

$x^2 - y^{-2} + 2\ln|y| = c$ gives the solution implicitly. Note that $y(x) = 0$ also satisfies the given D.E.

22. Multiplication of the given D.E. (which is not exact) by $\mu(x,y) = xe^x$ yields $(x^2 + 2x)e^x\sin y\, dx + x^2 e^x \cos y\, dy$, which is exact since $M_y(x,y) = N_x(x,y) = (x^2+2x)e^x\cos y$. To solve this exact equation it's easiest to integrate $N(x,y) = x^2 e^x \cos y$ with respect to y to get $\psi(x,y) = x^2 e^x \sin y + g(x)$. Finding ψ_x and setting that equal to $(x^2+2x)e^x\sin y$ yields $g'(x) = 0$.

23. This problem is similar to the derivation leading up to Eq.(26). Assuming that μ depends only on y, we find from Eq.(25) that $\mu' = Q\mu$, where $Q = (N_x - M_y)/M$ must depend on y alone. Solving this last D.E. yields $\mu(y)$ as given. This method provides an alternative approach to Problems 27 through 30.

25. The equation is not exact so we must attempt to find an integrating factor. Since $\dfrac{1}{N}(M_y-N_x) = \dfrac{3x^2 + 2x + 3y^2 - 2x}{x^2 + y^2} = 3$ is a function of x alone there is an integrating factor depending only on x, as shown in Eq.(26). Then $d\mu/dx = 3\mu$, and the integrating factor is $\mu(x) = e^{3x}$. Multiplying all terms in the given D.E. by e^{3x} will then yield an exact D.E.

26. An integrating factor can be found which is a function of x only, yielding $\mu(x) = e^{-x}$. Alternatively, you might recognize that $y' - y = e^{2x} - 1$ is a linear first order equation which can be solved as in Section 2.1.

27. Using the results of Prob. 23, it can be shown that $\mu(y) = y$ is an integrating factor. Thus multiplying the D.E. by y gives $y\,dx + (x - y\sin y)dy = 0$, which can be identified as an exact equation. Alternatively, one can rewrite the last equation as $(y\,dx + x\,dy) - y\sin y\, dy = 0$. The first term is $d(xy)$ and the last can be integrated by parts. Thus we have $xy + y\cos y - \sin y = c$.

29. Simplify the D.E. by multiplying by $\sin y$ to obtain

$e^x \sin y\, dx + e^x \cos y\, dy + 2y\, dy = 0$, which is exact. The first two terms are just $d(e^x \sin y)$ and thus, $e^x \sin y + y^2 = c$.

31. Using the results of Prob. 24, it can be shown that $\mu(xy) = xy$ is an integrating factor. Thus, multiplying by xy we have $(3x^2y + 6x)dx + (x^3 + 3y^2)dy = 0$, which can be identified as an exact equation. Alternatively, we can observe that the above equation can be written as $d(x^3y) + d(3x^2) + d(y^3) = 0$, so that $x^3y + 3x^2 + y^3 = c$.

Section 2.7, Page 107

1d. The exact solution to this I.V.P. is $y = \phi(t) = t + 2 - e^{-t}$.

3a. The Euler formula is $y_{n+1} = y_n + h(2y_n - t_n + 1/2)$ for $n = 0,1,2,3$ and with $t_0 = 0$ and $y_0 = 1$. Thus, for $h = .1$,
$y_1 = y_0 + .1(2y_0 - t_0 + 1/2) = 1.25$,
$y_2 = 1.25 + .1[2(1.25) - (.1) + 1/2] = 1.54$,
$y_3 = 1.54 + .1[2(1.54) - (.2) + 1/2] = 1.878$, and
$y_4 = 1.878 + .1[2(1.878) - (.3) + 1/2] = 2.2736$.

3b. Use the same formula as in Prob. 3a, except now $h = .05$ and $n = 0,1...7$. Notice that only results for $n = 1,3,5$ and 7 are needed to compare with part a.

3c. Again, use the same formula as above with $h = .025$ and $n = 0,1...15$. Notice that only results for $n = 3,7,11$ and 15 are needed to compare with parts a and b.

3d. $y' = 1/2 - t + 2y$ is a first order linear D.E. Rewrite the equation in the form $y' - 2y = 1/2 - t$ and multiply both sides by the integrating factor e^{-2t} to obtain $(e^{-2t}y)' = (1/2 - t)e^{-2t}$. Integrating the right side by parts and multiplying by e^{2t} we obtain $y = ce^{2t} + t/2$. The I.C. $y(0) = 1 \to c = 1$ and hence the solution of the I.V.P. is $y = \phi(x) = e^{2t} + t/2$. Thus $\phi(0.1) = 1.2714$, $\phi(0.2) = 1.59182$, $\phi(0.3) = 1.97212$, and $\phi(0.4) = 2.42554$.

4d. The exact solution to this I.V.P. is
$y = \phi(t) = (6\cos t + 3\sin t - 6e^{-2t})/5$.

6. For y(0) > 0 the
    solutions appear to
    converge to a number
    between 0 and 2. Note
    that y=0 is and
    equilibrium solution
    For y(0)<0 the
    solutions diverge.

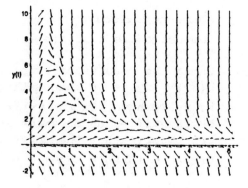

9. All solutions seem
    to diverge.

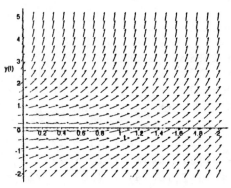

13a. The Euler formula is

$$y_{n+1} = y_n + h(\frac{4-t_n y_n}{1 + y_n^2}), \text{ where } t_0 = 0 \text{ and}$$

$y_0 = y(0) = -2$. Thus, for h = .1, we get

$y_1 = -2 + .1(4/5) = -1.92$

$$y_2 = -1.92 + .1(\frac{4-.1(-1.92)}{1 + (1.92)^2}) = -1.83055$$

$$y_3 = -1.83055 + .1(\frac{4-.2(-1.83055)}{1+(1.83055)^2}) = -1.7302$$

$$y_4 = -1.7302 + .1(\frac{4-.3(-1.7302)}{1 + (1.7302)^2}) = -1.617043$$

$$y_5 = -1.617043 + .1(\frac{4 - .4(-1.617043)}{1 + (1.617043)^2}) = -1.488494.$$

Thus, $y(.5) \cong -1.488494$.

15a. The Euler formula is

$$y_{n+1} = y_n + .1 (\frac{3t_n^2}{3y_n^2-4}), \text{ where } t_0 = 1 \text{ and } y_0 = 0. \text{ Thus}$$

$$y_1 = 0 + .1 (\frac{3}{-4}) = -.075 \text{ and}$$

$$y_2 = -.075 + .1 \left(\frac{3(1.1)^2}{3(-.075)^2 - 4} \right) = -.166134.$$

15c. There are two factors that explain the large differences. From the D.E., the slope of y, y', becomes very "large" for values of y near -1.155. Also, the slope changes sign at y = -1.155. Thus for part a, $y(1.7) \cong y_7 = -1.178$, which is close to -1.155 and the slope y' here is large and positive, creating the large change in $y_8 \cong y(1.8)$. For part b, $y(1.65) \cong -1.125$, resulting in a large negative slope, which yields $y(1.70) \cong -3.133$. The slope at this point is now positive and the remainder of the solutions "grow" to -3.098 for the approximation to y(1.8).

16. For the four step sizes given, the approximte values for y(.8) are 3.5078, 4.2013, 4.8004 and 5.3428. Thus, since these changes are still rather "large", it is hard to give an estimate other than y(.8) is at least 5.3428. By using h = .005, .0025 and .001, we find further approximate values of y(.8) to be 5.576, 5.707 and 5.790. Thus a better estimate now is for y(.8) to be between 5.8 and 6. No reliable estimate is obtainable for y(1), which is consistent with the direction field of Prob.9.

18. It is helpful, in understanding this problem, to also calculate $y'(t_n) = y_n(.1y_n^2 - t_n)$. For $\alpha = 2.38$ this term remains positive and grows very large for $t_n > 2$. On the other hand, for $\alpha = 2.37$ this term decreases and eventually becomes negative for $t_n \cong 1.6$ (for h = .01). For $\alpha = 2.37$ and h = .1, .05 and .01, y(2.00) has the approximations of 4.48, 4.01 and 3.50 respectively. A small step size must be used, due to the sensitivety of the slope field, given by $y_n(.1y_n^2 - t_n)$.

22. Using Eq.(8) we have $y_{n+1} = y_n + h(2y_n - 1) = (1+2h)y_n - h$. Setting n + 1 = k (and hence n = k-1) this becomes $y_k = (1 + 2h)y_{k-1} - h$, for k = 1,2,... . Since $y_0 = 1$, we have $y_1 = 1 + 2h - h = 1 + h = (1 + 2h)/2 + 1/2$, and hence $y_2 = (1 + 2h)y_1 - h = (1 + 2h)^2/2 + (1 + 2h)/2 - h$
    $= (1 + 2h)^2/2 + 1/2;$
$y_3 = (1 + 2h)y_2 - h = (1 + 2h)^3/2 + (1 + 2h)/2 - h$
    $= (1 + 2h)^3/2 + 1/2$. Continuing in this fashion (or using induction) we obtain $y_k = (1 + 2h)^k/2 + 1/2$. For fixed x > 0 choose h = x/k. Then substitute for h in the last formula to obtain $y_k = (1 + 2x/k)^k/2 + 1/2$. Letting

$k \to \infty$ we find (See hint for Problem 20d.)
$y(x) = y_k \to e^{2x}/2 + 1/2$, which is the exact solution.

Section 2.8, Page 117

1. Let $s = t-1$ and $w(s) = y(t(s)) - 2$, then when $t = 1$ and
    $y = 2$ we have $s = 0$ and $w(0) = 0$. Also,
    $$\frac{dw}{ds} = \frac{dw}{dt} \cdot \frac{dt}{ds} = \frac{d}{dt}(y-2)\frac{dt}{ds} = \frac{dy}{dt} \text{ (since } t = s+1) \text{ and hence}$$
    $$\frac{dw}{ds} = (s+1)^2 + (w+2)^2, \text{ upon substitution into the given}$$
    D.E.

4a. Following Ex. 1 of the text, from Eq.(7) we have
    $$\phi_{n+1}(t) = \int_0^t f(s,\phi(s))ds, \text{ where } f(t,\phi) = -1 - \phi. \quad \text{Thus if}$$
    $$\phi_0(t) = 0, \text{ then } \phi_1(t) = -\int_0^t ds = -t;$$
    $$\phi_2(t) = -\int_0^t (1-s)ds = -t + \frac{t^2}{2};$$
    $$\phi_3(t) = -\int_0^t (1-s + \frac{s^2}{2})ds = -t + \frac{t^2}{2} - \frac{t^3}{2\cdot3};$$
    $$\phi_4(t) = -\int_0^t (1 - s + \frac{s^2}{2} - \frac{s^3}{3!})ds = -t + \frac{t^2}{2} - \frac{t^3}{3!} + \frac{t^4}{4!}.$$
    Based upon these we hypothesize that $\phi_n(t) = \sum_{k=1}^{n} \frac{(-1)^k t^k}{k!}$

    and use mathematical induction to verify this form for
    $\phi_n(t)$. First, let $n = 1$, then $\phi_1(t) = -t$, so it is
    certainly true for $n = 1$. Then, using Eq.(7) again we
    have:
    $$\phi_{n+1}(t) = -\int_0^t [1 + \phi_n(s)]\, ds = -t - \sum_{k=1}^{n} \frac{(-1)^k t^{k+1}}{(k+1)!}$$
    $$= \sum_{k=0}^{n} \frac{(-1)^{k+1} t^{k+1}}{(k+1)!} = \sum_{i=1}^{n+1} \frac{(-1)^i t^i}{i!}, \text{ where } i = k+1. \quad \text{Since this}$$
    is the same form for $\phi_{n+1}(t)$ as derived from $\phi_n(t)$ above,
    we have verified by mathematical induction that $\phi_n(t)$ is
    as given.

4c. From part a, let $\phi(t) = \lim\limits_{n \to \infty} \phi_n(t) = \sum\limits_{k=1}^{\infty} \dfrac{(-1)^k t^k}{k!}$

$$= -t + \frac{t^2}{2} - \frac{t^3}{3!} + \ldots .$$

Since this is a power series, recall from calculus that:

$e^{at} = \sum\limits_{k=0}^{\infty} \dfrac{a^k t^k}{k!} = 1 + at + \dfrac{a^2 t^2}{2} + \dfrac{a^3 t^3}{3!} + \ldots .$ If we let

$a = -1$, then we have $e^{-t} = 1 - t + \dfrac{t^2}{2} - \dfrac{t^3}{3!} + \ldots = 1 + \phi(t)$.

Hence $\phi(t) = e^{-t} - 1$.

4b.

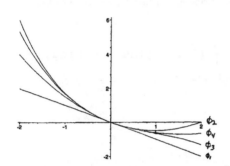

4d.

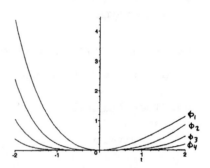

From the plot in 4d, it appears that ϕ_4 is a very good
estimate for $|t| < 1$.

7a. As in Prob.4,

$$\phi_1(t) = \int_0^t (s\phi_0(s) + 1)ds = s\Big|_0^t = t$$

$$\phi_2(t) = \int_0^t (s^2 + 1)ds = \left(\frac{s^3}{3} + s\right)\Big|_0^t = t + \frac{t^3}{3}$$

$$\phi_3(t) = \int_0^t \left(s^2 + \frac{s^4}{3} + 1\right)ds = \left(\frac{s^3}{3} + \frac{s^5}{3\cdot5} + s\right)\Big|_0^t = t + \frac{t^3}{3} + \frac{t^5}{3\cdot5}.$$

Based upon these we hypothesize that:

$$\phi_n(t) = \sum_{k=1}^{n} \frac{t^{2k-1}}{1\cdot3\cdot5\cdots(2k-1)} \quad \text{and use mathematical induction}$$

to verify this form for $\phi_n(t)$, which is clearly true for
n = 1. Using Eq.(7) again we have:

$$\phi_{n+1}(t) = \int_0^t \left(\sum_{k=1}^{n} \frac{s^{2k}}{1\cdot3\cdot5\cdots(2k-1)} + 1 \right)ds$$

$$= \sum_{k=1}^{n} \frac{t^{2k+1}}{1 \cdot 3 \cdot 5 \cdots (2k+1)} + t$$

$$= \sum_{k=0}^{n} \frac{t^{2k+1}}{1 \cdot 3 \cdot 5 \cdots (2k+1)}$$

$$= \sum_{i=1}^{n+1} \frac{t^{2i-1}}{1 \cdot 3 \cdot 5 \cdots (2i-1)}, \text{ where } i = k+1. \text{ Since this is}$$

the same form for $\phi_{n+1}(t)$ as derived from $\phi_n(t)$ above, we have verified by mathematical induction that $\phi_n(t)$ is as given.

7b. Your plot should show that the estimates appear to be converging.

10a. From Eq.(7) we have $\phi_{n+1} = \int_0^t [1-\phi_n^3(s)]ds$. Setting $\phi_0(t) = 0$ yields the desired iterates.

10b. The iterates
     appear
     to diverge.

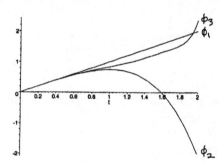

11. First, recall that $\sin x = x - \frac{x^3}{3!} + \frac{x^5}{5!} + O(x^7)$. Now, for this problem $\phi_1(t) \int_0^t [1-\sin\phi_0(s)]ds = t$ and hence

$$\phi_2(t) = \int_0^t [1-\sin(s)]ds = \int_0^t [1 - (s - \frac{s^3}{3!} + \frac{s^5}{5!} - O(s^7)]ds$$

$$= t - \frac{t^2}{2!} + \frac{t^4}{4!} - \frac{t^6}{6!} + O(t^8). \text{ For } \phi_3 \text{ we need to}$$

find $\sin[\phi_2(t)]$, which is given by

$$\sin[\phi_2(t)] = \phi_2(t) - \phi_2^3(t)/3! + \phi_2^5/5! + O(t^7)$$

$$= (t - \frac{t^2}{2!} + \frac{t^4}{4!} - \frac{t^6}{6!}) - \frac{(t - \frac{t^2}{2!})^3}{3!} + \frac{t^5}{5!} + O(t^7),$$

where we have retained only the terms less than $O(t^7)$.
Now use this in $\phi_3(t) = \int_0^t [1-\sin(\phi_2(s))]ds$, which gives
the desired answer up to $O(t^8)$.

Section 2.9, Page 129

2. Using the given difference equation we have for n=0,
    $y_1 = y_0/2$; for n=1, $y_2 = 2y_1/3 = y_0/3$; and for n=2,
    $y_3 = 3y_2/4 = y_0/4$. Thus we guess that $y_n = y_0/(n+1)$, and

    the given equation then gives $y_{n+1} = \dfrac{n+1}{n+2}y_n = y_0/(n+2)$,

    which, by mathematical induction, verifies $y_n = y_0/(n+1)$
    as the solution for all n.

5. From the given equation we have $y_1 = .5y_0+6$.

    $y_2 = .5y_1 + 6 = (.5)^2 y_0 + 6(1 + \dfrac{1}{2})$ and

    $y_3 = .5y_2 + 6 = (.5)^3 y_0 + 6(1 + \dfrac{1}{2} + \dfrac{1}{4})$. In general, then

    $y_n = (.5)^n y_0 + 6(1 + \dfrac{1}{2} + \cdots + \dfrac{1}{2^{n-1}})$

    $\quad = (.5)^n y_0 + 6(\dfrac{1 - (1/2)^n}{1 - 1/2})$

    $\quad = (.5)^n y_0 + 12 - (.5)^n 12$

    $\quad = (.5)^n(y_0-12) + 12$. Mathematical induction can now be
    used to prove that this is the correct solution.

10. As in Ex.(1), the governing equation is $y_{n+1} = \rho y_n - b$,

    which has the solution $y_n = \rho^n y_0 - \dfrac{1-\rho^n}{1-\rho}b$ (Eq.(14) with a

    negative b). Setting $y_{360} = 0$ and solving for b we
    obtain

    $b = \dfrac{(1-\rho)\rho^{360}y_0}{1-\rho^{360}}$, where $\rho = 1.0075$ for part a.

13. You must solve Eq.(14) numerically for ρ when n = 240,
    $y_{240} = 0$, b = -$900 and y_0 = $95,000.

14. Substituting Eq.(25), $u_n = \dfrac{\rho-1}{\rho} + v_n$, into Eq.(21) we get

$$\frac{\rho-1}{\rho} + v_{n+1} = \rho(\frac{\rho-1}{\rho} + v_n)(1 - \frac{\rho-1}{\rho} - v_n) \text{ or}$$

$$v_{n+1} = -\frac{\rho-1}{\rho} + (\rho-1 + \rho v_n)(\frac{1}{\rho} - v_n)$$

$$= \frac{1-\rho}{\rho} + \frac{\rho-1}{\rho} - (\rho-1)v_n + v_n - \rho v_n^2 = (2-\rho)v_n - \rho v_n^2$$

15a. For $u_0 = .2$ we have $u_1 = 3.2u_0(1-u_0) = .512$ and
$u_2 = 3.2u_1(1-u_1) = .7995392$. Likewise $u_3 = .51288406$,
$u_4 = .7994688$, $u_5 = .51301899$, $u_6 = .7994576$ and
$u_7 = .5130404$. Continuing in this fashion,
$u_{14} = u_{16} = .79945549$ and $u_{15} = u_{17} = .51304451$.

17. For both parts of this problem a computer spreadsheet
     was used and an initial value of $u_0 = .2$ was chosen.
     Different initial values or different computer programs
     may need a slightly different number of iterations to
     reach the limiting value.

17a. The limiting value of .65517 (to 5 decimal places) is
     reached after approximately 100 iterations for $\rho = 2.9$.
     The limiting value of .66102 (to 5 decimal places) is
     reached after approximately 200 iterations for $\rho = 2.95$.
     The limiting value of .66555 (to 5 decimal places) is
     reached after approximately 910 iterations for $\rho = 2.99$.

17b. The solution oscillates between .63285 and .69938 after
     approximately 400 iterations for $\rho = 3.01$. The solution
     oscillates between .59016 and .73770 after approximately
     130 iterations for $\rho = 3.05$. The solution oscillates
     between .55801 and .76457 after approximately 30
     iterations for $\rho = 3.1$. For each of these cases
     additional iterations verified the oscillations were
     correct to five decimal places.

18. For an initial value of .2 and $\rho = 3.448$ we have the
     solution oscillating between .4403086 and .8497146.
     After approximately 3570 iterations the eighth decimal
     place is still not fixed, though. For the same initial
     value and $\rho = 3.45$ the solution oscillates between the
     four values: .43399155, .84746795, .44596778 and
     .85242779 after 3700 iterations.. For $\rho = 3.449$, the
     solution is still varying in the fourth decimal place
     after 3570 iterations, but there appear to be four
     values.

Miscellaneous Problems, Page 131

Before trying to find the solution of a D.E. it is necessary
to know its type. The student should first classify the D.E.
before reading this section, which indentifies the type of
each equation in Problems 1 through 32.

    1. Linear   3. Exact

    2. Homgeneous (See Sect. 2.2, Prob.30)

    4. Rewrite the D.E. as $\dfrac{dx}{dy} - x = e^y$, which is linear
        in $x(y)$

    5. Exact   6. Linear

    7. Letting $u = x^2$ yields $\dfrac{dy}{dx} = 2x\dfrac{dy}{du}$ and thus
        $\dfrac{du}{dy} - 2yu = 2y^3$ which is linear in $u(y)$.

    8. Linear   9. Exact

  10. Integrating factor   11. Exact
      depends on x only

  12. Linear   13. Homogeneous

  14. Exact or homogeneous 15. Separable

  16. Homogeneous   17. Linear

  18. Linear or homogeneous 19. Integrating factor,
                               depends on x only

  20. Separable   21. Homogeneous

  22. Separable   24. Separable

  23. From Prob.23, Sect. 2.4, this is is a Bernoulli
      equation with $n = 2$. Let $v(x) = y$, then the linear
      equation $xv' - v = -e^{2x}$ is obtained.

  25. Exact. Your calculations might be simplified if you
      write the D.E. as $\dfrac{2xdx}{y} - \dfrac{x^2dy}{y^2} + \dfrac{-ydx}{x^2+y^2} + \dfrac{xdx}{x^2+y^2} = 0$

and show that the first two terms are exact and the last two terms are exact. Integrating each pair is then straight forward.

26. Integrating factor, depends on x only   27. Integrating factor, depends on x only

28. Exact   29. Homogeneous

30. Linear equation in x(y) 31. Separable

32. Integrating factor, depends on y only.

34b. For $y_1(t) = 1/t$ $y_1' = -1/t^2$ and substitution into the D.E. shows that $y_1(t)$ is indeed a solution. Comparing the D.E. with the equation in Prob.33 we see that $q_1(t) = -1/t^2$, $q_2(t) = -1/t$ and $q_3(t) = 1$. Hence, using the method suggested in Prob.33, we set $y = y_1(t) + \dfrac{1}{v(t)}$ in the D.E. to obtain $\dfrac{dv}{dt} = -(-1/t + 2y_1)v - 1$, or $\dfrac{dv}{dt} + \dfrac{1}{t}v = -1$. Thus $v(t) = \dfrac{c-t^2}{2t}$ and hence y $y_2(t) = \dfrac{1}{t} + \dfrac{2t}{c-t^2}$.

36. Let $v = y'$, then $v' = y''$ and thus the D.E. becomes $t^2 v' + 2tv - 1 = 0$ or $t^2 v' + 2tv = 1$. The left side is recognized as $(t^2 v)'$ and thus we may integrate to obtain $t^2 v = t + c$ (otherwise, divide both sides of the D.E. by t^2 and find the integrating factor, which is just t^2 in this case). Solving for $v = dy/dt$ we find $dy/dt = 1/t + c/t^2$ so that $y = \ln t + c_1/t + c_2$.

38. If $v = y'$, the D.E. becomes $v' + tv^2 = 0$. This equation is separable and has the solution $-v^{-1} + t^2/2 = c$ or $v = y' = -2/(c_1 - t^2)$ where $c_1 = 2c$. We must consider separately the cases $c_1 = 0$, $c_1 > 0$ and $c_1 < 0$. If $c_1 = 0$, then $y' = 2/t^2$ or $y = -2/t + c_2$. If $c_1 > 0$, let $c_1 = k^2$. Then $y' = -2/(k^2-t^2) = -(1/k)[1/(k-t) + 1/(k+t)]$, so that $y = (1/k)\ln|(k-t)/(k+t)|+c_2$. If $c_1 < 0$, let $c_1 = -k^2$.

Then $y' = 2/(k^2 + t^2)$ so that $y = (2/k)\tan^{-1}(t/k) + c_2$. Finally, we note that y = constant is also a solution of the D.E.

42. Following the procedure outlined, let $v = dy/dt$, then $y'' = \dfrac{dv}{dt} = v\dfrac{dv}{dy}$. Thus the D.E. becomes $yv\dfrac{dv}{dy} + v^2 = 0$, which is a separable equation with the solution $v = c/y$. Since $v = \dfrac{dy}{dt} = \dfrac{c}{y}$, we conclude that $y^2 = c_1 t + c_2$, by separating variables and integrating.

45. Again let $v = y'$ and $v' = v\,dv/dy$ to obtain $2y^2 v\dfrac{dv}{dy} + 2yv^2 = 1$, or $2y^2 v\,dv + 2yv^2\,dy = dy$. The left side can be written as $d(y^2 v^2)$ and thus integration gives $y^2 v^2 = y + c_1$ and thus $v = \pm y^{-1}(y + c_1)^{1/2}$. Setting $v = y'$ and separating variables gives $\pm y\,dy/(y+c_1)^{1/2} = dt$. On observing that the left side of the equation can be written as $\pm[(y+c_1) - c_1]dy/(y+c_1)^{1/2}$ we integrate and find $\pm (2/3)(y-2c_1)(y+c_1)^{1/2} = t + c_2$.

47. If $v = y'$, then $v' = v\,dv/dy$ and the D.E. becomes $v\,dv/dy + v^2 = 2e^{-y}$. Dividing by v we obtain $dv/dy + v = 2v^{-1}e^{-y}$, which is a Bernoulli equation with $n = -1$ (see Prob.27, Sect. 2.4). Let $w(y) = v^2$, then $dw/dy = 2v\,dv/dy$ and the D.E. then becomes $dw/dy + 2w = 4e^{-y}$, which is linear in w. Its solution is $w = v^2 = ce^{-2y} + 4e^{-y}$. Setting $v = dy/dt$ and separating variables gives $\dfrac{dy}{\pm\sqrt{4e^{-y}+ce^{-2y}}} = \dfrac{e^y dy}{\pm\sqrt{4e^y+c}} = dt$. Integrating and solving for e^y yields $e^y = (t+c_2)^2 + c_1$.

48. Since both t and y are missing, either approach used above will work. In this case it's easier to use the approach of Problems 36-41, so let $v = y'$ and thus $v' = y''$ and the D.E. becomes $v\,dv/dt = 2$.

51. The variable y is missing. Let $v = y'$, then $v' = y''$ and the D.E. becomes $vv' - t = 0$. The solution of this separable equation is $v^2 = t^2 + c_1$. Substituting $v = y'$ and applying the I.C. $y'(1) = 1$, we obtain $y' = t$. The

positive square root was chosen because $y' > 0$ at $t = 1$.
Solving this last equation and applying the I.C. $y(1) = 2$,
we obtain $y = t^2/2 + 3/2$.

CHAPTER 3

Section 3.1, Page 142

3.   Assume $y = e^{rt}$, which gives $y' = re^{rt}$ and $y'' = r^2 e^{rt}$. Substitution into the D.E. yields $(6r^2 - r - 1)e^{rt} = 0$. Since $e^{rt} \neq 0$, we have the characteristic equation $6r^2 - r - 1 = 0$, or $(3r+1)(2r-1) = 0$. Thus $r = -1/3, 1/2$ and $y = c_1 e^{t/2} + c_2 e^{-t/3}$.

5.   The characteristic equation is $r^2 + 5r = 0$, so the roots are $r_1 = 0$, and $r_2 = -5$. Thus
$$y = c_1 e^{0t} + c_2 e^{-5t} = c_1 + c_2 e^{-5t}.$$

7.   The characteristic equation is $r^2 - 9r + 9 = 0$ so that $r = (9 \pm \sqrt{81-36})/2 = (9 \pm 3\sqrt{5})/2$ using the quadratic formula. Hence
$$y = c_1 \exp[(9+3\sqrt{5})t/2] + c_2 \exp[(9-3\sqrt{5})t/2].$$

10.   Substituting $y = e^{rt}$ in the D.E. we obtain the characteristic equation $r^2 + 4r + 3 = 0$, which has the roots $r_1 = -1$, $r_2 = -3$. Thus $y = c_1 e^{-t} + c_2 e^{-3t}$ and $y' = -c_1 e^{-t} - 3c_2 e^{-3t}$.

Substituting $t = 0$ we then have $c_1 + c_2 = 2$ and $-c_1 - 3c_2 = -1$, yielding $c_1 = 5/2$ and $c_2 = -1/2$. Thus $y = \frac{5}{2}e^{-t} - \frac{1}{2}e^{-3t}$ and hence $y \to 0$ as $t \to \infty$.

15.   The characteristic equation is $r^2 + 8r - 9 = 0$, so that $r_1 = 1$ and $r_2 = -9$ and the general solution is $y = c_1 e^t + c_2 e^{-9t}$. Since the I.C. are given at $t = 1$, it is convenient to write the general solution in the form $y = k_1 e^{(t-1)} + k_2 e^{-9(t-1)}$. Note that $c_1 = k_1 e^{-1}$ and $c_2 = k_2 e^9$. The advantage of the latter form of the general solution becomes clear when we apply the I.C. $y(1) = 1$ and $y'(1) = 0$. This latter form of y

gives $y' = k_1 e^{(t-1)} - 9k_2 e^{-9(t-1)}$ and thus setting $t = 1$ in y and y' yields the equations $k_1 + k_2 = 1$ and $k_1 - 9k_2 = 0$. Solving for k_1 and k_2 we find that $y = (9e^{(t-1)} + e^{-9(t-1)})/10$. Since $e^{(t-1)}$ has a positive exponent for $t > 1$, $y \to \infty$ as $t \to \infty$.

17. Comparing the given solution to Eq(17), we see that $r_1 = 2$ and $r_2 = -3$ are the two roots of the characteristic equation. Thus we have $(r-2)(r+3) = 0$, or $r^2 + r - 6 = 0$ as the characteristic equation. Hence the given solution is for the D.E. $y'' + y' - 6y = 0$.

19. The roots of the characteristic equation are $r = 1, -1$ and thus the general solution is $y(t) = c_1 e^t + c_2 e^{-t}$. $y(0) = c_1 + c_2 = \frac{5}{4}$ and $y'(0) = c_1 - c_2 = -\frac{3}{4}$, yielding $y(t) = \frac{1}{4}e^t + e^{-t}$. From this $y'(t) = \frac{1}{4}e^t - e^{-t} = 0$ or $e^{2t} = 4$ or $t = \ln 2$. Since $y''(t) = y(t)$ is positive at $t = \ln 2$, this is a minimum point. Note that $y(t) \to \infty$ as $t \to \infty$.

21. The general solution is $y = c_1 e^{-t} + c_2 e^{2t}$. Using the I.C. we obtain $c_1 + c_2 = \alpha$ and $-c_1 + 2c_2 = 2$, so adding the two equations we find $3c_2 = \alpha + 2$. If y is to approach zero as $t \to \infty$, c_2 must be zero. Thus $\alpha = -2$.

24. The roots of the characteristic equation are given by $r = -2, \alpha - 1$ and thus $y(t) = c_1 e^{-2t} + c_2 e^{(\alpha-1)t}$. Hence, for $\alpha < 1$, all solutions tend to zero as $t \to \infty$. For $\alpha > 1$, the second term becomes unbounded, but not the first, so there are no values of α for which all solutions become unbounded.

25a. The characteristic equation is $2r^2 + 3r - 2 = 0$, so $r_1 = -2$ and $r_2 = 1/2$ and $y = c_1 e^{-2t} + c_2 e^{t/2}$. The I.C. yield $c_1 + c_2 = 1$ and $-2c_1 + \frac{1}{2}c_2 = -\beta$ so that $c_1 = (1 + 2\beta)/5$ and $c_2 = (4-2\beta)/5$.

25b. Setting $\beta = 1$ and differentiating we obtain

$y' = (-6e^{-2t} + e^{t/2})/5$. Setting this equal to zero and solving for t yields $t = \dfrac{2}{5}\ln 6$.

25c. From part (a), if $\beta = 2$ then $y(t) = e^{-2t}$ and the solution simply decays to zero. For $\beta > 2$, the solution becomes unbounded negatively, and again there is no minimum point. For $0 < \beta < 2$ there is always a minimum point, as found in part (b).

28a. The roots of the characteristic equation are given by $r = \dfrac{-b \pm \sqrt{b^2 - 4ac}}{2a}$. For the roots to be real and different we must have $b^2 - 4ac > 0$. If they are to be negative then we must have $b > 0$ (since we are given $a > 0$) and $c > 0$. This latter condition comes from the fact that if $c \le 0$ then $\sqrt{b^2 - 4ac} \ge b$ and hence the numerator of r would give both positive and negative values, or a zero if $c = 0$.

Section 3.2, Page 151

2. $W(\cos t, \sin t) = \begin{vmatrix} \cos t & \sin t \\ -\sin t & \cos t \end{vmatrix} = \cos^2 t + \sin^2 t = 1.$

4. $W(x, xe^x) = \begin{vmatrix} x & xe^x \\ 1 & e^x + xe^x \end{vmatrix} = xe^x + x^2 e^x - xe^x = x^2 e^x.$

8. Dividing by $(t-1)$ we have $p(t) = -3t/(t-1)$, $q(t) = 4/(t-1)$ and $g(t) = \sin t/(t-1)$, so the only point of discontinuity is $t = 1$. By Theorem 3.2.1, the largest interval is $-\infty < t < 1$, since the initial point is $t_0 = -2$.

12. $p(x) = 1/(x-2)$, $q(x) = \tan x$ and $g(x) = 0$, so $x = \pi/2$, 2, $3\pi/2$, $5\pi/2$ ... are points of discontinuity. Since $t_0 = 3$, the interval specified by Theorem 3.2.1 is $2 < x < 3\pi/2$.

14. For $y = t^{1/2}$, $y' = \dfrac{1}{2}t^{-1/2}$ and $y'' = -\dfrac{1}{4}t^{-3/2}$. Thus $yy'' + (y')^2 = -\dfrac{1}{4}t^{-1} + \dfrac{1}{4}t^{-1} = 0$. $y = 1$ is also a solution since $y' = y'' = 0$. If $y = c_1(1) + c_2 t^{1/2}$ is substituted in the D.E. you will get

$(c_1+c_2t^{1/2})(-\dfrac{c_2}{4}t^{-3/2}) + (\dfrac{c_2}{2}t^{-1/2})^2 = -\dfrac{c_1c_2}{4}t^{-3/2}$, which is
zero only if $c_1 = 0$ or $c_2 = 0$. Thus the linear
combination of two solutions is not, in general, a
solution. Theorem 3.2.2 is not contradicted however,
since the D.E. is not linear.

15. $y = \phi(t)$ is a solution of the D.E. so $L[\phi](t) = g(t)$.
Since L is a linear operator,
$L[c\phi](t) = cL[\phi](t) = cg(t)$. But, since $g(t) \neq 0$,
$cg(t) = g(t)$ if and only if $c = 1$. This is not a
contradiction of Theorem 3.2.2 since the linear D.E. is
not homogeneous.

18. $W(f,g) = W(t,g) = \begin{vmatrix} t & g \\ 1 & g' \end{vmatrix} = tg' - g = t^2e^t$, or $g' - \dfrac{1}{t}g = te^t$.

This has an integrating factor of $\dfrac{1}{t}$ and thus $\dfrac{1}{t}g' - \dfrac{1}{t^2}g$

$= e^t$ or $(\dfrac{1}{t}g)' = e^t$. Integrating and multiplying by t we

obtain $g(t) = te^t + ct$.

21. From Section 3.1, e^t and e^{-2t} are two solutions, and

since $W(e^t,e^{-2t}) = \begin{vmatrix} e^t & e^{-2t} \\ e^t & -2e^{-2t} \end{vmatrix} = -3e^{-t} \neq 0$ they form a

fundamental set of solutions. To find the fundamental
set specified by Theorem 3.2.5, let $y(t) = c_1e^t + c_2e^{-2t}$,
where c_1 and c_2 satisfy
$c_1 + c_2 = 1$ and $c_1 - 2c_2 = 0$ for y_1. Solving, we find
$y_1 = \dfrac{2}{3}e^t + \dfrac{1}{3}e^{-2t}$. Likewise, c_1 and c_2 satisfy
$c_1 + c_2 = 0$ and $c_1 - 2c_2 = 1$ for y_2, so that
$y_2 = \dfrac{1}{3}e^t - \dfrac{1}{3}e^{-2t}$.

25. For $y_1 = x$, we have $x^2(0) - x(x+2)(1) + (x+2)(x) = 0$ and
for $y_2 = xe^x$ we have $x^2(x+2)e^x - x(x+2)(x+1)e^x + (x+2)xe^x = 0$.
From Prob. 4, $W(x,xe^x) = x^2e^x \neq 0$ for $x > 0$, so y_1 and y_2
form a fundamental set of solutions for $x > 0$.

27a. From Sect. 3.1 the characteristic eq. is
$r^2 - r - 2 = (r-2)(r+1) = 0$ and thus e^{-t} and e^{2t} are

solutions. Since $\begin{vmatrix} e^{-t} & e^{2t} \\ -e^{-t} & 2e^{2t} \end{vmatrix} = 3e^t \neq 0$, e^{-t} and e^{2t} are

fundamental solutions.

27b. For particular choices of c_1 and c_2, all three are
solutions by Theorem 3.2.2.

27c. $W(y_2, y_3) = \begin{vmatrix} e^{2t} & -2e^{2t} \\ 2e^{2t} & -4e^{2t} \end{vmatrix} = -4e^{4t} + 4e^{4t} = 0$ and thus y_2 and

y_3 do not form a fundamental set of solutions. Similar
calculations show that $[y_1, y_3]$ and $[y_1, y_4]$ are
fundamental solutions but y_4 and y_5 do not form a
fundamental set of solutions.

28. Suppose that
$P(x)y'' + Q(x)y' + R(x)y = [P(x)y']' + [f(x)y]'$. On
expanding the right side and equating coefficients, we
find $f'(x) = R(x)$ and $P'(x) + f(x) = Q(x)$. These two
conditions on f can be satisfied if
$R(x) = Q'(x) - P''(x)$ which gives the necessary condition
$P''(x) - Q'(x) + R(x) = 0$.

31. We have $P(x) = x$, $Q(x) = -\cos x$, and $R(x) = \sin x$ and the
condition for exactness from Prob. 28 is satisfied. Also,
from Prob. 28, $f(x) = Q(x) - P'(x) = -\cos x - 1$, so the D.E.
becomes $(xy')' - [(1 + \cos x)y]' = 0$. Hence
$xy' - (1 + \cos x)y = c_1$. This is a first order linear
D.E. and the integrating factor (after dividing by x) is
$\mu(x) = \exp[-\int x^{-1}(1 + \cos x)dx]$. The general solution is
$$y = [\mu(x)]^{-1}[c_1\int_{x_0}^{x} t^{-1} \mu(t)dt + c_2].$$

33. We want to choose $\mu(x)$ and $f(x)$ so that $\mu(x)P(x)y'' +$
$\mu(x)Q(x)y' + \mu(x)R(x)y = [\mu(x)P(x)y']' + [f(x)y]'$.
Expand the right side and equate coefficients of y'', y'
and y. This gives $\mu'(x)P(x) + \mu(x)P'(x) + f(x) =$
$\mu(x)Q(x)$ and $f'(x) = \mu(x)R(x)$. Differentiate the first
equation and then eliminate $f'(x)$ to obtain the adjoint
equation $P\mu'' + (2P' - Q)\mu' + (P'' - Q' + R)\mu = 0$.

35. $P = 1-x^2$, $Q = -2x$ and $R = \alpha(\alpha+1)$. Thus
$2P' - Q = -4x + 2x = -2x$ and

$P'' - Q' + R = -2 + 2 + \alpha(\alpha+1) = \alpha(\alpha+1)$, and thus, from Prob. 33, $(1-x^2)\mu'' - 2x\mu' + \alpha(\alpha+1)\mu = 0$ is the adjoint of the given D.E.

37. Write the adjoint D.E. given in Problem 32 as $\hat{P}\mu'' + \hat{Q}\mu' + \hat{R}\mu = 0$ where $\hat{P} = P$, $\hat{Q} = 2P' - Q$, and $\hat{R} = P'' - Q' + R$. The adjoint of this equation, namely the adjoint of the adjoint, is $\hat{P}y'' + (2\hat{P}' - \hat{Q})y' + (\hat{P}'' - \hat{Q}' + \hat{R})y = 0$. After substituting for $\hat{P}$, $\hat{Q}$, and $\hat{R}$ and simplifying, we obtain $Py'' + Qy' + Ry = 0$. This is the same as the original equation.

38. From Prob. 33 the adjoint of $Py'' + Qy' + Ry = 0$ is $P\mu'' + (2P' - Q)\mu' + (P'' - Q' + R)\mu = 0$. The two equations are the same if $2P' - Q = Q$ and $P'' - Q' + R = R$. This will be true if $P' = Q$. Hence the original D.E. is self-adjoint if $P' = Q$. For Prob. 34, $P(x) = x^2$ so $P'(x) = 2x$ and $Q(x) = x$. Hence the Bessel equation of order ν is not self-adjoint. In a similar manner we find that Problems 35 and 36 are self-adjoint.

Section 3.3, Page 158

2. Since $\cos2\theta = 2\cos^2\theta - 1 = 1 - 2\sin^2\theta$ we have $f(\theta) + g(\theta) = -1 + 1 = 0$ for all θ. Thus $f(\theta)$ and $g(\theta)$ are linearly dependent.

6. $W(t, t^{-1}) = \begin{vmatrix} t & t^{-1} \\ 1 & -t^{-2} \end{vmatrix} = -\dfrac{2}{t} \neq 0.$

7. For $t > 0$ $g(t) = t$ and hence $f(t) - 3g(t) = 0$ for all t. Therefore f and g are linearly dependent on $0 < t$. For $t < 0$ $g(t) = -t$ and $f(t) + 3g(t) = 0$, so again f and g are linearly dependent on $t < 0$. For any interval that includes the origin, such as $-1 < t < 2$, there is no c for which $f(t) + cg(t) = 0$ for all t, and hence f and g are linearly independent on this interval.

12. The D.E. is linear and homogeneous. Hence, if y_1 and y_2 are solutions, then $y_3 = y_1 + y_2$ and $y_4 = y_1 - y_2$ are solutions. Now

$W(y_3, y_4) = y_3 y_4' - y_3' y_4$

$= (y_1 + y_2)(y_1' - y_2') - (y_1' + y_2')(y_1 - y_2) = -2(y_1 y_2' - y_1' y_2) = -2W(y_1, y_2)$

which is not zero since y_1 and y_2 are linearly independent solutions. Hence y_3 and y_4 form a fundamental set of solutions. Conversely, solving the first two equations for y_1 and y_2, we have

$y_1 = (y_3 + y_4)/2$ and $y_2 = (y_3 - y_4)/2$, so y_1 and y_2 are solutions. Finally, from above we have

$W(y_1, y_2) = -W(y_3, y_4)/2$.

15. Writing the D.E. in the form of Eq.(7), we have
$p(t) = -(t+2)/t$. Thus Eq.(8) yields

$W(t) = c \exp[-\int \dfrac{-(t+2)}{t}\ dt] = ct^2 e^t$.

20. From Eq.(8) we have $W(y_1, y_2) = c \exp[-\int p(t)dt]$, where

$p(t) = 2/t$ from the D.E. Thus $W(y_1, y_2) = c/t^2$. Since $W(y_1, y_2)(1) = 2$ we find $c = 2$ and thus $W(y_1, y_2)(5) = 2/25$.

24. Let c be the point in I at which both y_1 and y_2 vanish. Then $W(y_1, y_2)(c) = y_1(c)y_2'(c) - y_1'(c)y_2(c) = 0$. Hence, by Theorem 3.3.3 the functions y_1 and y_2 cannot form a fundamental set.

26. Suppose that y_1 and y_2 have a point of inflection at t_0 and either $p(t_0) \neq 0$ or $q(t_0) \neq 0$. Since $y_1''(t_0) = 0$ and $y_2''(t_0) = 0$ it follows from the D.E. that

$p(t_0)y_1'(t_0) + q(t_0)y_1(t_0) = 0$ and

$p(t_0)y_2'(t_0) + q(t_0)y_2(t_0) = 0$. If $p(t_0) = 0$ and $q(t_0) \neq 0$ then $y_1(t_0) = y_2(t_0) = 0$, and $W(y_1, y_2)(t_0) = 0$ so the solutions cannot form a fundamental set. If $p(t_0) \neq 0$ and $q(t_0) = 0$ then $y_1'(t_0) = y_2'(t_0) = 0$ and $W(y_1, y_2)(t_0) = 0$, so again the solutions cannot form a fundamental set. If $p(t_0) \neq 0$ and $q(t_0) \neq 0$ then $y_1'(t_0) = q(t_0)y_1(t_0)/p(t_0)$ and $y_2'(t_0) = q(t_0)y_2(t_0)/p(t_0)$ and thus

$$W(y_1,y_2)(t_0) = y_1(t_0)y_2'(t_0) - y_1'(t_0)y_2(t_0)$$
$$= y_1(t_0)[q(t_0)y_2(t_0)/p(t_0)] -$$
$$[q(t_0)y_1(t_0)/p(t_0)]y_2(t_0)$$
$$= 0.$$

27. Let $-1 < t_0$, $t_1 < 1$ and $t_0 \neq t_1$. If $y_1 = t$ and $y_2 = t^2$ are linearly dependent then $c_1t_1 + c_2t_1^2 = 0$ and $c_1t_0 + c_2t_0^2 = 0$ have a solution for c_1 and c_2 such that c_1 and c_2 are not both zero. But this system of equations has a non-zero solution only if $t_1 = 0$ or $t_0 = 0$ or $t_1 = t_0$. Hence, the only set c_1 and c_2 that satisfies the system for every choice of t_0 and t_1 in $-1 < t < 1$ is $c_1 = c_2 = 0$. Therefore t and t^2 are linearly independent on $-1 < t < 1$. Next, $W(t,t^2) = t^2$ clearly vanishes at $t = 0$. Since $W(t,t^2)$ vanishes at $t = 0$, but t and t^2 are linearly independent on $-1 < t < 1$, it follows that t and t^2 cannot be solutions of Eq.(7) on $-1 < t < 1$. To show that the functions $y_1 = t$ and $y_2 = t^2$ are solutions of $t^2y'' - 2ty' + 2y = 0$, substitute each of them in the equation. Clearly, they are solutions. There is no contradiction to Theorem 3.3.3 since $p(t) = -2/t$ and $q(t) = 2/t^2$ are discontinuous at $t = 0$, and hence the theorem does not apply on the interval $-1 < t < 1$.

28. On $0 < t < 1$, $f(t) = t^3$ and $g(t) = t^3$. Hence there are nonzero constants, $c_1 = 1$ and $c_2 = -1$, such that $c_1f(t) + c_2g(t) = 0$ for each t in $(0,1)$. On $-1 < t < 0$, $f(t) = -t^3$ and $g(t) = t^3$; thus $c_1 = c_2 = 1$ defines constants such that $c_1f(t) + c_2g(t) = 0$ for each t in $(-1,0)$. Thus f and g are linearly dependent on $0 < t < 1$ and on $-1 < t < 0$. We will show that $f(t)$ and $g(t)$ are linearly independent on $-1 < t < 1$ by demonstrating that it is impossible to find constants c_1 and c_2, not both zero, such that $c_1f(t) + c_2g(t) = 0$ for all t in $(-1,1)$. Assume that there are two such nonzero constants and choose two points t_0 and t_1 in $-1 < t < 1$ such that $t_0 < 0$ and $t_1 > 0$. Then $-c_1t_0^3 + c_2t_0^3 = 0$ and $c_1t_1^3 + c_2t_1^3 = 0$. These equations have a nontrivial solution for c_1 and c_2 only if the determinant of coefficients is

zero. But the determinant of coefficients is $-2t_0^3 t_1^3 \neq 0$ for t_0 and t_1 as specified. Hence f(t) and g(t) are linearly independent on $-1 < t < 1$.

Section 3.4, Page 164

1. $\exp(1+2i) = e^{1+2i} = ee^{2i} = e(\cos 2 + i\sin 2)$.

5. Recall that
   $2^{1-i} = e^{\ln(2^{1-i})} = e^{(1-i)\ln 2} = e^{\ln 2}e^{-i\ln 2} = 2(\cos\ln 2 - i\sin\ln 2)$.

7. As in Sect. 3.1, we seek solutions of the form $y = e^{rt}$. Substituting this into the D.E. yields the characteristic equation $r^2 - 2r + 2 = 0$, which has the roots $r_1 = 1 + i$ and $r_2 = 1 - i$, using the quadratic formula. Thus $\lambda = 1$ and $\mu = 1$ and from Eq.(17) the general solution is $y = c_1 e^t \cos t + c_2 e^t \sin t$.

11. The characteristic equation is $r^2 + 6r + 13 = 0$, which has the roots $r = \dfrac{-6 \pm \sqrt{-16}}{2} = -3 \pm 2i$. Thus $\lambda = -3$ and $\mu = 2$, so Eq.(17) becomes $y = c_1 e^{-3t}\cos 2t + c_2 e^{-3t}\sin 2t$.

14. The characteristic equation is $9r^2 + 9r - 4$, which has the real roots $-4/3$ and $1/3$. Thus the solution has the same form as in Section 3.1, $y(t) = c_1 e^{t/3} + c_2 e^{-4t/3}$.

18. The characteristic equation is $r^2 + 4r + 5 = 0$, which has the roots $r_1, r_2 = -2 \pm i$. Thus $y = c_1 e^{-2t}\cos t + c_2 e^{-2t}\sin t$ and $y' = (-2c_1+c_2)e^{-2t}\cos t + (-c_1-2c_2)e^{-2t}\sin t$, so that $y(0) = c_1 = 1$ and $y'(0) = -2c_1 + c_2 = 0$, or $c_2 = 2$. Hence $y = e^{-2t}(\cos t + 2\sin t)$. The oscillation is hard to see on this graph, but y(t) does cross the t axis at $t = \tan^{-1}(-.5) = 2.68$ and periodically after that.

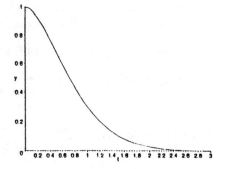

22. The characteristic equation is
$r^2 + 2r + 2 = 0$, so
$r_1, r_2 = -1 \pm i$. Since the I.C.
are given at $\pi/4$ we want to
alter Eq.(17) by letting
$c_1 = e^{\pi/4}d_1$ and $c_2 = e^{\pi/4}d_2$.
Thus, for $\lambda = -1$ and $\mu = 1$ we
have $y = e^{-(t-\pi/4)}(d_1\cos t + d_2\sin t)$;

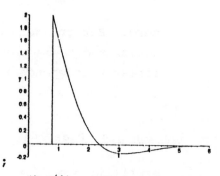

so $y' = -e^{-(t-\pi/4)}(d_1\cos t + d_2\sin t) + e^{-(t-\pi/4)}(-d_1\sin t + d_2\cos t)$.

Hence $y(\frac{\pi}{4}) = \sqrt{2}\, d_1/2 + \sqrt{2}\, d_2/2 = 2$ and

$y'(\frac{\pi}{4}) = -\sqrt{2}\, d_1 = -2$ and thus $y = \sqrt{2}\, e^{-(t-\pi/4)}(\cos t + \sin t)$.

23a. The characteristic equation is $3r^2 - r + 2 = 0$, which has
the roots $r_1, r_2 = \dfrac{1}{6} \pm \dfrac{\sqrt{23}}{6}i$. Thus

$u(t) = e^{t/6}(c_1\cos\dfrac{\sqrt{23}}{6}t + c_2\sin\dfrac{\sqrt{23}}{6}t)$ and we obtain

$u(0) = c_1 = 2$ and $u'(0) = \dfrac{1}{6}c_1 + \dfrac{\sqrt{23}}{6}c_2 = 0$. Solving

for c_2 we find $u(t) = e^{t/6}(2\cos\dfrac{\sqrt{23}}{6}t - \dfrac{2}{\sqrt{23}}\sin\dfrac{\sqrt{23}}{6}t)$.

23b. To estimate the first time that $|u(t)| = 10$ plot the graph
of $u(t)$ as found in part (a). Use this estimate in an
appropriate computer software program to find $t = 10.7598$.

25a. The characteristic equation is $r^2 + 2r + 6 = 0$, so
$r_1, r_2 = -1 \pm \sqrt{5}\, i$ and $y(t) = e^{-t}(c_1\cos\sqrt{5}\, t + c_2\sin\sqrt{5}\, t)$.
Thus $y(0) = c_1 = 2$ and $y'(0) = -c_1 + \sqrt{5}\, c_2 = \alpha$ and hence
$y(t) = e^{-t}(2\cos\sqrt{5}\, t + \dfrac{\alpha+2}{\sqrt{5}}\sin\sqrt{5}\, t)$.

25b. $y(1) = e^{-1}(2\cos\sqrt{5} + \dfrac{\alpha+2}{\sqrt{5}}\sin\sqrt{5}) = 0$ and hence

$\alpha = -2 - \dfrac{2\sqrt{5}}{\tan\sqrt{5}} = 1.50878$.

25c. For $y(t) = 0$ we must have $2\cos\sqrt{5}\, t + \dfrac{\alpha+2}{\sqrt{5}}\sin\sqrt{5}\, t = 0$ or

$$\tan\sqrt{5}\,t = \frac{-2\sqrt{5}}{\alpha+2}.$$ For $\alpha \geq 0$ (actually, for $\alpha > -2$) this yields $\sqrt{5}\,t = \pi - \arctan\dfrac{2\sqrt{5}}{\alpha+2}$ since arctanx is an odd function.

25d. From part (c) $\arctan\dfrac{2\sqrt{5}}{\alpha+2} \to 0$ as $\alpha \to \infty$, so $t \to \pi/\sqrt{5}$.

31. Let $r = \lambda+i\mu$, then $\dfrac{d}{dt}(e^{rt}) = \dfrac{d}{dt}[e^{\lambda t}(\cos\mu t + i\sin\mu t)]$

$= \lambda e^{\lambda t}(\cos\mu t + i\sin\mu t) + e^{\lambda t}(-\mu\sin\mu t + i\mu\cos\mu t)$

$= \lambda e^{\lambda t}(\cos\mu t + i\sin\mu t) + i\mu e^{\lambda t}(i\sin\mu t + \cos\mu t)$

$= e^{\lambda t}(\lambda+i\mu)(\cos\mu t + i\sin\mu t) = re^{rt}.$

33. Suppose that $t = a$ and $t = b$ $(b>a)$ are consecutive zeros of y_1. We must show that y_2 vanishes once and only once in the interval $a < t < b$. Assume that it does not vanish. Then we can form the quotient y_1/y_2 on the interval $a \leq t \leq b$. Note $y_2(a) \neq 0$ and $y_2(b) \neq 0$, otherwise y_1 and y_2 would not be linearly independent solutions. Next, y_1/y_2 vanishes at $t = a$ and $t = b$ and has a derivative in $a < t < b$. By Rolles theorem, the derivative must vanish at an interior point. But $(\dfrac{y_1}{y_2})' = \dfrac{y_1'y_2 - y_2'y_1}{y_2^{\,2}} = \dfrac{-W(y_1,y_2)}{y_2^{\,2}}$, which cannot be zero since y_1 and y_2 are linearly independent solutions. Hence we have a contradiction, and we conclude that y_2 must vanish at a point between a and b. Finally, we show that it can vanish at only one point between a and b. Suppose that it vanishes at two points c and d between a and b. By the argument we have just given we can show that y_1 must vanish between c and d. But this contradicts the hypothesis that a and b are consecutive zeros of y_1.

35. We use the result of Prob. 34. Note that $p(t) = t$ and $q(t) = e^{-t^2} > 0$ for $-\infty < t < \infty$. Thus $(q' + 2pq)/q^{3/2} = 0$ and the D.E. can be transformed into an equation with constant coefficients by letting $x = u(t) = \int e^{-t^2/2}dt$. Substituting $x = u(t)$ in the differential equation found in part (b) of Prob. 34 we obtain, after dividing by the coefficient of d^2y/dx^2, the D.E. $d^2y/dx^2 + y = 0$. Hence the

general solution of the original D.E. is

$y = c_1 \cos x + c_2 \sin x$, $x = \int e^{-t^2/2} dt$.

38. Rewrite the D.E. as $y'' + (\alpha/t)y' + (\beta/t^2)y = 0$ so that
    $p = \alpha/t$ and $q = \beta/t^2$, which satisfy the condition of
    part (d) of Prob. 34. Thus $x = \int (1/t^2)^{1/2} dt = \ln t$ will
    transform the D.E. into $dy^2/dx^2 + (\alpha-1)dy/dx + \beta y = 0$.
    Note that since β is constant, it can be neglected in
    defining x.

39. By direct substitution, or from Prob. 38, $x = \ln t$ will
    transform the D.E. into $d^2y/dx^2 + y = 0$, since $\alpha = 1$ and
    $\beta = 1$. Thus $y = c_1 \cos x + c_2 \sin x$, with $x = \ln t$, $t > 0$.

Section 3.5, Page 172

1. Substituting $y = e^{rt}$ into the D.E., we find that
    $r^2 - 2r + 1 = 0$, which gives $r_1 = 1$ and $r_2 = 1$. Since the
    roots are equal, the second linearly independent
    solution, see Eq.(26), is te^t and thus the general
    solution is $y = c_1 e^t + c_2 te^t$.

5. The characteristic equation is $r^2 - 2r + 10 = 0$, and thus
    $r = 1 \pm 3i$. The general solution, from Sect. 3.4, is
    $y(t) = e^t(c_1 \cos 3t + c_2 \sin 3t)$.

9. The characteristic equation is $25r^2 - 20r + 4 = 0$, which
    may be written as $(5r-2)^2 = 0$ and hence the roots are
    $r_1, r_2 = 2/5$. Thus $y = c_1 e^{2t/5} + c_2 te^{2t/5}$.

12. The characteristic equation is
    $r^2 - 6r + 9 = (r-3)^2$, which has
    the repeated root $r = 3$. Thus
    $y = c_1 e^{3t} + c_2 te^{3t}$, which gives
    $y(0) = c_1 = 0$, $y'(t) = c_2(e^{3t}+3te^{3t})$
    and $y'(0) = c_2 = 2$. Hence
    $y(t) = 2te^{3t}$.

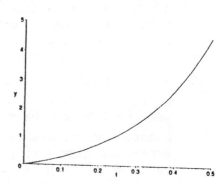

14. The characteristic equation is
    $r^2 + 4r + 4 = (r+2)^2 = 0$, which
    has the repeated root $r = -2$.
    Since the I.C. are given at
    $t = -1$, write the general
    solution as
    $y = d_1 e^{-2(t+1)} + d_2 t e^{-2(t+1)}$. Then

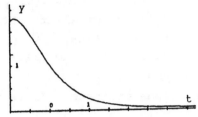

$y' = -2d_1 e^{-2(t+1)} + d_2 e^{-2(t+1)} - 2d_2 t e^{-2(t+1)}$ and hence
$d_1 - d_2 = 2$ and $-2d_1 + 3d_2 = 1$ which yield $d_1 = 7$ and $d_2 = 5$.
Thus $y = 7e^{-2(t+1)} + 5te^{-2(t+1)}$, a decaying exponential as
shown in the graph.

17a. The characteristic equation is $4r^2 + 4r + 1 = (2r+1)^2 = 0$,
    so we have $y(t) = (c_1 + c_2 t)e^{-t/2}$. Thus $y(0) = c_1 = 1$ and
    $y'(0) = -c_1/2 + c_2 = 2$ and hence $c_2 = 5/2$ and
    $y(t) = (1 + 5t/2)e^{-t/2}$.

17b. From part(a), $y'(t) = -\dfrac{1}{2}(1 + 5t/2)e^{-t/2} + \dfrac{5}{2}e^{-t/2} = 0$, when

    $-\dfrac{1}{2} - \dfrac{5t}{4} + \dfrac{5}{2} = 0$, or $t_M = \dfrac{8}{5}$ and $y_M = 5e^{-4/5}$.

17c. From part(a), c_1 is the same and $y'(0) = -\dfrac{1}{2} + c_2 = b$ or

    $c_2 = b + \dfrac{1}{2}$ and $y(t) = [1 + (b + \dfrac{1}{2})t]e^{-t/2}$.

17d. From part(c), $y'(t) = -\dfrac{1}{2}[1 + (b + \dfrac{1}{2})t]e^{-t/2} + (b + \dfrac{1}{2})e^{-t/2} = 0$

    which yields $t_M = \dfrac{4b}{2b+1} \to 2$ as $b \to \infty$ and

    $y_M = (1 + \dfrac{2b+1}{2} \cdot \dfrac{4b}{2b+1})e^{-2b/(2b+1)} = (1 + 2b)e^{-2b/(2b+1)}$. Since

    $e^{-2b/(2b+1)} = e^{-2/(2+b^{-1})} \to e^{-1}$ as $b \to \infty$, $y_M \to \infty$ as $b \to \infty$.

19. If $r_1 = r_2$ then $y(t) = (c_1 + c_2 t)e^{r_1 t}$. Since the exponential
    is never zero, $y(t)$ can be zero only if $c_1 + c_2 t = 0$, which
    yields at most one positive value of t if c_1 and c_2 differ
    in sign. If $r_2 > r_1$ then
    $y(t) = c_1 e^{r_1 t} + c_2 e^{r_2 t} = e^{r_1 t}(c_1 + c_2 e^{(r_2 - r_1)t})$. Again, this is
    zero only if c_1 and c_2 differ in sign, in which case
    $t = \dfrac{\ln(-c_1/c_2)}{(r_2 - r_1)}$.

21. If $r_2 \neq r_1$ then $\phi(t;r_1,r_2) = (e^{r_2 t} - e^{r_1 t})/(r_2 - r_1)$ is
     defined for all t. Note that ϕ is a linear combination of
     two solutions, $e^{r_1 t}$ and $e^{r_2 t}$, of the D.E. and thus ϕ is a
     solution of the differential equation. Think of r_1 as fixed
     and let $r_2 \to r_1$. The limit of ϕ as $r_2 \to r_1$ is
     indeterminate. If we use L'Hospital's rule, we find

$$\lim_{r_2 \to r_1} \frac{e^{r_2 t} - e^{r_1 t}}{r_2 - r_1} \cdot = \lim_{r_2 \to r_1} \frac{te^{r_2 t}}{1} = te^{r_1 t}. \quad \text{Hence, the}$$

     solution $\phi(t;r_1,r_2) \to te^{r_1 t}$ as $r_2 \to r_1$.

25. Let $y_2 = v/t$. Then $y_2' = v'/t - v/t^2$ and
     $y_2'' = v''/t - 2v'/t^2 + 2v/t^3$. Substituting in the D.E. we
     obtain
     $t^2(v''/t - 2v'/t^2 + 2v/t^3) + 3t(v'/t - v/t^2) + v/t = 0$.
     Simplifying the left side we get $tv'' + v' = 0$, which
     yields $v' = c_1/t$. Thus $v = c_1 \ln t + c_2$. Hence a second
     solution is $y_2(t) = (c_1 \ln t + c_2)/t$. However, we may set
     $c_2 = 0$ and $c_1 = 1$ without loss of generality and thus we
     have $y_2(t) = (\ln t)/t$ as a second solution. Note that in
     the form we actually calculated, $y_2(t)$ is a linear
     combination of $1/t$ and $\ln t/t$, and hence is the general
     solution.

27. In this case the calculations are somewhat easier if we
     do not use the explicit form for $y_1(x) = \sin x^2$ at the
     beginning but simply set $y_2(x) = y_1 v$. Substituting this
     form for y_2 in the D.E. gives $x(y_1 v)'' - (y_1 v)' + 4x^3(y_1 v) = 0$.
     On carrying out the differentiations and making use of
     the fact that y_1 is a solution, we obtain
     $xy_1 v'' + (2xy_1' - y_1)v' = 0$. Let $w = v'$ and separate

     variables to find $\dfrac{dw}{w} = (\dfrac{1}{x} - \dfrac{2y_1'}{y_1})dx$. Integrating and

     solving we find $w = v' = cx/(\sin x^2)^2$. Setting $u = x^2$
     allows integration of this to get $v = c_1 \cot x^2 + c_2$.
     Setting $c_1 = 1$, $c_2 = 0$ and multiplying by $y_1 = \sin x^2$ we
     obtain $y_2(x) = \cos x^2$ as the second solution of the D.E.

30. Substituting $y_2(x) = y_1(x)v(x)$ in the D.E. gives
$x^2(y_1v)'' + x(y_1v)' + (x^2 - \frac{1}{4})y_1v = 0$. On carrying out the differentiations and making use of the fact that y_1 is a solution, we obtain $x^2y_1v'' + (2x^2y_1' + xy_1)v' = 0$. This is a first order linear equation for $w = v'$,
$w' + (2y_1'/y_1 + 1/x)w = 0$, with solution (by separating variables)

$$w = v'(x) = c\exp[-\int(2\frac{y_1}{y_1} + \frac{1}{x})dx] = c\exp[-2\ln y_1 - \ln x]$$

$$= c\frac{1}{xy_1^2} = \frac{c}{x(x^{-1}\sin^2x)} = c\,\csc^2x, \text{ where } c \text{ is an}$$

arbitrary constant, which we will take to be one. Then $v(x) = \int\csc^2x\,dx = -\cot x + k$ where again k is an arbitrary constant which can be taken equal to zero. Thus $y_2(x) = y_1(x)v(x) = (x^{-1/2}\sin x)(-\cot x) = -x^{-1/2}\cos x$.
The second solution is usually taken to be $x^{-1/2}\cos x$. Note that $c = -1$ would have given this solution.

31b. Let $y_2(x) = e^xv(x)$, then $y_2' = e^xv' + e^xv$, and $y_2'' = e^xv'' + 2e^xv' + e^xv$. Substituting in the D.E. we obtain $xe^xv'' + (xe^x-Ne^x)v' = 0$, or $v'' + (1-N/x)v' = 0$. This is a first order linear D.E. for v' with integrating factor $\mu(x) = \exp[\int(1-N/x)dx] = x^{-N}e^x$. Hence $(x^{-N}e^xv')' = 0$, and $v' = cx^Ne^{-x}$ which gives $v(x) = c\int x^Ne^{-x}dx + k$. On taking $k = 0$ we obtain as the second solution $y_2(x) = ce^x\int x^Ne^{-x}dx$. The integral can be evaluated by using the method of integration by parts. At each stage let $u = x^N$ or x^{N-1}, or whatever the power of x that remains, and let $dv = e^{-x}$. Note that this dv is not related to the $v(x)$ in $y_2(x)$. For $N = 2$ we have

$$y_2(x) = ce^x\int x^2e^{-x}dx = ce^x[x^2\frac{e^{-x}}{-1} - \int 2x\frac{e^{-x}}{-1}dx]$$

$$= -cx^2 + ce^x[2x\frac{e^{-x}}{-1} - \int 2\frac{e^{-x}}{-1}dx]$$

$$= c(-x^2 - 2x - 2) = -2c(1 + x + x^2/2!).$$

Choosing $c = -1/2!$ gives the desired result. For the general case $c = -1/N!$

33. $(y_2/y_1)' = (y_1 y_2' - y_1' y_2)/y_1^2 = W(y_1, y_2)/y_1^2$. Abel's identity
is $W(y_1, y_2) = c \exp[-\int_{t_0}^{t} p(r)dr]$. Hence
$(y_2/y_1)' = c y_1^{-2} \exp[-\int_{t_0}^{t} p(r)dr]$. Integrating and setting
$c = 1$ (since a solution y_2 can be multiplied by any
constant) and taking the constant of integration to be
zero we obtain

$$y_2(t) = y_1(t) \int_{t_0}^{t} \frac{\exp[-\int_{s_0}^{s} p(r)dr]}{[y_1(s)]^2} \, ds.$$

35. From Prob. 33 and Abel's formula we have
$(\dfrac{y_2}{y_1})' = \dfrac{\exp[\int(1/t)dt]}{\sin^2(t^2)} = \dfrac{e^{\ln t}}{\sin^2(t^2)} = t\csc^2(t^2)$. Thus
$y_2/y_1 = -(1/2)\cot(t^2)$ and hence we can choose $y_2 = \cos(t^2)$
since $y_1 = \sin^2(t^2)$.

38. The general solution of the D.E. is $y = c_1 e^{r_1 t} + c_2 e^{r_2 t}$
where $r_1, r_2 = (-b \pm \sqrt{b^2 - 4ac})/2a$ provided $b^2 - 4ac \neq 0$.
In this case there are two possibilities. If $b^2 - 4ac > 0$
then $(b^2 - 4ac)^{1/2} < b$ and r_1 and r_2 are real and
<u>negative</u>. Consequently $e^{r_1 t} \to 0$ and $e^{r_2 t} \to 0$; and hence
$y \to 0$, as $t \to \infty$. If $b^2 - 4ac < 0$ then r_1 and r_2 are
complex conjugates with <u>negative</u> real part. Again
$e^{r_1 t} \to 0$ and $e^{r_2 t} \to 0$; and hence $y \to 0$, as $t \to \infty$.
Finally, if $b^2 - 4ac = 0$, then $y = c_1 e^{r_1 t} + c_2 t e^{r_1 t}$ where
$r_1 = -b/2a < 0$. Hence, again $y \to 0$ as $t \to \infty$. This
conclusion does not hold if either $b = 0$ (since, in this
case, $y(t) = c_1 \cos\omega t + c_2 \sin\omega t$) or $c = 0$ (since one of
the solutions would be $y_1(t) = c_1$).

42. Substituting $z = \ln t$ into the D.E. gives
$\dfrac{d^2 y}{dz^2} + \dfrac{dy}{dz} + 0.25y = 0$, which has the solution
$y(z) = c_1 e^{-z/2} + c_2 z e^{-z/2}$ so that $y(t) = c_1 t^{-1/2} + c_2 t^{-1/2} \ln t$.

Section 3.6, Page 184

1. First we find the solution of the homogeneous D.E., which has the characteristic equation $r^2 - 2r - 3 = (r-3)(r+1) = 0$. Hence $y_c = c_1 e^{3t} + c_2 e^{-t}$ and we can assume $Y = Ae^{2t}$ for the particular solution. Thus $Y' = 2Ae^{2t}$ and $Y'' = 4Ae^{2t}$ and substituting into the D.E. yields $4Ae^{2t} - 2(2Ae^{2t}) - 3(Ae^{2t}) = 3e^{2t}$. Thus $-3A = 3$ and $A = -1$, yielding $y = c_1 e^{3t} + c_2 e^{-t} - e^{2t}$.

4. Initially we might assume $Y = A + Bsin2t + Ccos2t$. However, since a constant is a solution of the related homogeneous D.E. we must modify Y by multiplying the constant A by t and thus the correct form is $Y = At + Bsin2t + Ccos2t$.

6. Since $y_c = c_1 e^{-t} + c_2 te^{-t}$ we must assume $Y = At^2 e^{-t}$, so that $Y' = 2Ate^{-t} - At^2 e^{-t}$ and $y'' = 2Ae^{-t} - 4Ate^{-t} + At^2 e^{-t}$. Substituting in the D.E. gives $(At^2 - 4At + 2A)e^{-t} + 2(-At^2 + 2At)e^{-t} + At^2 e^{-t} = 2e^{-t}$. Notice that all terms on the left involving t^2 and t add to zero and we are left with $2A = 2$, or $A = 1$. Hence $y = c_1 e^{-t} + c_2 te^{-t} + t^2 e^{-t}$.

8. The assumed form is $Y = (At + B)sin2t + (Ct + D)cos2t$, which is appropriate for both terms appearing on the right side of the D.E. Since none of the terms appearing in Y are solutions of the homogeneous equation, we do not need to modify Y. Calculating Y' and Y'' we have $Y' = Asin2t + Ccos2t + 2(At+B)cos2t - 2(Ct+D)sin2t$ and $Y'' = -4Csin2t - 4(At+B)sin2t - 4(Ct+D)cos2t$. Thus $Y'' + Y = -3Atsin2t - (3B+4C)sin2t - 3(Ct+D)cos2t$. Equating like coefficients yields $A = 0, -3B-4C = 3, -3C = 1$, and $-3D = 0$. Hence $Y(t) = -(5/9)sin2t - (1/3)tcos2t$.

11. First solve the homogeneous D.E. Substituting $y = e^{rt}$ gives $r^2 + r + 4 = 0$. Hence $y_c = e^{-t/2}[c_1 cos(\sqrt{15}\, t/2) + c_2 sin(\sqrt{15}\, t/2)]$. We replace sinht by $(e^t - e^{-t})/2$ and then assume $Y(t) = Ae^t + Be^{-t}$. Since neither e^t nor e^{-t} are solutions of the homogeneous equation, there is no

need to modify our assumption for Y. Substituting in the
D.E., we obtain $6Ae^t + 4Be^{-t} = e^t - e^{-t}$. Hence, $A = 1/6$
and $B = -1/4$. The general solution is
$$y = e^{-t/2}[c_1\cos(\sqrt{15}\,t/2) + c_2\sin(\sqrt{15}\,t/2)] + e^t/6 - e^{-t}/4.$$
[For this problem we could also have found a particular
solution as a linear combination of sinht and cosht:
$Y(t) = A\cosh t + B\sinh t$. Substituting this in the D.E.
gives $(5A + B)\cosh t + (A + 5B)\sinh t = 2\sinh t$. The
solution is $A = -1/12$ and $B = 5/12$. A simple calculation
shows that $-(1/12)\cosh t + (5/12)\sinh t = e^t/6 - e^{-t}/4$.]

13. $y_c = c_1e^{-2t} + c_2e^t$ so for the particular solution we
assume $Y = At + B$. Since neither At or B are solutions of
the homogeneous equation it is not necessary to modify
the original assumption. Substituting Y in the D.E. we
obtain $0 + A -2(At+B) = 2t$ or $-2A = 2$ and $A-2B = 0$.
Solving for A and B we obtain $y = c_1e^{-2t} + c_2e^t - t - 1/2$
as the general solution. $y(0) = 0 \Rightarrow c_1 + c_2 - 1/2 = 0$
and $y'(0) = 1 \Rightarrow -2c_1 + c_2 - 1 = 1$, which yield $c_1 = -1/2$
and $c_2 = 1$. Thus $y = e^t - (1/2)e^{-2t} - t - 1/2$.

16. Since the characteristic equation is $r^2 - 2r - 3 = 0$,
$y_c = c_1e^{3t} + c_2e^{-t}$. The nonhomogeneous term is the product
of a linear polynomial and an exponential, so assume Y of
the same form: $Y = (At+B)e^{2t}$, which we do not need to
modify since these terms are not in y_c. Thus
$Y' = Ae^{2t} + 2(At+B)e^{2t}$ and $Y'' = 4Ae^{2t}+4(At+B)e^{2t}$.
Substituting into the D.E. we find $-3At = 3t$ and
$2A - 3B = 0$, yielding $A = -1$ and $B = -2/3$. Thus, the
general solution is $y = c_1e^{3t} +c_2e^{-t} - \frac{2}{3}e^{2t} - te^{2t}$.

19a. The solution of the homogeneous D.E. is $y_c = c_1e^{-3t} + c_2$.
After inspection of the nonhomogeneous term, for $2t^4$ we
must assume a fourth order polynominial, for t^2e^{-3t} we
must assume a quadratic polynomial times the exponential,
and for sin3t we must assume $C\sin 3t + D\cos 3t$. Thus
$$Y(t) = (A_0t^4 +A_1t^3 +A_2t^2 +A_3t+A_4) + (B_0t^2 +B_1t+B_2)e^{-3t} + C\sin 3t + D\cos 3t.$$
However,since e^{-3t} and a constant are solutions of the
homogeneous D.E., we must multiply the coefficient of e^{-3t}

and the polynomial by t. The correct form, then, is

$$Y(t) = t(A_0t^4 + A_1t^3 + A_2t^2 + A_3t + A_4) +$$
$$t(B_0t^2 + B_1t + B_2)e^{-3t} + Csin3t + Dcos3t.$$

22a. The solution of the homgeneous D.E. is

$Y_c = e^{-t}[c_1cost + c_2sint]$. After inspection of the nonhomogeneous term, we assume

$$Y(t) = Ae^{-t} + (B_0t^2 + B_1t + B_2)e^{-t}cost + (C_0t^2 + C_1t + C_2)e^{-t}sint.$$

Since $e^{-t}cost$ and $e^{-t}sint$ are solutions of the homogeneous D.E., it is necessary to multiply both the last two terms by t. Hence the correct form is

$$Y(t) = Ae^{-t} + t(B_0t^2 + B_1t + B_2)e^{-t}cost + t(C_0t^2 + C_1t + C_2)e^{-t}sint.$$

27a. Calculating Y′ and Y″ and substituting into the D.E. we get $(v″ - 2v′ + v)e^{-t} - 3(v′ - v)e^{-t} - 4ve^{-t} = 2e^{-t}$. This reduces to $v″ - 5v′ = 2$, which is a 1st order D.E. for v′.

27b. The linear D.E. for w has an integrating factor of e^{-5t} and thus $(e^{-5t}w)′ = 2e^{-5t}$, which gives $w = v′ = \dfrac{-2}{5} + c_1e^{5t}$.

29. First solve the I.V.P. $y″ + y = t$, $y(0) = 0$, $y′(0) = 1$ for $0 \le t \le \pi$. The solution of the homogeneous D.E. is $y_c(t) = c_1cost + c_2sint$. The correct form for Y(t) is $y(t) = A_0t + A_1$. Substituting in the D.E. we find $A_0 = 1$ and $A_1 = 0$. Hence, $y = c_1cost + c_2sint + t$. Applying the I.C., we obtain $y = t$, for $0 \le t \le \pi$.

For $t > \pi$ we have $y″ + y = \pi e^{\pi-t}$ so the form, now, for Y(t) is $Y(t) = Ee^{\pi-t}$. Substituting Y(t) in the D.E., we obtain $Ee^{\pi-t} + Ee^{\pi-t} = \pi e^{\pi-t}$ so $E = \pi/2$. Hence the general solution for $t > \pi$ is $Y = D_1cost + D_2sint + (\pi/2)e^{\pi-t}$. If y and y′ are to be continuous at $t = \pi$, then the solutions and their derivatives for $t \le \pi$ and $t > \pi$ must have the same value at $t = \pi$. These conditions require $\pi = -D_1 + \pi/2$ and $1 = -D_2 - \pi/2$. Hence $D_1 = -\pi/2$, $D_2 = -(1 + \pi/2)$, and

$$y = \phi(t) = \begin{cases} t, & 0 \le t \le \pi \\ -(\pi/2)cost - (1 + \pi/2)sint + (\pi/2)e^{\pi-t}, & t > \pi. \end{cases}$$

The graphs of the nonhomogeneous term and ϕ follow.

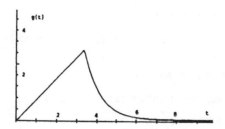
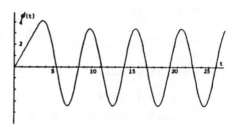

31. According to Theorem 3.6.1, the difference of any two
    solutions of the linear second order nonhomogeneous D.E.
    is a solution of the corresponding homogeneous D.E.
    Hence $Y_1 - Y_2$ is a solution of $ay'' + by' + cy = 0$. In
    Problem 38 of Section 3.5 we showed that if $a > 0$, $b > 0$,
    and $c > 0$ then every solution of this D.E. goes to zero
    as $t \to \infty$. If $b = 0$, then y_c involves only sines and
    cosines, so $Y_1 - Y_2$ does not approach zero as $t \to \infty$.

34. From Problem 33 we write the D.E. as $(D-4)(D+1)y = 3e^{2t}$
    Thus let $(D+1)y = u$ and then $(D-4)u = 3e^{2t}$. This last
    equation is the same as $du/dt - 4u = 3e^{2t}$, which may be
    solved by multiplying both sides by e^{-4t} and integrating
    (see Sect. 2.1). This yields $u = (-3/2)e^{2t} + Ce^{4t}$.
    Substituting this form of u into $(D+1)y = u$ we obtain
    $dy/dt + y = (-3/2)e^{2t} + Ce^{4t}$. Again, multiplying by e^{t}
    and integrating gives $y = (-1/2)e^{2t} + C_1 e^{4t} + C_2 e^{-t}$, where
    $C_1 = C/5$.

Section 3.7, Page 190

2. Two linearly independent solutions of the homogeneous
    D.E. are $y_1(t) = e^{2t}$ and $y_2(t) = e^{-t}$. Assume
    $Y = u_1(t)e^{2t} + u_2(t)e^{-t}$, then
    $Y'(t) = [2u_1(t)e^{2t} - u_2(t)e^{-t}] + [u_1'(t)e^{2t} + u_2'(t)e^{-t}]$.
    We set $u_1'(t)e^{2t} + u_2'(t)e^{-t} = 0$.   Then
    $Y'' = 4u_1 e^{2t} + u_2 e^{-t} + 2u_1' e^{2t} - u_2' e^{-t}$ and substituting in

the D.E. gives $2u_1'(t)e^{2t} - u_2'(t)e^{-t} = 2e^{-t}$. Thus we have two algebraic equations for $u_1'(t)$ and $u_2'(t)$ with the solution $u_1'(t) = 2e^{-3t}/3$ and $u_2'(t) = -2/3$. Hence $u_1(t) = -2e^{-3t}/9$ and $u_2(t) = -2t/3$. Substituting in the expression for Y(t) we obtain
$Y(t) = (-2e^{-3t}/9)e^{2t} + (-2t/3)e^{-t} = (-2e^{-t}/9) - (2te^{-t}/3)$.
Since e^{-t} is a solution of the homogeneous D.E., we can choose $Y(t) = -2te^{-t}/3$.

5.   Since cost and sint are solutions of the homogeneous
     D.E., we assume $Y = u_1(t)\cos t + u_2(t)\sin t$. Thus
     $Y' = -u_1(t)\sin t + u_2(t)\cos t$, after setting
     $u_1'(t)\cos t + u_2'(t)\sin t = 0$. Finding Y'' and substituting
     into the D.E. then yields $-u_1'(t)\sin t + u_2'(t)\cos t = \tan t$.
     The two equations for $u_1'(t)$ and $u_2'(t)$ have the solution:
     $u_1'(t) = -\sin^2 t/\cos t = -\sec t + \cos t$ and
     $u_2'(t) = \sin t$. Thus $u_1(t) = \sin t - \ln(\tan t + \sec t)$ and
     $u_2(t) = -\cos t$, which when substituted into the assumed
     form for Y, simplified, and added to the homogeneous
     solution yields
     $y = c_1\cos t + c_2\sin t - (\cos t)\ln(\tan t + \sec t)$.

11. Two linearly independent solutions of the homogeneous
     D.E. are $y_1(t) = e^{3t}$ and $y_2(t) = e^{2t}$. Applying Theorem
     3.7.1 with $W(y_1,y_2)(t) = -e^{5t}$, we obtain
     $$Y(t) = -e^{3t}\int\frac{e^{2s}g(s)}{-e^{5s}}\,ds + e^{2t}\int\frac{e^{3s}g(s)}{-e^{5s}}\,ds$$
     $$= \int[e^{3(t-s)} - e^{2(t-s)}]g(s)\,ds.$$
     The complete solution is then obtained by adding
     $c_1e^{3t} + c_2e^{2t}$ to Y(t). Note: since we are taking e^{3t} and e^{2t}
     under the integral the integration variable can't be t.

14. That t and te^t are solutions of the homogeneous D.E. can
     be verified by direction substitution. Thus we assume
     $Y = tu_1(t) + te^t u_2(t)$. Following the pattern of earlier
     problems we find $tu_1'(t) + te^t u_2'(t) = 0$, Eq.(21), and
     $u_1'(t) + (t+1)e^t u_2' = 2t$, Eq.(25). [Note that $g(t) = 2t$,

since the D.E. must be put into the form of Eq.(16)].
The solution of these equations gives $u_1'(t) = -2$ and
$u_2'(t) = 2e^{-t}$. Hence, $u_1(t) = -2t$ and $u_2(t) = -2e^{-t}$, and
$Y(t) = t(-2t) + te^t(-2e^{-t}) = -2t^2 - 2t$. However, since t
is a solution of the homogeneous D.E. we can choose as
our particular solution $Y(t) = -2t^2$.

18. For this problem, and for many others, it is probably
easier to rederive Eqs.(26) without using the explicit
form for $y_1(x)$ and $y_2(x)$ and then to substitute for $y_1(x)$
and $y_2(x)$ in Eqs.(26). In this case if we take
$y_1 = x^{-1/2} \sin x$ and $y_2 = x^{-1/2} \cos x$, then $W(y_1,y_2) = -1/x$.
If the D.E. is put in the form of Eq.(16), then
$g(x) = 3x^{-1/2} \sin x$ and thus $u_1'(x) = 3\sin x \cos x$ and
$u_2'(x) = -3\sin^2 x = 3(-1 + \cos 2x)/2$. Hence
$u_1(x) = (3\sin^2 x)/2$ and $u_2(x) = -3x/2 + 3(\sin 2x)/4$, and

$$Y(x) = \frac{3\sin^2 x}{2} \frac{\sin x}{\sqrt{x}} + \left(-\frac{3x}{2} + \frac{3\sin 2x}{4} \right) \frac{\cos x}{\sqrt{x}}$$

$$= \frac{3\sin^2 x}{2} \frac{\sin x}{\sqrt{x}} + \left(-\frac{3x}{2} + \frac{3\sin x \cos x}{2} \right) \frac{\cos x}{\sqrt{x}}$$

$$= \frac{3\sin x}{2\sqrt{x}} - \frac{3\sqrt{x}\cos x}{2}.$$

The first term is a multiple of $y_1(x)$ and thus can be
neglected for $Y(x)$.

22. Putting limits on the integrals of Eq.(28) and changing
the integration variable to s yields

$$Y(t) = -y_1(t)\int_{t_0}^t \frac{y_2(s)g(s)ds}{W(y_1,y_2)(s)} + y_2(t)\int_{t_0}^t \frac{y_1(s)g(s)ds}{W(y_1,y_2)(s)}$$

$$= \int_{t_0}^t \frac{-y_1(t)y_2(s)g(s)ds}{W(y_1,y_2)(s)} + \int_{t_0}^t \frac{y_2(t)y_1(s)g(s)ds}{W(y_1,y_2)(s)}$$

$$= \int_{t_0}^t \frac{[y_1(s)y_2(t) - y_1(t)y_2(s)]g(s)ds}{y_1(s)y_2'(s) - y_1'(s)y_2(s)}.$$ To show that

$Y(t)$ satisfies $L[y] = g(t)$ we must take the derivative of
Y using Leibnitz's rule, which says that if

$$Y(t) = \int_{t_0}^t G(t,s)ds, \text{ then } Y'(t) = G(t,t) + \int_{t_0}^t \frac{\partial G}{\partial t}(t,s)ds.$$

Letting $G(t,s)$ be the above integrand, then $G(t,t) = 0$

and $\dfrac{\partial G}{\partial t} = \dfrac{y_1(s)y_2'(t) - y_1'(t)y_2(s)}{W(y_1,y_2)(s)} \, g(s)$.   Likewise

$$Y'' = \dfrac{\partial G(t,t)}{\partial t} + \int_{t_0}^{t} \dfrac{\partial^2 G}{\partial t^2}(t,s)ds$$

$$= g(t) + \int_{t_0}^{t} \dfrac{y_1(s)y_2''(t) - y_1''(t)y_2(s)}{W(y_1,y_2)(s)} g(s)ds.$$

Since y_1 amd y_2 are solutions of $L[y] = 0$, we have
$L[Y] = g(t)$ since all the terms involving the integral
will add to zero. Clearly $Y(t_0) = 0$ and $Y'(t_0) = 0$.

25. Note that $y_1 = e^{\lambda t}\cos\mu t$ and $y_2 = e^{\lambda t}\sin\mu t$ and thus

$W(y_1,y_2) = \mu e^{2\lambda t}$. From Prob. 22 we then have:

$$Y(t) = \int_{t_0}^{t} \dfrac{e^{\lambda s}\cos\mu s\, e^{\lambda t}\sin\mu t - e^{\lambda t}\cos\mu t\, e^{\lambda s}\sin\mu s}{\mu e^{2\lambda s}} \, g(s)ds$$

$$= \mu^{-1} \int_{t_0}^{t} e^{\lambda(t-s)}[\cos\mu s\, \sin\mu t - \cos\mu t\, \sin\mu s]g(s)ds$$

$$= \mu^{-1} \int_{t_0}^{t} e^{\lambda(t-s)}[\sin\mu(t-s)]g(s)ds.$$

29. First, we put the D.E. in standard form by dividing by
t^2: $y'' - 2y'/t + 2y/t^2 = 4$. Assuming that $y = tv(t)$ and
substituting in the D.E. we obtain $tv'' = 4$. Hence
$v'(t) = 4\ln t + c_2$ and $v(t) = 4\int\ln t\, dt + c_2 t + c_1 =$
$4(t\ln t - t) + c_2 t + c_1$, using integration by parts. Thus
$y = 4t^2\ln t + c_3 t^2 + c_1 t$, where $c_3 = c_2 - 4$. Since $y_1 = c_1 t$,
we can take $y_2 = 4t^2\ln t + c_3 t^2$, where $c_3 t^2$ represents the
second fundamental solution of the related homogeneous
equation and $4t^2\ln t$ is the particular solution.

Section 3.8, Page 203

2.   From Eq.(15) we have $R\cos\delta = -1$, and $R\sin\delta = \sqrt{3}$. Thus
$R = \sqrt{1+3} = 2$ and $\delta = \tan^{-1}(-\sqrt{3}) + \pi = 2\pi/3 \approx 2.09440$.   Note
that we have to "add" π to the inverse tangent value since δ
must be a second quadrant angle. Thus $u = 2\cos(t-2\pi/3)$.

6.   The motion is an undamped free vibration. The units are
in the CGS system. The spring constant
$k = (100 \text{ gm})(980\text{cm/sec}^2)/5\text{cm}$. Hence the D.E. for the
motion is $100u'' + [(100 \cdot 980)/5]u = 0$ where u is measured

in cm and time in sec. We obtain u" + 196u = 0 so
u = Acos14t + Bsin14t. The I.C. are u(0) = 0 ⇒ A = 0
and u'(0) = 10 cm/sec ⇒ B = 10/14 = 5/7. Hence
u(t) = (5/7)sin14t, which first reaches equilibrium when
14t = π, or t = π/14.

8. We use Eq.(33) without R and E(t) (there is no resistor or
impressed voltage) and with L = 1 henry and 1/C = 4x10^6 since
C = .25x10^{-6} farads. Thus the I.V.P. is Q" + 4x10^6 Q = 0,
Q(0) = 10^{-6} coulombs and Q'(0) = 0.

9. The spring constant is k = (20)(980)/5 = 3920 dyne/cm.
The I.V.P. for the motion is 20u" + 400u' + 3920u = 0 or
u" + 20u' + 196u = 0 and u(0) = 2, u'(0) = 0. Here u is
measured in cm and t in sec. The general solution of the
D.E. is u = Ae^{-10t}cos4$\sqrt{6}$ t + Be^{-10t}sin4$\sqrt{6}$ t. The I.C.
u(0) = 2 ⇒ A = 2 and u'(0) = 0 ⇒ −10A + 4$\sqrt{6}$ B = 0. The
solution is u = e^{-10t}[2cos4$\sqrt{6}$ t + 5(sin4$\sqrt{6}$ t)/$\sqrt{6}$]cm.
The quasi frequency is μ = 4$\sqrt{6}$, the quasi period is
T$_d$ = 2πμ = π/2$\sqrt{6}$ and T$_d$/T = 7/2$\sqrt{6}$ since
T = 2π/14 = π/7. To find an upper bound for τ, write u
in the form of Eq.(26): u(t) = $\sqrt{4+25/6}$ e^{-10t}cos(4$\sqrt{6}$ t−δ).
Now, since |cos(4$\sqrt{6}$ t−δ)| ≤ 1, we have |u(t)| < .05 ⇒
$\sqrt{4+25/6}$ e^{-10t} < .05, which

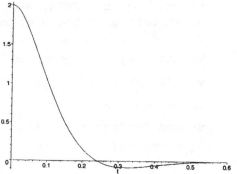

yields τ = .4046. A more
precise answer can be
obtained with a computer
algebra system, which in
this case yields τ = .4045.
The original estimate was
unusually close for this
problem since
cos(4$\sqrt{6}$ t−δ) = −0.9996
for t = .4046.

12. Substituting the given values for L, C and R in Eq.(33),
we obtain the D.E. .2Q" + 3x10^2 Q' + 10^5 Q = 0. The I.C.
are Q(0) = 10^{-6} and Q'(0) = I(0) = 0. The roots of the
characteristic equation are r$_1$ = −500 and r$_2$ = −1000.
Thus Q = c$_1$e^{-500t} + c$_2$e^{-1000t} and hence Q(0) = 10^{-6} ⇒
c$_1$ + c$_2$ = 10^{-6} and Q'(0) = 0 ⇒ −500c$_1$ − 1000c$_2$ = 0.
Solving for c$_1$ and c$_2$ yields the solution.

17. The mass is 8/32 lb-sec^2/ft, and the spring constant is
8/(1/8) = 64 lb/ft. Hence (1/4)u" + γu' + 64u = 0 or

$u'' + 4\gamma u' + 256u = 0$, where u is measured in ft, t in sec
and the units of γ are lb-sec/ft. The characteristic
equation is $r^2 + 4\gamma r + 256 = 0$, so
$r_1, r_2 = [-4\gamma \pm \sqrt{16\gamma^2 - 1024}\,]/2$. The system will be
overdamped, critically damped or underdamped as
$(16\gamma^2 - 1024)$ is > 0, $= 0$, or < 0, respectively. Thus
the system is critically damped when $\gamma = 8$ lb-sec/ft.

19. The general solution of the D.E. is $u = Ae^{r_1 t} + Be^{r_2 t}$
where $r_1, r_2 = [-\gamma \pm (\gamma^2 - 4km)^{1/2}]/2m$ provided
$\gamma^2 - 4km \neq 0$, and where A and B are determined by the
I.C. When the motion is overdamped, $\gamma^2 - 4km > 0$ and
$r_1 > r_2$. Setting $u = 0$, we obtain $Ae^{r_1 t} = - Be^{r_2 t}$ or
$e^{(r_1 - r_2)t} = - B/A$. Since the exponential function is a
monotone function, there is at most one value of t
(when $B/A < 0$) for which this equation can be satisfied.
Hence u can vanish at most once. If the system is
critically damped, the general solution is
$u(t) = (A + Bt)e^{-\gamma t/2m}$. The exponential function is never
zero; hence u can vanish only if $A + Bt = 0$. If $B = 0$
then u never vanishes; if $B \neq 0$ then u vanishes once at
$t = - A/B$ provided $A/B < 0$.

20. The general solution of Eq.(21) for the case of critical
damping is $u = (A + Bt)e^{-\gamma t/2m}$. The I.C. $u(0) = u_0 \Rightarrow$
$A = u_0$ and $u'(0) = v_0 \Rightarrow A(-\gamma/2m) + B = v_0$. Hence
$u = [u_0 + (v_0 + \gamma u_0/2m)t]e^{-\gamma t/2m}$. If $v_0 = 0$, then
$u = u_0(1 + \gamma t/2m)e^{-\gamma t/2m}$, which is never zero since γ and
m are postive. By L'Hospital's Rule $u \to 0$ as $t \to \infty$.
Finally for $u_0 > 0$, we want the condition which will
insure that $v = 0$ at least once. Since the exponential
function is never zero we require
$u_0 + (v_0 + \gamma u_0/2m)t = 0$ at a positive value of t. This
requires that $v_0 + \gamma u_0/2m \neq 0$ and that
$t = -u_0(v_0 + u_0\gamma/2m)^{-1} > 0$. We know that $u_0 > 0$ so we must
have $v_0 + \gamma u_0/2m < 0$ or $v_0 < -\gamma u_0/2m$.

23. From Prob. 21: $\Delta = \dfrac{2\pi\gamma}{\mu(2m)} = T_d\gamma/2m$. Substituting the
given values (and $m = 1/4$ from Prob.17) we find
$\gamma = \dfrac{(1/2)\,(3)}{.3} = 5$ lb sec/ft.

24. From Eq.(13) $\omega_0^2 = \dfrac{2k}{3}$ so $P = 2\pi/\sqrt{2k/3} = \pi \Rightarrow k = 6$.

Thus $u(t) = c_1\cos 2t + c_2\sin 2t$ and $u(0) = 2 \Rightarrow c_1 = 2$ and $u'(0) = v \Rightarrow c_2 = v/2$. Hence

$$u(t) = 2\cos 2t + \frac{v}{2}\sin 2t = \sqrt{4+\frac{v^2}{4}}\,\cos(2t-\gamma).$$

Thus $\sqrt{4+\dfrac{v^2}{4}} = 3$ and $v = \pm 2\sqrt{5}$.

27. First, consider the static case. Let Δl denote the length of the block below the surface of the water. The weight of the block, which is a downward force, is $w = \rho l^3 g$. This is balanced by an equal and opposite buoyancy force B, which is equal to the weight of the

displaced water. Thus $B = (\rho_0 l^2 \Delta l)g = \rho l^3 g$. Now let x be the displacement of the block from its equilibrium position. We take downward as the positive direction. In a displaced position the forces acting on the block are its weight, which acts downward and is unchanged, and the buoyancy force which is now $\rho_0 l^2(\Delta l + x)g$ and acts upward. The resultant force must be equal to the mass of the block times the acceleration, namely $\rho l^3 x''$. Hence $\rho l^3 g - \rho_0 l^2(\Delta l + x)g = \rho l^3 x''$. Hence the D.E. for the motion of the block is $\rho l^3 x'' + \rho_0 l^2 gx = 0$ or

$x'' + \dfrac{\rho_0 g}{\rho l}x = 0$. This gives a simple harmonic motion with frequency $(\rho_0 g/\rho l)^{1/2}$ and natural period $T = 2\pi(\rho l/\rho_0 g)^{1/2}$.

29a. The characteristic equation is $4r^2 + r + 8 = 0$, so $r = (-1\pm\sqrt{127})/8$ and hence

$$u(t) = e^{-t/8}(c_1\cos\frac{\sqrt{127}}{8}t + c_2\sin\frac{\sqrt{127}}{8}t. \quad u(0) = 0 \Rightarrow c_1 = 0$$

and $u'(0) = 2 \Rightarrow \dfrac{\sqrt{127}}{8}c_2 = 2$. Thus

$$u(t) = \frac{16}{\sqrt{127}}e^{-t/8}\sin\frac{\sqrt{127}}{8}t.$$

29c. The phase plot is the
     spiral shown and the
     direction of motion is
     clockwise since the
     graph starts at (0,2) and
     u increases initially.

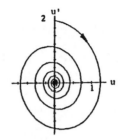

30a. The kinetic energy of a mass is given by $\frac{1}{2}mv^2$, so at t = 0

     we have v = u'(0) = b and thus $\frac{1}{2}mb^2$ is the initial kenetic

     energy. The work done deforming a spring an amount y from
     its undeformed state is stored in the spring and is known as
     the elastic potential energy. For our example, then, the

     potential energy is given by $\int_0^x F\,dy = \int_0^x ky\,dy = \frac{1}{2}kx^2$. For

     x = u(0) = a, this becomes $\frac{1}{2}ka^2$, as the initial potential

     energy.

30c. From part (a), the total energy in the system is
     $ku^2/2 + m(u')^2/2$. Using u(t) as found in part(b), calculate
     u' and show that $ku^2/2 + m(u')^2/2 = (ka^2 + mb^2)/2$ for all t.
     This confirms the principle of conservation of energy when
     there is no damping.

Section 3.9, Page 214

1.   We use the trigonometric identities
         cos(A + B) = cosA cosB - sinA sinB
         cos(A - B) = cosA cosB + sinA sinB
     to obtain cos(A + B) - cos(A - B) = -2sinA sinB. If we
     choose A + B = 9t and A - B = 7t, then A = 8t and B = t.
     Substituting in the formula just derived, we obtain
     cos9t - cos7t = -2sin8t sint.

5.   The mass m = 4/32 = 1/8 lb-sec²/ft and the spring
     constant k = 4/(1/8) = 32 lb/ft. Since there is no
     damping, the I.V.P. is (1/8)u" + 32u = 2cos3t,
     u(0) = 1/6, u'(0) = 0 where u is measured in ft and t in
     sec.

7a. From the solution to Prob. 5, we have m = 1/8, F_0 = 2,
     ω_0^2 = 256, and ω^2 = 9, so Eq.(10) becomes

$$u = c_1\cos 16t + c_2\sin 16t + \frac{16}{247}\cos 3t. \quad \text{The I.C.}$$

$u(0) = 1/6 \Rightarrow c_1 + 16/247 = 1/6$ and $u'(0) = 0 \Rightarrow 16c_2 = 0$,
and thus $u = (151/1482)\cos 16t + (16/247)\cos 3t$ ft.

7b.

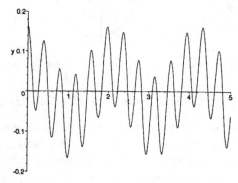

7c. Resonance occurs when the frequency ω of the forcing
     function $4\sin\omega t$ is the same as the natural frequency ω_0
     of the system. Since $\omega_0 = 16$, the system will resonate
     when $\omega = 16$ rad/sec.

10. The I.V.P. is $.25u'' + 16u = 8\sin 8t$, or $u'' + 64u = 32\sin 8t$,
     $u(0) = 3$ and $u'(0) = 0$. Since $u_c = c_1\cos 8t + c_2\sin 8t$, we
     assume the form $U(t) = t(A\cos 8t + B\sin 8t)$ and resonance
     occurs. Substituting U into the D.E. we find
     $U'' + 64U = -16A\sin 8t + 16B\cos 8t$ and thus $A = -2$, $B = 0$.
     Therefore $u(t) = c_1\cos 8t + c_2\sin 8t - 2t\cos 8t$. The I.C.
     yield $c_1 = 1/4$ and $8c_2 - 2 = 0$ and hence
     $u(t) = (\cos 8t + \sin 8t - 8t\cos 8t)/4$. The velocity will be
     zero when $u'(t) = 8\sin 8t(8t-1) = 0$.

11a. For this problem the mass $m = 8/32$ lb-sec^2/ft and the
     spring constant $k = 8/(1/2) = 16$ lb/ft, so the D.E. is
     $0.25u'' + 0.25u' + 16u = 4\cos 2t$ where u is measured in ft
     and t in sec. To determine the steady state response we
     need only compute a particular solution of the
     nonhomogeneous D.E. since the solutions of the
     homogeneous D.E. decay to zero as $t \to \infty$. We assume
     $u(t) = A\cos 2t + B\sin 2t$, and substitute in the D.E.:
     $-A\cos 2t - B\sin 2t + (1/2)(-A\sin 2t + B\cos 2t) + 16(A\cos 2t +$
     $B\sin 2t) = 4\cos 2t$. Hence $15A + (1/2)B = 4$ and
     $-(1/2)A + 15B = 0$, from which we obtain $A = 240/901$ and
     $B = 8/901$. Thus $U(t) = (240\cos 2t + 8\sin 2t)/901$.

11b. In order to determine the value of m that maximizes the
     steady state response, we note that the present problem

has exactly the form of Eq.(1) considered in the text. Referring to Eqs.(3) and (4), the magnitude of the response, R, is a maximum when Δ is a minimum since F_0 is constant. Δ, as given in Eq.(5), will be a minimum when $f(m) = m^2(\omega_0^2 - \omega^2)^2 + \gamma^2\omega^2$, where $\omega_0^2 = k/m$, is a minimum. We calculate df/dm and set this quantity equal to zero to obtain $m = k/\omega^2$. We verify that this value of m gives a minimum of f(m) by the second derivative test. For this problem k = 16 lb/ft and ω = 2 rad/sec so the value of m that maximizes the response of the system is m = 4 slugs.

14. Since $U(t) = R\cos(\omega t-\delta)$ we have $U'(t) = \dfrac{-F_0\omega}{\Delta}\sin(\omega t-\delta)$, where Δ is given by Eq.(5). Since F_0 is a constant, differentiate $\dfrac{\omega}{\Delta}$ with respect to ω and set it equal to zero. Alternatively, you can minimize $(\Delta/\omega)^2$, which simplifies the differentiation.

15. We must solve the three I.V.P.:   (1)$u_1'' + u_1 = F_0 t$, $0 < t < \pi$, $u_1(0) = u_1'(0) = 0$; (2) $u_2'' + u_2 = F_0(2\pi-t)$, $\pi < t < 2\pi$, $u_2(\pi) = u_1(\pi)$, $u_2'(\pi) = u_1'(\pi)$; and (3) $u_3'' + u_3 = 0$, $2\pi < t$, $u_3(2\pi) = u_2(2\pi)$, $u_3'(2\pi) = u_2'(2\pi)$. The conditions at π and 2π insure the continuity of u and u' at those points. The general solutions of the D.E. are $u_1 = b_1\cos t + b_2\sin t + F_0 t$, $u_2 = c_1\cos t + c_2\sin t + F_0(2\pi-t)$, and $u_3 = d_1\cos t + d_2\sin t$. The I.C. and matching conditions, in order, give $b_1 = 0$, $b_2 + F_0 = 0$, $-b_1 + \pi F_0 = -c_1 + \pi F_0$, $-b_2 + F_0 = -c_2 - F_0$, $c_1 = d_1$, and $c_2 - F_0 = d_2$. Solving these equations we obtain

$$u = F_0 \begin{cases} t - \sin t & , \ 0 \le t \le \pi \\ (2\pi - t) - 3\sin t & , \ \pi < t \le 2\pi \\ -4\sin t & , \ 2\pi < t. \end{cases}$$

16. From Eq.(33) of Sect. 3.8, the I.V.P. is $Q'' + 5\times10^3 Q' + 4\times10^6 Q = 12$, $Q(0) = 0$, and $Q'(0) = 0$. The particular solution is of the form Q = A, so that upon substitution into the D.E. we obtain $4\times10^6 A = 12$ or $A = 3\times10^{-6}$. The general solution of the D.E. is $Q = c_1 e^{r_1 t} + c_2 e^{r_2 t} + 3\times10^{-6}$, where r_1 and r_2 satisfy $r^2 + 5\times10^3 r + 4\times10^6 = 0$ and thus $r_1 = -1000$ and

$r_2 = -4000$. The I.C. yield $c_1 = -4 \times 10^{-6}$ and $c_2 = 10^{-6}$ and thus $Q = 10^{-6}(e^{-4000t} - 4e^{-1000t} + 3)$ coulombs. Substituting $t = .001$ sec we obtain
$Q(.001) = 10^{-6}(e^{-4} - 4e^{-1} + 3) = 1.5468 \times 10^{-6}$ coulombs.
Since exponentials are to a negative power $Q(t) \to 3 \times 10^{-6}$ coulombs as $t \to \infty$, which is the steady state charge.

22. The steady-state response is 12sin2t and thus the amplitude of the steady state response is four times the amplitude of the forcing term. This large an increase is due to the fact that the forcing function has the same frequency as the natural frequency, $\omega_0 = 2$, of the system. The graph also shows a phase lag of approximately 1/4 of a period. That is, the maximum of the response occurs 1/4 of a period after the maximum of the forcing function. Both these results are substantially different than those of either Probs. 21 or 23.

24.

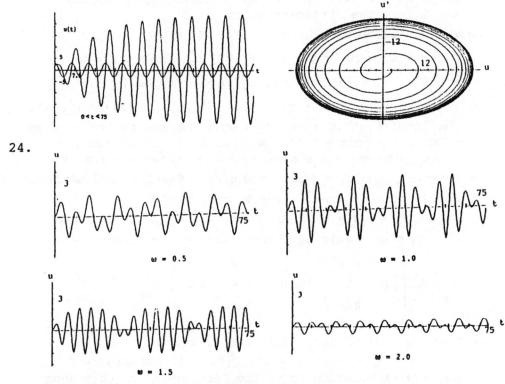

From viewing the above graphs, it appears that the system exhibits a beat near $\omega = 1.5$, while the pattern for $\omega = 1.0$ is more irregular. However, the system exhibits the resonance characteristic of the linear system for ω near 1, as the amplitude of the response is the largest here.

CHAPTER 4

Section 4.1, Page 222

2. Writing the equation in standard form of Eq.(2), we obtain $y''' + [(\sin t)/t]y'' + (3/t)y = \cos t/t$. The functions $p_1(t) = \sin t/t$, $p_3(t) = 3/t$ and $g(t) = \cos t/t$ each have a discontinuity at $t = 0$. Hence Theorem 4.1.1 guarantees that a solution exists for $t < 0$ or for $t > 0$.

4. The equation is in standard form with $p_1(t)$, $p_2(t)$ and $p_3(t)$ being continuous for all t. However, $g(t) = \ln t$ is defined and continuous only for $t > 0$.

8. We have $W(f_1, f_2, f_3) = \begin{vmatrix} 2t-3 & 2t^2+1 & 3t^2+t \\ 2 & 4t & 6t+1 \\ 0 & 4 & 6 \end{vmatrix} = 0$ for all t.

   Thus by the extension of Theorem 3.3.1 (or by the discussion following Eq.(8), the given functions are linearly dependent. To find a linear relation we have $c_1(2t-3) + c_2(2t^2+1) + c_3(3t^2+t) =$ $(2c_2+3c_3)t^2 + (2c_1+c_3)t + (-3c_1+c_2)$ which is zero when $(2c_2+3c_3) = 0$, $2c_1+c_3 = 0$ and $-3c_1+c_2 = 0$. Solving, we find $c_1 = 1$, $c_2 = 3$ and $c_3 = -2$ and hence $(2t-3)+3(2t^2+1)-2(3t^2+t) = 0$.

13. That e^t, e^{-t}, and e^{-2t} are solutions can be verified by direct substitution. For $y = e^{-2t}$, $y' = -2e^{-2t}$, $y'' = 4e^{-2t}$, and $y''' = -8e^{-2t}$ and thus $(-8e^{-2t}) + 2(4e^{-2t}) - (-2e^{-2t}) - 2(e^{-2t}) = 0$. Computing the Wronskian we obtain,

$$W(e^t, e^{-t}, e^{-2t}) = \begin{vmatrix} e^t & e^{-t} & e^{-2t} \\ e^t & -e^{-t} & -2e^{-2t} \\ e^t & e^{-t} & 4e^{-2t} \end{vmatrix} = e^{-2t}\begin{vmatrix} 1 & 1 & 1 \\ 1 & -1 & -2 \\ 1 & 1 & 4 \end{vmatrix} = -6e^{-2t}$$

17. To show that the given Wronskian is zero, it is helpful, in evaluating the Wronskian, to note that $(\sin^2 t)' = 2\sin t \cos t = \sin 2t$. The result can also be obtained directly since $\sin^2 t = (1 - \cos 2t)/2 = \dfrac{1}{10}(5) + (-1/2)\cos 2t$ and hence

$\sin^2 t$ is a linear combination of 5 and $\cos 2t$. Thus the functions are linearly dependent and their Wronskian is zero.

19a. The $(n-1)$st derivative of t^n is $[n(n-1)(n-2) \cdots 2]t$, so $y^{(n)} = [n(n-1)(n-2) \cdots (2)(1)]$.

19b. The nth derivative of e^{rt} is $y^{(n)} = r^n e^{rt}$.

19c. If we let $L[y] = y^{(4)} - 5y'' + 4y$ and if we use the result of Prob. 19b, we have $L[e^{rt}] = (r^4 - 5r^2 + 4)e^{rt}$. Thus e^{rt} will be a solution of the D.E. provided $(r^2-4)(r^2-1) = 0$. Solving for r, we obtain the four solutions e^t, e^{-t}, e^{2t} and e^{-2t}. Since $W(e^t, e^{-t}, e^{2t}, e^{-2t}) \neq 0$, the four functions form a fundamental set of solutions.

23. Writing the equation in the form of Eq.(2), we have

$p_1(t) = \dfrac{2}{t}$ and from Prob. 20, $W = ce^{-\int \frac{2dt}{t}} = \dfrac{c}{t^2}$.

27. Differentiating e^t and substituting in the D.E. we verify that $y = e^t$ is a solution: $(2-t)e^t + (2t-3)e^t - te^t + e^t = 0$. Now, as in Prob. 26, we let $y = v(t)e^t$. Differentiating three times and substituting into the D.E. yields $(2-t)e^t v''' + (3-t)e^t v'' = 0$. Dividing by $(2-t)e^t$ and letting $w = v''$ we obtain the first order separable equation $w' = -\dfrac{t-3}{t-2}w = (-1 + \dfrac{1}{t-2})w$. Separating t and w, integrating, and then solving for w yields $w = v'' = c_1(t-2)e^{-t}$. Integrating this twice then gives $v = c_1 te^{-t} + c_2 t + c_3$ so that $y = ve^t = c_1 t + c_2 te^t + c_3 e^t$, which is the complete solution, since it contains the given $y_1(t)$ and three constants.

Section 4.2, Page 230

2. If $-1 + i\sqrt{3} = R(\cos\theta + i\sin\theta) = Re^{i\theta}$, then $R\cos\theta = -1$ and $R\sin\theta = \sqrt{3}$. Thus $R^2 = (-1)^2 + (\sqrt{3})^2 = 4$ and the angle θ is given by $R\cos\theta = 2\cos\theta = -1$ and $R\sin\theta = 2\sin\theta = \sqrt{3}$. Hence $\cos\theta = -1/2$ and $\sin\theta = \sqrt{3}/2$ which has the solution $\theta = 2\pi/3$. The angle

θ is only determined up to an additive integer multiple of $\pm 2\pi$.

8. Writing $(1-i)$ in the form $Re^{i\theta}$, we have $R\cos\theta = 1$ and $R\sin\theta = -1$, which yield $R = \sqrt{2}$ and $\theta = -\pi/4$. Thus $(1-i) = \sqrt{2}\,e^{i(-\pi/4+2m\pi)}$ (where m is any integer) and hence $(1-i)^{1/2} = [2^{1/2}e^{i(-\pi/4+2m\pi)}]^{1/2} = 2^{1/4}e^{i(-\pi/8+m\pi)}$. We obtain the two square roots by setting $m = 0,1$. They are $2^{1/4}e^{-i\pi/8}$ and $2^{1/4}e^{i7\pi/8}$. Note that any other integer value of m gives one of these two values. Also note that $1-i$ could be written as $1-i = \sqrt{2}\,e^{i(7\pi/4 + 2m\pi)}$.

12. We look for solutions of the form $y = e^{rt}$. Substituting in the D.E., we obtain the characteristic equation $r^3 - 3r^2 + 3r - 1 = 0$ which has roots $r = 1,1,1$. Since the roots are repeated, the general solution is $y = c_1e^t + c_2te^t + c_3t^2e^t$.

15. We look for solutions of the form $y = e^{rt}$. Substituting in the D.E. we obtain the characteristic equation $r^6 + 1 = 0$. The six roots of $-1 = e^{i\pi}$ are obtained by setting $m = 0,1,2,3,4,5$ in $(-1)^{1/6} = e^{i(\pi+2m\pi)/6}$. They are $e^{i\pi/6} = (\sqrt{3} + i)/2$, $e^{i\pi/2} = i$, $e^{i5\pi/6} = (-\sqrt{3} + i)/2$, $e^{i7\pi/6} = (-\sqrt{3} - i)/2$, $e^{i3\pi/2} = -i$, and $e^{i11\pi/6} = (\sqrt{3} - i)/2$. Note that there are three pairs of conjugate roots. The general solution is
$$y = e^{\sqrt{3}\,t/2}[c_1\cos(t/2) + c_2\sin(t/2)]$$
$$+ e^{-\sqrt{3}\,t/2}[c_3\cos(t/2) + c_4\sin(t/2)] + c_5\cos t + c_6\sin t.$$

23. The characteristic equation is $r^3 - 5r^2 + 3r + 1 = 0$. Using the procedure suggested following Eq. (12) we try $r = 1$ as a root and find that indeed it is. Factoring out $(r-1)$ we are then left with $r^2 - 4r - 1 = 0$, which has the roots $2 \pm \sqrt{5}$.

27. The characteristic equation in this case is $12r^4 + 31r^3 + 75r^2 + 37r + 5 = 0$. Using an equation solver we find $r = -\dfrac{1}{4}, -\dfrac{1}{3}, -1 \pm 2i$. Thus $y = c_1e^{-t/4} + c_2e^{-t/3} + e^{-t}(c_3\cos 2t + c_4\sin 2t)$. As in Prob. 23, it is possible to find the first two of these

roots without using an equation solver. Factoring then
reduces the characteristic equation to a quadratic, which
can be solved for the other two roots.

29. The characteristic equation is $r^3 + r = 0$ and hence
$r = 0, +i, -i$ are the roots and the general solution is
$y(t) = c_1 + c_2\cos t + c_3\sin t$. $y(0) = 0$ implies
$c_1 + c_2 = 0$, $y'(0) = 1$ implies $c_3 = 1$ and $y''(0) = 2$
implies $-c_2 = 2$. Use this last equation in the first to
find $c_1 = 2$ and thus $y(t) = 2 - 2\cos t + \sin t$, which
continues to oscillate about $y = 2$ as $t \to \infty$.

29.   30.

30. The general solution is given by Eq. (21).

31. The characteristic equation is $r^4 - 4r^3 + 4r^2 = 0$, which
has the roots $r = 0,0,2,2$. Thus the general solution
would normally be written $y(t) = c_1 + c_2t + c_3e^{2t} + c_4te^{2t}$.
However, in order to evaluate the c's when the initial
conditions are given at $t = 1$, it is advantageous to
rewrite this as $y(t) = c_1 + c_2t + c_5e^{2(t-1)} + c_6(t-1)e^{2(t-1)}$,
which also satisfies the given D.E.

31.   34.

34. The characteristic equation is $4r^3 + r + 5 = 0$, which has

roots -1, $\dfrac{1}{2} \pm i$.

Thus $y(t) = c_1 e^{-t} + e^{t/2}(c_2\cos t + c_3\sin t)$,

$y'(t) = -c_1 e^{-t} + e^{t/2}[(c_2/2 + c_3)\cos t + (-c_2 + c_3/2)\sin t]$

and

$y''(t) = c_1 e^{-t} + e^{t/2}[(-3c_2/4 + c_3)\cos t + (-c_2 - 3c_3/4)\sin t]$.
The I.C. then yield $c_1 + c_2 = 2$, $-c_1 + c_2/2 + c_3 = 1$ and
$c_1 - 3c_2/4 + c_3 = -1$. Solving these last three equations
give $c_1 = 2/13$, $c_2 = 24/13$ and $c_3 = 3/13$.

37. The approach for solving the D.E. would normally yield
$y(t) = c_1\cos t + c_2\sin t + c_5 e^t + c_6 e^{-t}$ as the solution.
Since $\cosh t = (e^t + e^{-t})/2$ and $\sinh t = (e^t - e^{-t})/2$, $y(t)$ can
be written as $y(t) = c_1\cos t + c_2\sin t + c_3\cosh t + c_4\sinh t$,
where c_3 and c_4 can be written in terms of c_5 and c_6. It
is convenient to use $\cosh t$ and $\sinh t$ rather than e^t and
e^{-t} because the I.C. are given at $t = 0$. Since $\cosh t$ and
$\sinh t$ and all of their derivatives are either 0 or 1 at
$t = 0$, the algebra in satisfying the I.C. is greatly
simplified.

38a. Since $p_1(t) = 0$, $W = ce^{-\int 0\,dt} = c$.

39a. As in Section 3.8, the force that the spring designated
by k_1 exerts on mass m_1 is $-3u_1$. By an analysis similar
to that shown in Section 3.8, the middle spring exerts a
force of $-2(u_1-u_2)$ on mass m_1 and a force of $-2(u_2-u_1)$ on
mass m_2. Thus Newton's Law gives $m_1 u_1'' = -3u_1 - 2(u_1-u_2)$
and $m_2 u_2'' = -2(u_2-u_1)$. Setting the masses equal to 1 and
rewriting each equation yields Eqs.(i). In all cases the
positive direction is taken in the direction shown in
Figure 4.2.4.

39b. The characteristic equation for (ii) is $r^4 + 7r^2 + 6 = 0$,
or $(r^2+1)(r^2+6) = 0$. Thus the general solution of Eq.(ii)
is $u_1(t) = c_1\cos t + c_2\sin t + c_3\cos\sqrt{6}\,t + c_4\sin\sqrt{6}\,t$.

39c. From Eq.(iii) and u_1 from 39b we have $u_1(0) = c_1 + c_3 = 1$
and $u_1'(0) = c_2 + \sqrt{6}\,c_4 = 0$. From Eq.(i) we have
$u_1''(0) = 2u_2(0) - 5u_1(0) = -1$ and

$u_1'''(0) = 2u_2'(0) - 5u_1'(0) = 0$, hence $-c_1 - 6c_3 = -1$ and
$-c_2 - 6\sqrt{6}\,c_4 = 0$. Solving the four equations for the c_i
we find $c_1 = 1$ and $c_2 = c_3 = c_4 = 0$, so that $u_1 = \cos t$.
The first of Eqs.(i) then gives $2u_2 = u_1'' + 5u_1 = 4\cos t$
and thus $u_2 = 2\cos t$.

Section 4.3, Page 235

1. First solve the homogeneous D.E. The characteristic
    equation is $r^3 - r^2 - r + 1 = 0$, and the roots are $r = -1$,
    1, 1; hence $y_c(t) = c_1 e^{-t} + c_2 e^t + c_3 t e^t$. Using the
    superposition principle, we can write a particular
    solution as the sum of particular solutions corresponding
    to the D.E. $y''' - y'' - y' + y = 2e^{-t}$ and $y''' - y'' - y' + y = 3$. Our
    initial choice for a particular solution, Y_1, of the
    first equation is Ae^{-t}; but e^{-t} is a solution of the
    homogeneous equation so we multiply by t. Thus,
    $Y_1(t) = Ate^{-t}$. For the second equation we choose
    $Y_2(t) = B$, and there is no need to modify this choice.
    The constants are determined by substituting into the
    individual equations. We obtain $A = 1/2$, $B = 3$. Thus,
    the general solution is
    $y = c_1 e^{-t} + c_2 e^t + c_3 t e^t + 3 + (te^{-t})/2$.

5. The characteristic equation is $r^4 - 4r^2 = r^2(r^2 - 4) = 0$,
    so $y_c(t) = c_1 + c_2 t + c_3 e^{-2t} + c_4 e^{2t}$. For the particular
    solution correspnding to t^2 we assume
    $Y_1 = t^2(At^2 + Bt + C)$ and for the particular solution
    corresponding to e^t we assume $Y_2 = De^t$. Substituting Y_1,
    in the D.E. yields $-48A = 1$, $B = 0$ and $24A - 8C = 0$ and
    substituting Y_2 yields $-3D = 1$. Solving for A, B, C and
    D gives the desired solution.

9. The characteristic equation for the related homogeneous
    D.E. is $r^3 + 4r = 0$ with roots $r = 0$, $+2i$, $-2i$. Hence
    $y_c(t) = c_1 + c_2\cos 2t + c_3\sin 2t$. The initial choice for
    $Y(t)$ is $At + B$, but since B is a solution of the
    homogeneous equation we must multiply by t and assume
    $Y(t) = t(At+B)$. A and B are found by substituting in the
    D.E., which gives $A = 1/8$, $B = 0$, and thus the general

solution is $y(t) = c_1 + c_2\cos 2t + c_3\sin 2t + (1/8)t^2$.
Applying the I.C. we have $y(0) = 0 \Rightarrow c_1 + c_2 = 0$,
$y'(0) = 0 \Rightarrow 2c_3 = 0$, and $y''(0) = 1 \Rightarrow -4c_2 + 1/4 = 1$,
which have the solution $c_1 = 3/16$, $c_2 = -3/16$, $c_3 = 0$.
For small t the graph will approximate $3(1-\cos 2t)/16$ and
for large t it will be approximated by $t^2/8$.

13. The characteristic equation for the homogeneous D.E. is
$r^3 - 2r^2 + r = 0$ with roots $r = 0,1,1$. Hence the
complementary solution is $y_c(t) = c_1 + c_2e^t + c_3te^t$. We
consider the differential equations $y''' - 2y'' + y' = t^3$
and $y''' - 2y'' + y' = 2e^t$ separately. Our initial choice
for a particular solution, Y_1, of the first equation is
$A_0t^3 + A_1t^2 + A_2t + A_3$; but since a constant is a solution
of the homogeneous equation we must multiply by t. Thus
$Y_1(t) = t(A_0t^3 + A_1t^2 + A_2t + A_3)$. For the second equation
we first choose $Y_2(t) = Be^t$, but since both e^t and te^t
are solutions of the homogeneous equation, we multiply by
t^2 to obtain $Y_2(t) = Bt^2e^t$. Then $Y(t) = Y_1(t) + Y_2(t)$ by
the superposition principle and $y(t) = y_c(t) + Y(t)$.

17. The characteristic equation is
$r^4 - r^3 - r^2 + r = r(r-1)(r^2-1) = 0$, so the complementary
solution is $y_c(t) = c_1 + c_2e^{-t} + c_3e^t + c_4te^t$. The
superposition principle allows us to consider separately
the D.E. $y^{(4)} - y''' - y'' + y' = t^2 + 4$ and
$y^{(4)}s - y''' - y'' + y' = t\sin t$. For the first equation our
initial choice is $Y_1(t) = A_0t^2 + A_1t + A_2$; but this must
be multiplied by t since a constant is a solution of the
homogeneous D.E. Hence $Y_1(t) = t(A_0t^2 + A_1t + A_2)$. For
the second equation our initial choice that
$Y_2 = (B_0t + B_1)\cos t + + (C_0t + C_1)\sin t$ does not need to be
modified. Hence
$Y(t) = t(A_0t^2 + A_1t + A_2) + (B_0t + B_1)\cos t + (C_0t + C_1)\sin t$.

20. $(D-a)(D-b)f = (D-a)(Df-bf) = D^2f - (a+b)Df + abf$ and
$(D-b)(D-a)f = (D-b)(Df-af) = D^2f - (b+a)Df + baf$. Since
$a+b = b+a$ and $ab = ba$, we find the given equation holds
for any function f.

22a. The D.E. of Prob.13 can be written as
$D(D-1)^2 y = t^3 + 2e^t$. Since D^4 annihilates t^3 and $(D-1)$
annihilates $2e^t$, we have $D^5(D-1)^3 y = 0$, which corresponds
to Eq.(ii) of Prob. 21. The solution of this equation is
$y(t) = A_1 t^4 + A_2 t^3 + A_3 t^2 + A_4 t + A_5 + (B_1 t^2 + B_2 t + B_3)e^{-t}$.
Since A_5 and $(B_2 t + B_3)e^{-t}$ are solutions of the
homogeneous equation related to the original D.E., they
may be deleted and thus
$Y(t) = A_1 t^4 + A_2 t^3 + A_3 t^2 + A_4 t + B_1 t^2 e^{-t}$.

22b. $(D+1)^2(D^2+1)$ annihilates the right side of the D.E. of
Prob.14.

22e. $D^3(D^2+1)^2$ annihilates the right side of the D.E.of Prob.17.

Section 4.4, Page 240

1. The complementary solution is $y_c = c_1 + c_2\cos t + c_3\sin t$
    and thus we assume a particular solution of the form
    $Y = u_1(t) + u_2(t)\cos t + u_3(t)\sin t$. Differentiating and
    assuming Eq.(5), we obtain $Y' = -u_2\sin t + u_3\cos t$ and
    $$u_1' + u_2'\cos t + u_3'\sin t = 0 \qquad (a).$$
    Continuing this process we obtain $Y'' = -u_2\cos t - u_3\sin t$,
    $Y''' = u_2\sin t - u_3\cos t - u_2'\cos t - u_3'\sin t$ and
    $$-u_2'\sin t + u_3'\cos t = 0 \qquad\qquad (b).$$
    Substituting Y and its derivatives, as given above, into
    the D.E. we obtain the third equation:
    $$-u_2'\cos t - u_3'\sin t = \tan t \qquad (c).$$
    Equations (a), (b) and (c) constitute Eqs.(10) of the
    text for this problem and may be solved to give
    $u_1' = \tan t$, $u_2' = -\sin t$, and $u_3' = -\sin^2 t/\cos t$. Thus
    $u_1 = -\ln\cos t$, $u_2 = \cos t$ and $u_3 = \sin t - \ln(\sec t + \tan t)$
    and substitution into Y above gives
    $Y = -\ln\cos t + 1 - (\sin t)\ln(\sec t + \tan t)$, since
    $\sin^2 t + \cos^2 t = 1$ Note that the constant 1 can be
    absorbed in c_1.

4. Replace tant in Eq.(c) of Prob. 1 by sect and use

Eqs.(a) and (b) as in Prob.1 to obtain $u_1' = \sec t$, $u_2' = -1$ and $u_3' = -\sin t/\cos t$.

5.   Replace $\sec t$ in Prob. 7 with $e^{-t}\sin t$.

7.   Since e^t, $\cos t$ and $\sin t$ are solutions of the related homogenous equation we have

$Y(t) = u_1 e^t + u_2 \cos t + u_3 \sin t$. Eqs. (10) then are

$u_1' e^t + u_2' \cos t + u_3' \sin t = 0$

$u_1' e^t - u_2' \sin t + u_3' \cos t = 0$

$u_1' e^t - u_2' \cos t - u_3' \sin t = \sec t$.

Using Abel's identity, $W(t) = c\exp(-\int p_1(t)dt) = ce^t$.

Using the above equations, $W(0) = \begin{vmatrix} 1 & 1 & 0 \\ 1 & 0 & 1 \\ 1 & -1 & 0 \end{vmatrix} = 2$, so

$c = 2$ and $W(t) = 2e^t$. From Eq.(11), we have

$u'_1(t) = \dfrac{\sec t\, W_1(t)}{2e^t}$, where $W_1 = \begin{vmatrix} 0 & \cos t & \sin t \\ 0 & -\sin t & \cos t \\ 1 & -\cos t & -\sin t \end{vmatrix} = 1$

and thus

$u_1'(t) = \dfrac{1}{2}e^{-t}/\cos t$. Likewise

$u_2' = \dfrac{\sec t\, W_2(t)}{2e^t} = -\dfrac{1}{2}\sec t(\cos t - \sin t)$ and

$u_3' = \dfrac{\sec t\, W_3(t)}{2e^t} = -\dfrac{1}{2}\sec t(\sin t + \cos t)$. Thus

$u_1 = \dfrac{1}{2}\displaystyle\int_{t_0}^{t}\dfrac{e^{-s}ds}{\cos s}$, $u_2 = -\dfrac{1}{2}t - \dfrac{1}{2}\ln(\cos t)$ and $u_3 = -\dfrac{1}{2}t + \dfrac{1}{2}\ln(\cos t)$

which, when substituted into the assumed form for Y, yields the desired solution.

11. Since the D.E. is the same as in Prob. 7, we may use the complete solution from that, with $t_0 = 0$. Thus

$y(0) = c_1 + c_2 = 2$, $y'(0) = c_1 + c_3 - \dfrac{1}{2} + \dfrac{1}{2} = -1$ and

$y''(0) = c_1 - c_2 + \dfrac{1}{2} - 1 + \dfrac{1}{2} = 1$. A computer algebra

system may be used to find the respective derivatives.

11.

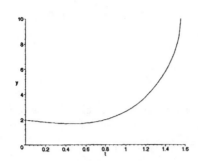

14. Since a fundamental set of solutions of the homogeneous
    D.E. is $y_1 = e^t$, $y_2 = \cos t$, $y_3 = \sin t$, a particular
    solution is of the form $Y(t) = e^t u_1(t) + (\cos t)u_2(t) +$
    $(\sin t)u_3(t)$. Differentiating and making the same
    assumptions that lead to Eqs.(10), we obtain

$$u_1' e^t + u_2' \cos t + u_3' \sin t = 0$$
$$u_1' e^t - u_2' \sin t + u_3' \cos t = 0$$
$$u_1' e^t - u_2' \cos t - u_3' \sin t = g(t)$$

Solving these equations using either determinants or by
elimination, we obtain $u_1' = (1/2)e^{-t}g(t)$,

$u_2' = (1/2)(\sin t - \cos t)g(t), u_3' = -(1/2)(\sin t + \cos t)g(t)$.
Integrating these and substituting into Y yields

$$Y(t) = \frac{1}{2}\{e^t\int_{t_0}^t e^{-s}g(s)ds + \cos t \int_{t_0}^t (\sin s - \cos s)g(s)ds$$

$$-\sin t\int_{t_0}^t (\sin s + \cos s)g(s)ds\}.$$

Putting e^t, $\cos t$ and $\sin t$ inside the respective integrals
yields

$$Y(t) = (1/2)\int_{t_0}^t (e^{t-s} + \cos t\sin s - \cos t\cos s -$$
$$\sin t\sin s - \sin t\cos s)g(s)ds.$$

If we use the trigonometric identities
$\sin(A-B) = \sin A\cos B - \cos A\sin B$ and
$\cos(A-B) = \cos A\cos B + \sin A\sin B$, we obtain the desired
result. Note: Eqs.(11) and (12) of this section give the
same result, but it is not recommended to memorize these
equations.

16. The characteristic equation has the repeated roots
    $r = 1,1,1$ and thus the particular solution has the form
    $Y = e^t u_1(t) + te^t u_2(t) + t^2 e^t u_3(t)$. Differentiating,
    making the same assumptions as in the earlier problems,
    and solving the three linear equations for u_1', u_2', and u_3'

yields
$u_1' =(1/2)t^2e^{-t}g(t)$, $u_2' = -te^{-t}g(t)$ and $u_3' = (1/2)e^{-t}g(t)$.
Integrating and substituting into Y yields the desired
solution.   For instance

$$te^tu_2 = -te^t\int_{t_0}^t se^{-s}g(s)ds = -\frac{1}{2}\int_{t_0}^t 2tse^{(t-s)}g(s)ds,\text{ and}$$

likewise for u_1 and u_3. If $g(t) = t^{-2}e^t$ then $g(s) = e^s/s^2$,

and thus $e^{(t-s)}g(s) = \dfrac{e^t}{s^2}$ and the integration with repect

to s is accomplished using the power rule.   Note that
terms involving t_0 become part of the complimentary
solution.

84

CHAPTER 5

Section 5.1, Page 249

2. Use the ratio test:
$$\lim_{n \to \infty} \frac{\left|(n+1)x^{n+1}/2^{n+1}\right|}{\left|nx^n/2^n\right|} = \lim_{n \to \infty} \frac{n+1}{n} \frac{1}{2} |x| = \frac{|x|}{2}.$$
Therefore the series converges absolutely for $|x| < 2$.
For $x = 2$ and $x = -2$ the nth term does not approach zero
as $n \to \infty$ so the series diverge. Hence the radius of
convergence is $\rho = 2$.

5. Use the ratio test:
$$\lim_{n \to \infty} \frac{\left|(2x+1)^{n+1}/(n+1)^2\right|}{\left|(2x+1)^n/n^2\right|} = \lim_{n \to \infty} \frac{n^2}{(n+1)^2} |2x+1| = |2x+1|.$$
Therefore the series converges absolutely for $|2x+1| < 1$,
or $|x+1/2| < 1/2$. At $x = 0$ and $x = -1$ the series also
converge absolutely. However, for $|x+1/2| > 1/2$ the
series diverges by the ratio test. The radius of
convergence is $\rho = 1/2$.

9. For this problem $f(x) = \sin x$, so $f'(x) = \cos x$,
$f''(x) = -\sin x$, $f'''(x) = -\cos x$, $f^{iv}(x) = \sin x...$, and thus
$f(0) = 0$, $f'(0) = 1$, $f''(0) = 0$, $f'''(0) = -1,...$. The even
terms in the series will vanish and the odd terms will
alternate in sign. We obtain $\sin x = \sum_{n=0}^{\infty} (-1)^n x^{2n+1}/(2n+1)!$.

Now, $\lim_{n \to \infty} \frac{\left|(-1)^{n+1}x^{2n+3}/(2n+3)!\right|}{\left|(-1)^n x^{2n+1}/(2n+1)!\right|} = \lim_{n \to \infty} x^2 \frac{1}{(2n+3)(2n+2)} = 0,$
so the series converges for all x and hence $\rho = \infty$.

12. For this problem $f(x) = x^2$. Hence $f'(x) = 2x$, $f''(x) = 2$,
and $f^{(n)}(x) = 0$ for $n > 2$. Then $f(-1) = 1$, $f'(-1) = -2$,
$f''(-1) = 2$ and $x^2 = 1 - 2(x+1) + 2(x+1)^2/2! = 1 - 2(x+1) + (x+1)^2$. Since the series terminates after a finite
number of terms, it converges for all x. Thus $\rho = \infty$.

13. For this problem $f(x) = \ln x$. Hence $f'(x) = 1/x$,
$f''(x) = -1/x^2$, $f'''(x) = 1 \cdot 2/x^3,...$, and $f^{(n)}(x) = (-1)^{n+1}(n-1)!/x^n$. Then $f(1) = 0$, $f'(1) = 1$, $f''(1) = -1$,
$f'''(1) = 1 \cdot 2,...$, $f^{(n)}(1) = (-1)^{n+1}(n-1)!$ The Taylor

series is $\ln x = (x-1) - (x-1)^2/2 + (x-1)^3/3 - \ldots =$

$\sum_{n=1}^{\infty} (-1)^{n+1}(x-1)^n/n$. It follows from the ratio test that

the series converges absolutely for $|x-1| < 1$. However,
the series diverges at $x = 0$ so $\rho = 1$.

18. Writing the individual terms of y, we have
$y = a_0 + a_1 x + a_2 x^2 + \ldots + a_n x^n + \ldots$, so
$y' = a_1 + 2a_2 x + 3a_3 x^2 + \ldots + (n+1)a_{n+1}x^n + \ldots$, and
$y'' = 2a_2 + 3 \cdot 2a_3 x + 4 \cdot 3a_4 x^2 + \ldots + (n+2)(n+1)a_{n+2}x^n + \ldots$.
If $y'' = y$, we then equate coefficients of like powers of x to
obtain $2a_2 = a_0$, $3 \cdot 2a_3 = a_1$, $4 \cdot 3a_4 = a_2$, $\ldots (n+2)(n+1)a_{n+2} = a_n$,
which yields the desired result for $n = 0,1,2,3 \ldots$.

19. Set $m = n-1$ on the right hand side of the equation. Then
$n = m+1$ and when $n = 1$, $m = 0$. Thus the right hand side

becomes $\sum_{m=0}^{\infty} a_m(x-1)^{m+1}$, which is the same as the left hand

side when m is replaced by n.

23. Multiplying each term of the first series by x yields

$x\sum_{n=1}^{\infty} na_n x^{n-1} = \sum_{n=1}^{\infty} na_n x^n = \sum_{n=0}^{\infty} na_n x^n$, where the last

equality can be verified by writing out the first few
terms (or noting that $na_n = 0$ for $n = 0$). Changing the
index from k to n (n=k) in the second series then yields

$\sum_{n=0}^{\infty} na_n x^n + \sum_{n=0}^{\infty} a_n x^n = \sum_{n=0}^{\infty} (n+1)a_n x^n.$

25. $\sum_{m=2}^{\infty} m(m-1)a_m x^{m-2} + x\sum_{k=1}^{\infty} ka_k x^{k-1} =$

$\sum_{n=0}^{\infty} (n+2)(n+1)a_{n+2}x^n + \sum_{k=1}^{\infty} ka_k x^k =$

$\sum_{n=0}^{\infty} [(n+2)(n+1)a_{n+2} + na_n]x^n.$ In the first case we have

let $n = m - 2$ in the first summation and multiplied each
term of the second summation by x. In the second case we
have let $n = k$ and noted that for $n = 0$, $na_n = 0$.

28. If we shift the index of summation in the first sum by
     letting m = n-1, we have

$$\sum_{n=1}^{\infty} n a_n x^{n-1} = \sum_{m=0}^{\infty} (m+1) a_{m+1} x^m.$$ Substituting this into the

given equation and letting m = n again, we obtain:

$$\sum_{n=0}^{\infty} (n+1) a_{n+1} x^n + 2 \sum_{n=0}^{\infty} a_n x^n = 0, \text{ or}$$

$$\sum_{n=0}^{\infty} [(n+1) a_{n+1} + 2a_n] x^n = 0.$$

Hence $a_{n+1} = -2a_n/(n+1)$ for n = 0,1,2,3,... . Thus
$a_1 = -2a_0$, $a_2 = -2a_1/2 = 2^2 a_0/2$, $a_3 = -2a_2/3 = -2^3 a_0/2 \cdot 3 =$
$-2^3 a_0/3!...$ and $a_n = (-1)^n 2^n a_0/n!$. Notice that for n = 0
this formula reduces to a_0 so we can write

$$\sum_{n=0}^{\infty} a_n x^n = \sum_{n=0}^{\infty} (-1)^n 2^n a_0 x^n/n! = a_0 \sum_{n=0}^{\infty} (-2x)^n/n! = a_0 e^{-2x}.$$

Section 5.2, Page 259

2. $y = \displaystyle\sum_{n=0}^{\infty} a_n x^n$; $y' = \displaystyle\sum_{n=1}^{\infty} n a_n x^{n-1}$ and since we must multiply

y' by x in the D.E. we do not shift the index; and

$$y'' = \sum_{n=2}^{\infty} n(n-1) a_n x^{n-2} = \sum_{n=0}^{\infty} (n+2)(n+1) a_{n+2} x^n.$$ Substituting

in the D.E., we obtain

$$\sum_{n=0}^{\infty} (n+2)(n+1) a_{n+2} x^n - \sum_{n=1}^{\infty} n a_n x^n - \sum_{n=0}^{\infty} a_n x^n = 0.$$ In order to

have the starting point the same in all three summations,
we let n = 0 in the first and third terms to obtain the
following

$$(2 \cdot 1 \ a_2 - a_0) x^0 + \sum_{n=1}^{\infty} [(n+2)(n+1) a_{n+2} - (n+1) a_n] x^n = 0.$$

Thus $a_{n+2} = a_n/(n+2)$ for n = 1,2,3,... . Note that the
recurrence relation is also correct for n = 0. We show

how to calculate the odd a_n's:

$a_3 = a_1/3$, $a_5 = a_3/5 = a_1/5 \cdot 3$, $a_7 = a_5/7 = a_1/7 \cdot 5 \cdot 3, \ldots$.

Now notice that $a_3 = 2a_1/(2 \cdot 3) = 2a_1/3!$, that

$a_5 = 2 \cdot 4 a_1/(2 \cdot 3 \cdot 4 \cdot 5) = 2^2 \cdot 2 a_1/5!$, and that

$a_7 = 2 \cdot 4 \cdot 6 a_1/(2 \cdot 3 \cdot 4 \cdot 5 \cdot 6 \cdot 7) = 2^3 \cdot 3! \ a_1/7!$. Likewise

$a_9 = a_7/9 = 2^3 \cdot 3! \ a_1/(7!)9 = 2^3 \cdot 3! \ 8a_1/9! = 2^4 \cdot 4! \ a_1/9!$.

Continuing we have $a_{2m+1} = 2^m m! \ a_1/(2m+1)!$. In the same

way we find that the even a's are given by $a_{2m} = a_0/2^m \ m!$.

Thus

$$y = a_0 \sum_{m=0}^{\infty} \frac{x^{2m}}{2^m m!} + a_1 \sum_{m=0}^{\infty} \frac{2^m m! \ x^{2m+1}}{(2m+1)!}.$$

3. $\displaystyle y = \sum_{n=0}^{\infty} a_n(x-1)^n$; $\displaystyle y' = \sum_{n=1}^{\infty} n a_n(x-1)^{n-1} = \sum_{n=0}^{\infty} (n+1)a_{n+1}(x-1)^n$,

and

$$y'' = \sum_{n=2}^{\infty} n(n-1)a_n(x-1)^{n-2} = \sum_{n=0}^{\infty} (n+2)(n+1)a_{n+2}(x-1)^n.$$

Substituting in the D.E. and setting $x = 1 + (x-1)$ we
obtain

$$\sum_{n=0}^{\infty} (n+2)(n+1)a_{n+2}(x-1)^n - \sum_{n=0}^{\infty} (n+1)a_{n+1}(x-1)^n - \sum_{n=1}^{\infty} n a_n(x-1)^n$$

$$- \sum_{n=0}^{\infty} a_n(x-1)^n = 0,$$

where the third term comes from:

$$-(x-1)y' = -\sum_{n=0}^{\infty} (n+1)a_{n+1}(x-1)^{n+1} = -\sum_{n=1}^{\infty} n a_n(x-1)^n.$$

Letting $n = 0$ in the first, second, and the fourth sums,
we obtain

$$(2 \cdot 1 \cdot a_2 - 1 \cdot a_1 - a_0)(x-1)^0 + \sum_{n=1}^{\infty} \big[(n+2)(n+1)a_{n+2}$$

$$- (n+1)a_{n+1} - (n+1)a_n \big](x-1)^n = 0. \text{ Setting the}$$

terms in the square brackets equal to zero and dividing
by $(n-1)$ yields $(n+2)a_{n+2} - a_{n+1} - a_n = 0$ for $n = 1, 2, 3, \ldots,$
(which also holds for $n = 0$). This recurrance relation
can be used to solve for a_2 in terms of a_0 and a_1, then

for a_3 in terms of a_0 and a_1, etc. In many cases it is easier to first take $a_0 = 0$ and generate one solution and then take $a_1 = 0$ and generate the second linearly independent solution. Thus, choosing $a_0 = 0$ we find that $a_2 = a_1/2$, $a_3 = (a_2+a_1)/3 = a_1/2$, $a_4 = (a_3+a_2)/4 = a_1/4$, $a_5 = (a_4+a_3)/5 = 3a_1/20,\ldots$. This yields the solution $y_2(x) = a_1[(x-1) + (x-1)^2/2 + (x-1)^3/2 + (x-1)^4/4 + 3(x-1)^5/20 + \ldots]$. The second independent solution may be obtained by choosing $a_1 = 0$. Then $a_2 = a_0/2$, $a_3 = (a_2+a_1)/3 = a_0/6$, $a_4 = (a_3+a_2)/4 = a_0/6$, $a_5 = (a_4+a_3)/5 = a_0/15,\ldots$. This yields the solution $y_1(x) = a_0[1+(x-1)^2/2+(x-1)^3/6+(x-1)^4/6+(x-1)^5/15+\ldots]$.

5. $y = \sum\limits_{n=0}^{\infty} a_n x^n$; $y' = \sum\limits_{n=1}^{\infty} n a_n x^{n-1}$; and $y'' = \sum\limits_{n=2}^{\infty} n(n-1) a_n x^{n-2}$.

Substituting in the D.E. and shifting the index in both summations for y'' gives

$$\sum_{n=0}^{\infty} (n+2)(n+1)a_{n+2}x^n - \sum_{n=1}^{\infty} (n+1)n\, a_{n+1}x^n + \sum_{n=0}^{\infty} a_n x^n =$$

$$(2\cdot 1\cdot a_2 + a_0)x^0 + \sum_{n=1}^{\infty}[(n+2)(n+1)a_{n+2} - (n+1)na_{n+1} + a_n]x^n = 0.$$

Thus $a_2 = -a_0/2$ and $a_{n+2} = na_{n+1}/(n+2) - a_n/(n+2)(n+1)$, $n = 1,2,\ldots$. Choosing $a_0 = 0$ yields $a_2 = 0$, $a_3 = -a_1/6$, $a_4 = 2a_3/4 = -a_1/12$, $a_5 = 3a_4/5 - a_3/20 = -a_1/24 \ldots$, and hence $y_2(x) = a_1(x - x^3/6 - x^4/12 - x^5/24 + \ldots)$. A second linearly independent solution is obtained by choosing $a_1 = 0$. Then $a_2 = -a_0/2$, $a_3 = a_2/3 = -a_0/6$, $a_4 = 2a_3/4 - a_2/12 = -a_0/24,\ldots$ which gives $y_1(x) = a_0(1 - x^2/2 - x^3/6 - x^4/24 + \ldots)$.

8. If $y = \sum\limits_{n=1}^{\infty} a_n(x-1)^n$ then

$$xy = [1+(x-1)]y = \sum_{n=1}^{\infty} a_n(x-1)^n + \sum_{n=1}^{\infty} a_n(x-1)^{n+1},$$

$$y' = \sum_{n=1}^{\infty} na_n(x-1)^{n-1}, \text{ and}$$

$$xy'' = [1+(x-1)]y''$$

$$= \sum_{n=1}^{\infty} n(n-1)a_n(x-1)^{n-2} + \sum_{n=1}^{\infty} n(n-1)a_n(x-1)^{n-1}.$$

14. You will need to rewrite x+1 as 3 + (x-2) in order to multiply x+1 times y' as a power series about $x_0 = 2$.

16a. From Prob. 6 we have

$$y(x) = c_1(1 - x^2 + \frac{1}{6}x^4+...) + c_2(x - \frac{1}{4}x^3 + \frac{7}{160}x^5 + ...).$$

Now $y(0) = c_1 = -1$ and $y'(0) = c_2 = 3$ and thus

$$y(x) = -1 + x^2 - \frac{1}{6}x^4 + ... + 3x - \frac{3}{4}x^3 + \cdots$$

$$= -1 + 3x + x^2 - \frac{3}{4}x^3 - \frac{1}{6}x^4 + ...$$

16b.

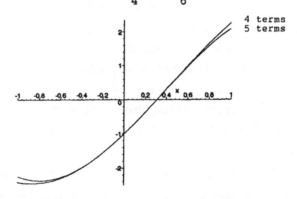

16c. It appears that f is a reasonable approximation for $|x| < 0.7$. In fact, the magnitude of the difference in the two graphs is .02 for $|x| = .6$ and .04 for $|x| = .7$

19. Letting t = x-1 yields $(x-1)^2 = t^2$ and $(x^2-1) = t^2+2t$. Now let u(t) = y(t+1) and hence $u' = y'$ and $u'' = y''$. Thus the D.E. transforms into $u''(t) + t^2u'(t) + (t^2+2t)u(t) = 0$.

Assuming that $u(t) = \sum_{n=0}^{\infty} a_n t^n$, we have $u'(t) = \sum_{n=1}^{\infty} na_n t^{n-1}$ and

$u''(t) = \sum_{n=2}^{\infty} n(n-1)a_n t^{n-2}$. Substituting in the D.E. and shifting indices yields

$$\sum_{n=0}^{\infty} (n+2)(n+1)a_{n+2}t^n + \sum_{n=2}^{\infty} (n-1)a_{n-1}t^n + \sum_{n=2}^{\infty} a_{n-2}t^n$$

$$+ \sum_{n=1}^{\infty} 2a_{n-1}t^n = 0,$$

$$2 \cdot 1 \cdot a_2 t^0 + (3 \cdot 2 \cdot a_3 + 2 \cdot a_0)t^1 + \sum_{n=2}^{\infty} [(n+2)(n+1)a_{n+2}$$

$$+ (n+1)a_{n-1} + a_{n-2}]t^n = 0.$$

It follows that $a_2 = 0$, $a_3 = -a_0/3$ and
$a_{n+2} = -a_{n-1}/(n+2) - a_{n-2}/[(n+2)(n+1)]$, $n = 2,3,4\ldots$. We
obtain one solution by choosing $a_1 = 0$. Then $a_4 = -a_0/12$,
$a_5 = -a_2/5 - a_1/20 = 0$, $a_6 = -a_3/6 - a_2/30 = a_0/18,\ldots$. Thus
one solution is $u_1(t) = a_0(1 - t^3/3 - t^4/12 + t^6/18 + \ldots)$ so
$y_1(x) = u_1(x-1) = a_0[1 - (x-1)^3/3 - (x-1)^4/12 + (x-1)^6/18 + \ldots]$.
We obtain a second solution by choosing $a_0 = 0$. Then
$a_4 = -a_1/4$, $a_5 = -a_2/5 - a_1/20 = -a_1/20$,
$a_6 = -a_3/6 - a_2/30 = 0$, $a_7 = -a_4/7 - a_3/42 = a_1/28,\ldots$.
Thus a second linearly independent solution is
$u_2(t) = a_1[t - t^4/4 - t^5/20 + t^7/28 + \ldots]$ or
$y_2(x) = u_2(x-1)$
$$= a_1[(x-1) - (x-1)^4/4 - (x-1)^5/20 + (x-1)^7/28 + \ldots].$$
The Taylor series for $x^2 - 1$ about $x = 1$ may be obtained by
writing $x = (x-1) + 1$ so $x^2 = (x-1)^2 + 2(x-1) + 1$ and
$x^2 - 1 = (x-1)^2 + 2(x-1)$. The D.E. now appears as
$y'' + (x-1)^2 y' + [(x-1)^2 + 2(x-1)]y = 0$ which is identical to
the transformed equation with $t = x - 1$.

22b. $y = a_0 + a_1 x + a_2 x^2 + \ldots$, $y^2 = a_0^2 + 2a_0 a_1 x + (2a_0 a_2 + a_1^2)x^2$
$+ \ldots$, $y' = a_1 + 2a_2 x + 3a_3 x^2 + \ldots$, and
$(y')^2 = a_1^2 + 4a_1 a_2 x + (6a_1 a_3 + 4a_2{}^2)x^2 + \ldots$. Substituting
these into $(y')^2 = 1 - y^2$ and collecting coefficients of
like powers of x yields $(a_1^2 + a_0^2 - 1) + (4a_1 a_2 + 2a_0 a_1)x +$
$(6a_1 a_3 + 4a_2^2 + 2a_0 a_2 + a_1^2)x^2 + \ldots = 0$. As in the earlier
problems, each coefficient must be zero. The I.C. $y(0) = 0$
requires that $a_0 = 0$, and thus $a_1^2 + a_0^2 - 1 = 0$ gives $a_1^2 = 1$.
However, the D.E. indicates that y' is always positive, so

$y'(0) = a_1 > 0$ implies $a_1 = 1$. Then $4a_1a_2 + 2a_0a_1 = 0$ implies that $a_2 = 0$; and $6a_1a_3 + 4a_2^2 + 2a_0a_2 + a_1^2 = 6a_1a_3 + a_1^2 = 0$ implies that $a_3 = -1/6$. Thus $y = x - x^3/3! + \dots$.

23. We have $y(x) = a_0y_1 + a_1y_2$, where y_1 and y_2 are found in Prob.2. Now, $y(0) = a_0 = 1$ and $y'(0) = a_1 = 0$ and thus $y(x) = 1 + x^2/2! + x^4/4! + x^6/6! + \cdots$.

23.   26.

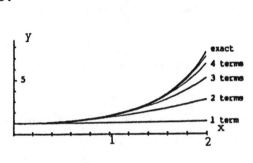
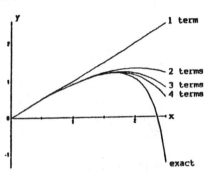

26. Again, $y(x) = a_0y_1 + a_1y_2$, where y_1 and y_2 are found in Prob.10, and $y(0) = a_0 = 0$ and $y'(0) = a_1 = 1$.

Thus $y(x) = x - \dfrac{x^3}{12} - \dfrac{x^5}{240} - \dfrac{x^7}{2240}$.

Section 5.3, Page 265

1. The D.E. can be solved for y'' to yield $y'' = -xy' - y$. If $y = \phi(x)$ is a solution, then $\phi''(x) = -x\phi'(x) - \phi(x)$ and thus setting $x = 0$ we obtain $\phi''(0) = -0 - 1 = -1$. Differentiating the equation for y'' yields $y''' = -xy'' - 2y'$ and hence setting $y = \phi(x)$ again yields $\phi'''(0) = -0 - 0 = 0$. In a similar fashion $y^{(4)} = -xy''' - 3y''$ and thus $\phi^{(4)}(0) = -0 - 3(-1) = 3$. The process can be continued to calculate higher derivatives of $\phi(x)$.

6. The zeros of $P(x) = x^2 - 2x - 3$ are $x = -1$ and $x = 3$. For $x_0 = 4$, $x_0 = -4$, and $x_0 = 0$ the distance to the nearest zero of $P(x)$ is 1,3, and 1, respectively. Thus a lower bound for the radius of convergence for series solutions in powers of $(x-4)$, $(x+4)$, and x is $\rho = 1$, $\rho = 3$, and $\rho = 1$, respectively.

9a. Since $P(x) = 1$ has no zeros, the radius of convergence about $x_0 = 0$ is $\rho = \infty$.

9f. Since $P(x) = 2 + x^2$ has zeros at $x = \pm\sqrt{2}\,i$, the lower bound for the radius of convergence of the series solution about $x_0 = 0$ is $\rho = \sqrt{2}$.

9h. $P(x) = x$ has a zero at $x = 0$ and since $x_0 = 1$, $\rho = 1$.

10a. If we assume that $y = \sum\limits_{n=0}^{\infty} a_n x^n$, then $y' = \sum\limits_{n=1}^{\infty} n a_n x^{n-1}$ and

$y'' = \sum\limits_{n=2}^{\infty} n(n-1) a_n x^{n-2}$. Substituting in the D.E. gives:

$$\sum\limits_{n=2}^{\infty} n(n-1) a_n x^{n-2} - \sum\limits_{n=2}^{\infty} n(n-1) a_n x^n - \sum\limits_{n=1}^{\infty} n a_n x^n + \alpha^2 \sum\limits_{n=0}^{\infty} a_n x^n = 0.$$

Shifting indices of summation and collecting coefficients of like powers of x yields the equation:

$(2\cdot1\cdot a_2 + \alpha^2 a_0)x^0 + [3\cdot2\cdot a_3 + (\alpha^2-1)a_1]x^1$

$$+ \sum\limits_{n=2}^{\infty} [(n+2)(n+1)a_{n+2} + (\alpha^2-n^2)a_n]x^n = 0.$$

Hence the recurrence relation is
$a_{n+2} = (n^2-\alpha^2)a_n/(n+2)(n+1)$, $n = 0, 1, 2, \ldots$. For the first solution we choose $a_1 = 0$. We find that
$a_2 = -\alpha^2 a_0/2\cdot1$, $a_3 = 0$, $a_4 = (2^2-\alpha^2)a_2/4\cdot3 = -(2^2-\alpha^2)\alpha^2 a_0/4!$
$\ldots$, $a_{2m} = -[(2m-2)^2 - \alpha^2]\ldots (2^2-\alpha^2)\alpha^2 a_0/(2m)!$,

and $a_{2m+1} = 0$, so $y_1(x) = 1 - \dfrac{\alpha^2}{2!}x^2 - \dfrac{(2^2-\alpha^2)\alpha^2}{4!}x^4 - \ldots$

$$-\dfrac{[(2m-2)^2-\alpha^2]\ldots(2^2-\alpha^2)\alpha^2}{(2m)!}x^{2m} - \ldots,$$

where we have set $a_0 = 1$. For the second solution we take $a_0 = 0$ and $a_1 = 1$ in the recurrence relation to obtain the desired solution.

10b. If α is an even integer $2k$ then $(2m-2)^2-\alpha^2 = 4(m-1)^2-4k^2$. Thus when $m = k+1$ all terms in the series for $y_1(x)$ are zero after the x^{2k} term. A similar argument shows that if $\alpha = 2k+1$ then all terms in $y_2(x)$ are zero after the x^{2k+1} term.

11. The Taylor series about $x = 0$ for sinx is

$\sin x = x - x^3/3! + x^5/5! - \dots$. Assuming that

$y = \displaystyle\sum_{n=2}^{\infty} a_n x^n$ we find $y'' + (\sin x)y = 2a_2 + 6a_3 x + 12a_4 x^2$

$+ 20a_5 x^3 + 30a_6 x^4 + 42a_7 x^5 + \dots$

$+ (x - x^3/3! + x^5/5! - \dots)(a_0 + a_1 x + a_2 x^2 + a_3 x^3 + a_4 x^4 + \dots)$

$= 2a_2 + (6a_3 + a_0)x + (12a_4 + a_1)x^2 + (20a_5 + a_2 - a_0/6)x^3 +$

$(30a_6 + a_3 - a_1/6)x^4 + (42a_7 + a_4 - a_2/3! + a_0/5!)x^5 + \dots = 0.$

Hence $a_2 = 0$, $a_3 = -a_0/6$, $a_4 = -a_1/12$, $a_5 = a_0/120$,

$a_6 = (a_1 + a_0)/180$, $a_7 = -a_0/7! + a_1/504$, $\dots$. We set

$a_0 = 1$ and $a_1 = 0$ and obtain

$y_1(x) = (1 - x^3/6 + x^5/120 + x^6/180 + \dots)$. Next we set

$a_0 = 0$ and $a_1 = 1$ and obtain

$y_2(x) = (x - x^4/12 + x^6/180 + x^7/504 + \dots)$. Since

$p(x) = 1$ and $q(x) = \sin x$ both have $\rho = \infty$, the solution in this case converges for all x, that is, $\rho = \infty$

18. We know that $e^x = 1 + x + x^2/2! + x^3/3! + \dots$, and

therefore $e^{x^2} = 1 + x^2 + x^4/2! + x^6/3! + \dots$. Hence, if

$y = \displaystyle\sum_{n=0}^{\infty} a_n x^n$, we have

$a_1 + 2a_2 x + 3a_3 x^2 + \dots = (1 + x^2 + x^4/2 + \dots)(a_0 + a_1 x + a_2 x^2 + \dots)$

$= a_0 + a_1 x + (a_0 + a_2)x^2 + \dots$.

Thus, $a_1 = a_0$ $2a_2 = a_1$ and $3a_3 = a_0 + a_2$, which yield the desired solution.

20. Substituting $y = \displaystyle\sum_{n=0}^{\infty} a_n x^n$ into the D.E. we obtain

$\displaystyle\sum_{n=1}^{\infty} na_n x^{n-1} - \sum_{n=0}^{\infty} a_n x^n = x^2.$ Shifting indices in the summation

yields $\displaystyle\sum_{n=0}^{\infty}[(n+1)a_{n+1} - a_n]x^n = x^2$. Equating coefficients of
both sides then gives: $a_1 - a_0 = 0$, $2a_2 - a_1 = 0$, $3a_3 - a_2 = 1$
and $(n+1)a_{n+1} = a_n$ for $n = 3,4,\ldots$. Thus $a_1 = a_0$,
$a_2 = a_1/2 = a_0/2$, $a_3 = 1/3 + a_2/3 = 1/3 + a_0/2\cdot3$,
$a_4 = a_3/4 = 1/3\cdot4 + a_0/2\cdot3\cdot4 = 2/4! + a_0/4!$, and in general
$a_n = a_{n-1}/n = 2/n! + a_0/n!$. Hence

$$y(x) = a_0(1 + x + \frac{x^2}{2!}+\ldots+\frac{x^n}{n!}\ldots) + 2(\frac{x^3}{3!}+\frac{x^4}{4!}+\ldots+\frac{x^n}{n!}+\ldots).$$

Using the power series for e^x, the first and second sums
can be rewritten as $a_0 e^x + 2(e^x - 1 - x - x^2/2)$.

22. Substituting $y = \displaystyle\sum_{n=0}^{\infty}a_n x^n$ into the Legendre equation,

shifting indices, and collecting coefficients of like
powers of x yields
$[2\cdot1\cdot a_2 + \alpha(\alpha+1)a_0]x^0 + \{3\cdot2\cdot a_3 - [2\cdot1 - \alpha(\alpha+1)]a_1\}x^1 +$

$\displaystyle\sum_{n=2}^{\infty}\{(n+2)(n+1)a_{n+2} - [n(n+1) - \alpha(\alpha+1)]a_n\}x^n = 0$. Thus

$a_2 = -\alpha(\alpha+1)a_0/2!$, $a_3 = [2\cdot1 - \alpha(\alpha+1)]a_1/3! =$
$-(\alpha-1)(\alpha+2)a_1/3!$ and the recurrence relation is
$(n+2)(n+1)a_{n+2} = -[\alpha(\alpha+1) - n(n+1)]a_n = -(\alpha-n)(\alpha+n+1)a_n$,
$n = 2,3,\ldots$. Setting $a_1 = 0$, $a_0 = 1$ yields a solution
with $a_3 = a_5 = a_7 = \ldots = 0$ and
$a_4 = \alpha(\alpha-2)(\alpha+1)(\alpha+3)/4!,\ldots$, and

$a_{2m} = (-1)^m[\alpha(\alpha-2)\ldots(\alpha-2m+2)][(\alpha+1)\ldots(\alpha+2m-1)]/(2m)!$.
The second linearly independent solution is obtained by
setting $a_0 = 0$ and $a_1 = 1$. The coefficients are
$a_2 = a_4 = a_6 = \ldots = 0$ and $a_3 = -(\alpha-1)(\alpha+2)/3!$, and
$a_5 = -(\alpha-3)(\alpha+4)a_3/5\cdot4 = (\alpha-1)(\alpha-3)(\alpha+2)(\alpha+4)/5!$.

26. Using the chain rule we have:
$$\frac{dF(\phi)}{d\phi} = \frac{dF[\phi(x)]}{dx}\frac{dx}{d\phi} = -f'(x)\sin\phi(x) = -f'(x)\sqrt{1-x^2},$$
$$\frac{d^2F(\phi)}{d\phi^2} = \frac{d}{dx}[-f'(x)\sqrt{1-x^2}]\frac{dx}{d\phi} = (1-x^2)f''(x) - xf'(x),$$
which when substituted into the D.E. yields the desired
result.

28. Since $[(1-x^2)y']' = (1-x^2)y'' - 2xy'$, the Legendre
Equation, from Prob. 22, can be written as shown. Thus,
carrying out the multiplications indicated yields the two
equations:

$$P_m[(1-x^2)P_n']' = -n(n+1)P_nP_m$$

$$P_n[(1-x^2)P_m']' = -m(m+1)P_nP_m.$$

As long as $n \neq m$ the second equation can be subtracted
from the first and the result integrated from -1 to 1 to
obtain

$$\int_{-1}^{1}\{P_m[(1-x^2)P_n']'-P_n[(1-x^2)P_m']'\}dx = [m(m+1)-n(n+1)]\int_{-1}^{1}P_nP_mdx$$

The left side may be integrated by parts to yield

$$[P_m(1-x^2)P_n' - P_n(1-x^2)P_m']_{-1}^{1} + \int_{-1}^{1}[P_m'(1-x^2)P_n' - P_n'(1-x^2)P_m']dx,$$

which is zero. Thus $\int_{-1}^{1}P_n(x)P_m(x)dx = 0$ for $n \neq m$.

Section 5.4, Page 271

1. Since the coefficients of y, y' and y'' have no common
factors and since $P(x)$ vanishes only at $x = 0$ we conclude
that $x = 0$ is a singular point. Writing the D.E. in the
form $y'' + p(x)y' + q(x)y = 0$, we obtain $p(x) = (1-x)/x$
and $q(x) = 1$. Thus for the singular point we have
$$\lim_{x\to 0} x\, p(x) = \lim_{x\to 0} 1-x = 1, \quad \lim_{x\to 0} x^2 q(x) = 0 \text{ and thus } x = 0$$
is a regular singular point.

5. Writing the D.E. in the form $y'' + p(x)y' + q(x)y = 0$, we
find $p(x) = x/(1-x)(1+x)^2$ and $q(x) = 1/(1-x^2)(1+x)$.
Therefore $x = \pm 1$ are singular points. Since
$$\lim_{x\to 1} (x-1)p(x) \text{ and } \lim_{x\to 1} (x-1)^2 q(x) \text{ both exist, we conclude}$$
$x = 1$ is a regular singular point. Finally, since
$\lim_{x\to -1} (x+1)p(x)$ does not exist, we conclude that $x = -1$ is
an irregular singular point.

12. Writing the D.E. in the form $y + p(x)y' + q(x)y = 0$, we
see that $p(x) = e^x/x$ and $q(x) = (3\cos x)/x$. Thus $x = 0$ is
a singular point. Since $xp(x) = e^x$ is analytic at $x = 0$
and $x^2 q(x) = 3x\cos x$ is analytic at $x = 0$ the point $x = 0$
is a regular singular point.

17. Writing the D.E. in the form $y'' + p(x)y' + q(x)y = 0$, we see that $p(x) = \dfrac{x}{\sin x}$ and $q(x) = \dfrac{4}{\sin x}$. Since $\lim\limits_{x \to 0} q(x)$ does not exist, the point $x_0 = 0$ is a singular point and since neither $\lim\limits_{x \to \pm n\pi} p(x)$ nor $\lim\limits_{x \to \pm n\pi} q(x)$ exist, either, the points $x_0 = \pm n\pi$ are also singular points. To determine whether the singular points are regular or irregular we must use Eq.(9) and the result #7 of multiplication and division of power series from Section 5.1. For $x_0 = 0$, we have

$$xp(x) = \frac{x^2}{\sin x} = \frac{x^2}{x - \dfrac{x^3}{6} + \dots} = x[1 + \frac{x^2}{6} + \dots]$$

$$= x + \frac{x^3}{6} + \dots,$$

which converges about $x_0 = 0$ and thus $xp(x)$ is analytic at $x_0 = 0$. $x^2 q(x)$, by similar steps, is also analytic at $x_0 = 0$ and thus $x_0 = 0$ is a regular singular point. For $x_0 = n\pi$, we have

$$(x-n\pi)p(x) = \frac{(x-n\pi)x}{\sin x} = \frac{(x-n\pi)[(x-n\pi) + n\pi]}{\pm(x-n\pi) + \dfrac{-(x-n\pi)^3}{6} \pm \dots}$$

$$= \pm[(x-n\pi) + n\pi][1 + \frac{(x-n\pi)^2}{6} + \dots], \text{ which}$$

converges about $x_0 = n\pi$ and thus $(x-n\pi)p(x)$ is analytic at $x = n\pi$. Similarly $(x+n\pi)p(x)$ and $(x \pm n\pi)^2 q(x)$ are analytic and thus $x_0 = \pm n\pi$ are regular singular points.

19. Substituting $y = \sum\limits_{n=0}^{\infty} a_n x^n$ into the D.E. yields

$$2\sum_{n=2}^{\infty} n(n-1)a_n x^{n-1} + 3\sum_{n=1}^{\infty} na_n x^{n-1} + \sum_{n=0}^{\infty} a_n x^{n+1} = 0. \text{ The last sum}$$

becomes $\sum\limits_{n=2}^{\infty} a_{n-2} x^{n-1}$ (let $m = n+2$ and then replace m by n), the first term of the middle sum is $3a_1$, and thus we have

$$3a_1 + \sum_{n=2}^{\infty} \{[2n(n-1)+3n]a_n + a_{n-2}\}x^{n-1} = 0. \text{ Hence } a_1 = 0 \text{ and}$$

$$a_n = \frac{-a_{n-2}}{n(2n+1)},$$ which is the desired recurrance relation.
Thus all even coefficients are found in terms of a_0 and all odd coefficients are zero, thereby yielding only one solution of the desired form.

21. If $\xi = 1/x$ then
$$\frac{dy}{dx} = \frac{dy}{d\xi}\frac{d\xi}{dx} = -\frac{1}{x^2}\frac{dy}{d\xi} = -\xi^2\frac{dy}{d\xi},$$

$$\frac{d^2y}{dx^2} = \frac{d}{d\xi}(-\xi^2\frac{dy}{d\xi})\frac{d\xi}{dx} = (-2\xi\frac{dy}{d\xi} - \xi^2\frac{d^2y}{d\xi^2})(-\frac{1}{x^2})$$

$$= \xi^4\frac{d^2y}{d\xi^2} + 2\xi^3\frac{dy}{d\xi}.$$

Substituting in the D.E. we have

$$P(1/\xi)[\xi^4\frac{d^2y}{d\xi^2} + 2\xi^3\frac{dy}{d\xi}] + Q(1/\xi)[-\xi^2\frac{dy}{d\xi}] + R(1/\xi)y = 0, \text{ or}$$

$$\xi^4 P(1/\xi)\frac{d^2y}{d\xi^2} + [2\xi^3 P(1/\xi) - \xi^2 Q(1/\xi)]\frac{dy}{d\xi} + R(1/\xi)y = 0.$$

The result then follows from the theory of singular points at $\xi = 0$.

23. Since $P(x) = x^2$, $Q(x) = x$ and $R(x) = -4$ we have
$f(\xi) = [2P(1/\xi)/\xi - Q(1/\xi)/\xi^2]/P(1/\xi) = 2/\xi - 1/\xi = 1/\xi$
and $g(\xi) = R(1/\xi)/\xi^4 P(1/\xi) = -4/\xi^2$. Thus the point at infinity is a singular point. Since both $\xi f(\xi)$ and $\xi^2 g(\xi)$ are analytic at $\xi = 0$, the point at infinity is a regular singular point.

25. Since $P(x) = x^2$, $Q(x) = x$, and $R(x) = x^2 - \upsilon^2$,
$f(\xi) = [2P(1/\xi)/\xi - Q(1/\xi)/\xi^2]/P(1/\xi) = 2/\xi - 1/\xi = 1/\xi$
and $g(\xi) = R(1/\xi)/\xi^4 P(1/\xi) = (1/\xi^2 - \upsilon^2)/\xi^2 = 1/\xi^4 - \upsilon^2/\xi^2$.
Thus the point at infinity is a singular point. Although $\xi f(\xi) = 1$ is analytic at $\xi = 0$, $\xi^2 g(\xi) = 1/\xi^2 - \upsilon^2$ is not, so the point at infinity is an irregular singular point.

Section 5.5, Page 278

2. Comparing the D.E. to Eq.(27), we seek solutions of the form $y = (x+1)^r$ for $x + 1 > 0$. Substitution of y into the D.E. yields $F(r) = [r(r-1) + 3r + 3/4](x+1)^r = 0$. Thus $r^2 + 2r + 3/4 = 0$, which yields $r = -3/2, -1/2$. The

general solution of the D.E. is then

$$y = c_1|x+1|^{-1/2} + c_2|x+1|^{-3/2}, \ x \neq -1.$$

4.   If $y = x^r$ then $F(r) = r(r-1) + 3r + 5 = 0$.
     So $r^2 + 2r + 5 = 0$ and $r = (-2 \pm \sqrt{4-20})/2 = -1 \pm 2i$.
     Thus the general solution of the D.E. is
     $$y = c_1x^{-1}\cos(2\ln|x|) + c_2x^{-1}\sin(2\ln|x|), \ x \neq 0.$$

9.   Again let $y = x^r$ to obtain $F(r) = r(r-1) - 5r + 9 = 0$, or
     $(r-3)^2 = 0$. Thus the roots are $x = 3,3$ and
     $y = c_1x^3 + c_2x^3\ln|x|, \ x \neq 0$, is the solution of the D.E.

13. In this csae $F(r) = 2r(r-1) + r - 3 = 2r^2 - r - 3 =$
     $(2r-3)(r+1) = 0$, so $y = c_1x^{3/2} + c_2x^{-1}$ (since $y_0 = 1$, we
     don't need $|x|$) and $y' = \frac{3}{2}c_1x^{1/2} - c_2x^{-2}$. Setting $x = 1$
     in y and y' we obtain $c_1 + c_2 = 1$ and $\frac{3}{2}c_1 - c_2 = 4$, which
     yield $c_1 = 2$ and
     $c_2 = -1$. Hence $y = 2x^{3/2} - x^{-1}$. As $x \to 0^+$ we have
     $y \to -\infty$ due to the second term.

16. We have $F(r) = r(r-1) + 3r + 5 = r^2 + 2r + 5 = 0$. Thus
     $r_1,r_2 = -1 \pm 2i$ and $y = x^{-1}[c_1\cos(2\ln x) + c_2\sin(2\ln x)]$.
     Then $y(1) = c_1 = 1$ and $y' = -x^{-2}[\cos(2\ln x) + c_2\sin(2\ln x)]$
     $+ x^{-1}[-\sin(2\ln x)2/x + c_2\cos(2\ln x)2/x]$ so that
     $y'(1) = -1-2c_2 = -1$, or $c_2 = 0$. Hence $y = x^{-1}\cos(2\ln x)$
     for $x > 0$. As $x \to 0^+$ this will oscillate rapidly, with
     large amplitudes.

17. Substituting $y = x^r$, we find that $r(r-1) + \alpha r + 5/2 = 0$
     or $r^2 + (\alpha-1)r + 5/2 = 0$. Thus
     $r_1,r_2 = [-(\alpha-1) \pm \sqrt{(\alpha-1)^2-10}]/2$. In order for
     solutions to approach zero as $x \to 0$ it is necessary that
     the real parts of r_1 and r_2 be positive. Suppose that α
     > 1, then
     $\sqrt{(\alpha-1)^2-10}$ is either imaginary or real and less than
     $\alpha - 1$; hence the real parts of r_1 and r_2 will be

negative. Suppose that $\alpha = 1$, then $r_1, r_2 = \pm i\sqrt{10}$ and the solutions are oscillatory. Suppose that $\alpha < 1$, then $\sqrt{(\alpha-1)^2-10}$ is either imaginary or real and less than $|\alpha-1| = 1 - \alpha$; hence the real parts of r_1 and r_2 will be positive. Thus if $\alpha < 1$ the solutions of the D.E. will approach zero as $x \to 0$.

21. In all cases the roots of $F(r) = 0$ are given by Eq.(5) and the forms of the solution are given in Theorem 5.5.1.

21a. The real part of the root must be positive so, from Eq.(5), $\alpha < 1$. Also $\beta > 0$, since the $\sqrt{(\alpha-1)^2-4\beta}$ term must be less than $|\alpha-1|$.

21d. The real part of the root must be negative, so $\alpha > 1$, with $\beta \geq 0$ (for $\beta = 0$ one root is zero, which is bounded as $x \to \infty$). If $\alpha = 1$, then the roots are $\pm\sqrt{-4\beta}$, so $\beta > 0$ will yield oscillatory solutions as $x \to \infty$, which are bounded.

22. Assume that $y = v(x)x^{r_1}$. Then $y' = v(x)r_1 x^{r_1-1} + v'(x)x^{r_1}$ and $y'' = v(x)r_1(r_1-1)x^{r_1-2} + 2v'(x)r_1 x^{r_1-1} + v''(x)x^{r_1}$. Substituting in the D.E. and collecting terms yields $x^{r_1+2} v'' + (\alpha + 2r_1)x^{r_1+1} v' + [r_1(r_1-1) + \alpha r_1 + \beta]x^{r_1} v = 0$. Now we make use of the fact that r_1 is a double root of $f(r) = r(r-1) + \alpha r + \beta$. This means that $f(r_1) = 0$ and $f'(r_1) = 2r_1 - 1 + \alpha = 0$. Hence the D.E. for v reduces to $x^{r_1+2} v'' + x^{r_1+1} v'$. Since $x > 0$ we may divide by x^{r_1+1} to obtain $xv'' + v' = 0$. Thus $v(x) = \ln x$ and a second solution is $y = x^{r_1}\ln x$.

25. From Prob.23b, the change of variable $x = e^z$ transforms the D.E. into $u'' - 4u' + 4u = \ln(e^z) = z$, which has the solution $u(z) = c_1 e^{2z} + c_2 z e^{2z} + (1/4)z + 1/4$ (using the Method of Undetermined Coefficients). Hence $y(x) = c_1 x^2 + c_2 x^2 \ln x + (1/4)\ln x + 1/4$.

31. If $x > 0$, then $|x| = x$ and $|x|^{r_1} = x^{r_1}$ so we can choose $c_1 = k_1$. If $x < 0$, then $|x| = -x$ and $|x|^{r_1} = (-x)^{r_1} = (-1)^{r_1}x^{r_1}$ and we can choose $c_1 = (-1)^{r_1}k_1$, or $k_1 = (-1)^{-r_1}c_1 = (-1)^{r_1}c_1$. In both cases we have $c_2 = k_2$.

Section 5.6, Page 284

2. If the D.E. is put in the standard form
$y'' + p(x)y + q(x)y = 0$, then $p(x) = x^{-1}$ and
$q(x) = 1 - 1/9x^2$. Thus $x = 0$ is a singular point. Since
$xp(x) \to 1$ and $x^2 q(x) \to -1/9$ as $x \to 0$ it follows that
$x = 0$ is a regular singular point. In determining a
series solution of the D.E. it is more convenient to
leave the equation in the form given rather than divide
by x^2, the coefficient of y''. If we substitute

$$y = \sum_{n=0}^{\infty} a_n x^{n+r}, \text{ we have}$$

$$\sum_{n=0}^{\infty}(n+r)(n+r-1)a_n x^{n+r} + \sum_{n=0}^{\infty}(n+r)a_n x^{n+r} + (x^2 - \frac{1}{9})\sum_{n=0}^{\infty} a_n x^{n+r} = 0.$$

Note that $x^2 \sum_{n=0}^{\infty} a_n x^{n+r} = \sum_{n=0}^{\infty} a_n x^{n+r+2} = \sum_{n=2}^{\infty} a_{n-2} x^{n+r}$. Thus we

have $[r(r-1) + r - \frac{1}{9}]a_0 x^r + [(r+1)r + (r+1) - \frac{1}{9}]a_1 x^{r+1} +$

$$\sum_{n=2}^{\infty}\{[(n+r)(n+r-1) + (n+r) - \frac{1}{9}]a_n + a_{n-2}\} x^{n+r} = 0. \text{ From}$$

the first term, the indicial equation is $r^2 - 1/9 = 0$
with roots $r_1 = 1/3$ and $r_2 = -1/3$. For either value of
r it is necessary to take $a_1 = 0$ in order that the
coefficient of x^{r+1} be zero. The recurrence relation is
$[(n+r)^2 - 1/9]a_n = -a_{n-2}$. For $r = 1/3$ we have

$$a_n = \frac{-a_{n-2}}{(n + \frac{1}{3})^2 - (\frac{1}{3})^2} = -\frac{a_{n-2}}{(n + \frac{2}{3})n}, \ n = 2,3,4,\ldots .$$

Since $a_1 = 0$ it follows from the recurrence relation that
$a_3 = a_5 = a_7 = \ldots = 0$. For the even coefficients it is
convenient to let $n = 2m$, $m = 1,2,3,\ldots .$ Then
$a_{2m} = -a_{2m-2}/2^2 m(m + \frac{1}{3})$. The first few coefficients are
given by

$$a_2 = \frac{(-1)a_0}{2^2(1 + \frac{1}{3})1}, \ a_4 = \frac{(-1)a_2}{2^2(2 + \frac{1}{3})2} = \frac{a_0}{2^4(1 + \frac{1}{3})(2 + \frac{1}{3})2!}$$

$$a_6 = \frac{(-1)a_4}{2^2(3 + \frac{1}{3})3} = \frac{(-1)a_0}{2^6(1 + \frac{1}{3})(2 + \frac{1}{3})(3 + \frac{1}{3})3!} , \text{ and the}$$

coefficent of x^{2m} for $m = 1, 2, \ldots$ is

$$a_{2m} = \frac{(-1)^m a_0}{2^{2m} m!(1 + \frac{1}{3})(2 + \frac{1}{3}) \ldots (m + \frac{1}{3})} . \quad \text{Thus one}$$

solution (on setting $a_0 = 1$) is

$$y_1(x) = x^{1/3}[1 + \sum_{m=1}^{\infty} \frac{(-1)^m}{m! \ (1 + \frac{1}{3})(2 + \frac{1}{3}) \ldots (m + \frac{1}{3})} (\frac{x}{2})^{2m}] .$$

Since $r_2 = -1/3 \neq r_1$ and $r_1 - r_2 = 2/3$ is not an integer, we can calculate a second series solution corresponding to $r = -1/3$. The recurrence relation is $n(n-2/3)a_n = -a_{n-2}$, which yields the desired solution following the steps just outlined. Note that $a_1 = 0$, as in the first solution, and thus all the odd coefficients are zero.

4.   Putting the D.E. in the form $y'' + p(x)y' + q(x)y = 0$, we see that $p(x) = 1/x$ and $q(x) = -1/x$. Thus $x = 0$ is a singular point, and since $xp(x) \to 1$ and $x^2 q(x) \to 0$, as $x \to 0$, $x = 0$ is a regular singular point. Substituting

$$y = \sum_{n=0}^{\infty} a_n x^{n+r} \text{ in } xy'' + y' - y = 0 \text{ and shifting indices we}$$

obtain

$$\sum_{n=-1}^{\infty} a_{n+1}(r+n+1)(r+n)x^{n+r} + \sum_{n=-1}^{\infty} a_{n+1}(r+n+1)x^{n+r} - \sum_{n=0}^{\infty} a_n x^{n+r} = 0, \text{ or}$$

$$[r(r-1) + r]a_0 x^{-1+r} + \sum_{n=0}^{\infty} [(r+n+1)^2 a_{n+1} - a_n]x^{n+r} = 0. \quad \text{The}$$

indicial equation is $r^2 = 0$ so $r = 0$ is a double root. Thus we will obtain only one series of the form

$$y = x^r \sum_{n=0}^{\infty} a_n x^n. \quad \text{Setting } r = 0 \text{ in the coefficient of } x^{n+r},$$

we find that the recurrence relation is $(n+1)^2 a_{n+1} = a_n$, $n = 0, 1, 2, \ldots$. The coefficients are

$a_1 = a_0$, $a_2 = a_1/2^2 = a_0/2^2$, $a_3 = a_2/3^2 = a_0/3^2 \cdot 2^2$,
$a_4 = a_3/4^2 = a_0/4^2 \cdot 3^2 \cdot 2^2, \ldots$ and $a_n = a_0/(n!)^2$. Thus one

solution (on setting $a_0 = 1$) is $y = \displaystyle\sum_{n=0}^{\infty} x^n/(n!)^2$.

11. If we make the change of variable $t = x-1$ and let
$y = u(t)$, then the Legendre equation transforms to
$(t^2 + 2t)u''(t) + 2(t+1)u'(t) - \alpha(\alpha+1)u(t) = 0$. Since
$x = 1$ is a regular singular point of the original
equation, we know that $t = 0$ is a regular singular point

of the transformed equation. Substituting $u = \displaystyle\sum_{n=0}^{\infty} a_n t^{n+r}$

in the transformed equation and shifting indices, we
obtain

$$\sum_{n=0}^{\infty}(n+r)(n+r-1)a_n t^{n+r} + 2\sum_{n=-1}^{\infty}(n+r+1)(n+r)a_{n+1}t^{n+r}$$

$$+ 2\sum_{n=0}^{\infty}(n+r)a_n t^{n+r} + 2\sum_{n=-1}^{\infty}(n+r+1)a_{n+1}t^{n+r} - \alpha(\alpha+1)\sum_{n=0}^{\infty}a_n t^{n+r} = 0,$$

or $[2r(r-1) + 2r]a_0 t^{r-1} + \displaystyle\sum_{n=0}^{\infty}\{2(n+r+1)^2 a_{n+1}$

$$+ [(n+r)(n+r+1) - \alpha(\alpha+1)]a_n\}t^{n+r} = 0.$$

The indicial equation is $2r^2 = 0$ so $r = 0$ is a double
root. Thus there will be only one series solution of the

form $y = \displaystyle\sum_{n=0}^{\infty} a_n t^{n+r}$. The recurrence relation is

$2(n+1)^2 a_{n+1} = [\alpha(\alpha+1) - n(n+1)]a_n, n = 0,1,2,\ldots$. We have
$a_1 = [\alpha(\alpha+1)]a_0/2 \cdot 1^2$, $a_2 = [\alpha(\alpha+1)][\alpha(\alpha+1) - 1 \cdot 2]a_0/2^2 \cdot 2^2 \cdot 1^2$,
$a_3 = [\alpha(\alpha+1)][\alpha(\alpha+1) - 1 \cdot 2][\alpha(\alpha+1) - 2 \cdot 3]a_0/2^3 \cdot 3^2 \cdot 2^2 \cdot 1^2,\ldots,$
and $a_n = [\alpha(\alpha+1)][\alpha(\alpha+1)-1 \cdot 2]..[\alpha(\alpha+1)-(n-1)n]a_0/2^n(n!)^2$.
Reverting to the variable x it follows that one solution
of the Legendre equation in powers of $x-1$ is

$$y_1(x) = \sum_{n=0}^{\infty} [\alpha(\alpha+1)][\alpha(\alpha+1) - 1 \cdot 2] \ldots$$

$[\alpha(\alpha+1) - (n-1)n](x-1)^n/2^n(n!)^2$ where we have set $a_0 = 1$, which is equivalent to the answer in the text if a (-1) is taken out of each square bracket.

14. The standard form is $y'' + p(x)y' + q(x)y = 0$, with $p(x) = 1/x$ and $q(x) = 1$. Thus $x = 0$ is a singular point; and since $xp(x) \to 1$ and $x^2q(x) \to 0$ as $x \to 0$, $x = 0$ is a regular singular point. Substituting $y = \sum\limits_{n=0}^{\infty} a_n x^{n+r}$ into $x^2y'' + xy' + x^2y = 0$ and shifting indices appropriately, we obtain

$$\sum_{n=0}^{\infty}(n+r)(n+r-1)a_n x^{n+r} + \sum_{n=0}^{\infty}(n+r)a_n x^{n+r} + \sum_{n=2}^{\infty}a_{n-2}x^{n+r} = 0, \text{ or}$$

$$[r(r-1)+r]a_0 x^r + [(1+r)r+1+r]a_1 x^{r+1} + \sum_{n=2}^{\infty}[(n+r)^2 a_n + a_{n-2}]x^{n+r} = 0.$$

The indicial equation is $r^2 = 0$ so $r = 0$ is a double root. It is necessary to take $a_1 = 0$ in order that the coefficient of x^{r+1} be zero. The recurrence relation in $n^2 a_n = -a_{n-2}$, $n = 2,3,\ldots$. Since $a_1 = 0$ it follows that $a_3 = a_5 = a_7 = \ldots = 0$. For the even coefficients we let $n = 2m$, $m = 1,2,\ldots$. Then $a_{2m} = -a_{2m-2}/2^2 m^2$ so $a_2 = -a_0/2^2 \cdot 1^2$, $a_4 = a_0/2^2 \cdot 2^2 \cdot 1^2 \cdot 2^2, \ldots$, and $a_{2m} = (-1)^m a_0/2^{2m}(m!)^2$. Thus one solution of the Bessel equation of order zero is

$$J_0(x) = 1 + \sum_{m=1}^{\infty}(-1)^m x^{2m}/2^{2m}(m!)^2 \text{ where we have set } a_0 = 1.$$

Using the ratio test it can be shown that the series converges for all x. Also note that $J_0(x) \to 1$ as $x \to 0$.

15. In order to determine the form of the integral for x near zero we must study the integrand for x small. Using the above series for J_0, we have

$$\frac{1}{x[J_0(x)]^2} = \frac{1}{x[1 - x^2/2 +\ldots]^2} = \frac{1}{x[1 - x^2 +\ldots]}$$

$$= \frac{1}{x}[1 + x^2 + \ldots] \text{ for } x \text{ small.}$$

Thus

$$y_2(x) = J_0(x)\int\frac{dx}{x[J_0(x)]^2} = J_0(x)\int[\frac{1}{x} + x + \dots]dx$$

$$= J_0(x)[\ln x + \frac{x^2}{x} + \dots], \text{ and it is clear that } y_2(x)$$

will contain a logarithmic term.

16a. Putting the D.E. in the standard form
$y'' + p(x)y' + q(x)y = 0$ we see that $p(x) = 1/x$ and
$q(x) = (x^2-1)/x^2$. Thus $x = 0$ is a singular point and
since $xp(x) \to 1$ and $x^2q(x) \to -1$ as $x \to 0$, $x = 0$ is a

regular singular point. Substituting $y = \sum_{n=0}^{\infty} a_n x^{n+r}$ into

$x^2y'' + xy' + (x^2-1)y = 0$, shifting indices appropriately,
and collecting coefficients of common powers of x we
obtain $[r(r-1) + r - 1]a_0 x^r + [(1+r)r + 1 + r -1]a_1 x^{r+1}$

$$+ \sum_{n=2}^{\infty} \{[(n+r)^2 - 1]a_n + a_{n-2}\}x^{n+r} = 0.$$

The indicial equation is $r^2-1 = 0$ so the roots are $r_1 = 1$
and $r_2 = -1$. For either value of r it is necessary to
take $a_1 = 0$ in order that the coefficient of x^{r+1} be zero.
The recurrence relation is $[(n+r)^2 - 1]a_n = -a_{n-2}$,
$n = 2,3,4\dots$. For $r = 1$ we have $a_n = -a_{n-2}/[n(n+2)]$,
$n = 2,3,4,\dots$. Since $a_1 = 0$ it follows that $a_3 = a_5 = a_7$
$= \dots = 0$. Let $n = 2m$. Then $a_{2m} = -a_{2m-2}/2^2 m(m+1)$, $m =$
$1,2,\dots$, so $a_2 = -a_0/2^2 \cdot 1 \cdot 2$, $a_4 = -a_2/2^2 \cdot 1 \cdot 2 \cdot 3 =$
$a_0/2^2 \cdot 2^2 \cdot 1 \cdot 2 \cdot 2 \cdot 3, \dots$, and $a_{2m} = (-1)^m a_0/2^{2m} m!(m+1)!$. Thus one
solution (set $a_0 = 1/2$) of the Bessel equation of order

one is $J_1(x) = (x/2) \sum_{n=0}^{\infty} (-1)^n x^{2n}/(n+1)!n!2^{2n}$. The ratio

test shows that the series converges for all x. Also
note that $J_1(x) \to 0$ as $x \to 0$.

16b. For $r = -1$ the recurrence relation is
$[(n-1)^2 - 1]a_n = -a_{n-2}$, $n = 2,3,\dots$. Substituting $n = 2$
into the relation yields $[(2-1)^2 - 1]a_2 = 0$ $a_2 = -a_0$.
Hence it is impossible to determine a_2 and consequently

impossible to find a series solution of the form

$$x^{-1} \sum_{n=0}^{\infty} b_n x^n.$$

Secton 5.7, Page 292

1.   The D.E. has the form $P(x)y'' + Q(x)y' + R(x)y = 0$ with
     $P(x) = x$, $Q(x) = 2x$, and $R(x) = 6e^x$. From this we find
     $p(x) = Q(x)/P(x) = 2$ and $q(x) = R(x)/P(x) = 6e^x/x$ and
     thus $x = 0$ is a singular point. Since $xp(x) = 2x$ and
     $x^2q(x) = 6xe^x$ are analytic at $x = 0$ we conclude that
     $x = 0$ is a regular singular point. Next, we have
     $xp(x) \to 0 = p_0$ and $x^2q(x) \to 0 = q_0$ as $x \to 0$ and thus
     Eq.(7), the indicial equation, is $F(r) = r(r-1) + 0 \cdot r + 0$
     $= r^2 - r = 0$, which has the roots $r_1 = 1$ and $r_2 = 0$. These
     are the exponents of the singularity at $x = 0$.

3.   The equation has the form $P(x)y'' + Q(x)y' + R(x)y = 0$
     with $P(x) = x(x-1)$, $Q(x) = 6x^2$ and $R(x) = 3$. Since $P(x)$,
     $Q(x)$, and $R(x)$ are polynomials with no common factors and
     $P(0) = 0$ and $P(1) = 0$, we conclude that $x = 0$ and $x = 1$
     are singular points. The first point, $x = 0$, can be shown
     to be a regular singular point using steps similar to
     those to shown in Prob. 1. For $x = 1$, we must put the
     D.E. in a form similar to Eq.(1) for this case. To do
     this, divide the D.E. by x and multiply by $(x-1)$ to
     obtain $(x-1)^2y'' + 6x(x-1)y + \dfrac{3}{x}(x-1)y = 0$. Comparing this
     to Eq.(1) we find that $(x-1)p(x) = 6x$ and
     $(x-1)^2q(x) = 3(x-1)/x$ which are both analytic at
     $x = 1$ and hence $x = 1$ is a regular singular point. These
     last two expressions approach $p_0 = 6$ and $q_0 = 0$
     respectively as $x \to 1$, and thus the indicial equation is
     $F(r) = r(r-1) + 6r + 0 = r(r+5) = 0$.

9.   For this D.E., $p(x) = \dfrac{-(1+x)}{x^2(1-x)}$ and $q(x) = \dfrac{2}{x(1-x)}$ and thus
     $x = 0$, -1 are singular points. Since $xp(x)$ is not
     analytic at $x = 0$, $x = 0$ is not a regular singular point.
     Looking at $(x-1)p(x) = \dfrac{1+x}{x^2}$ and $(x-1)^2q(x) = \dfrac{2(1-x)}{x}$ we
     see that $x = 1$ is a regular singular point and that
     $p_0 = 2$ and $q_0 = 0$. Thus $F(r) = r^2 + r$.

17a. We have $p(x) = \dfrac{\sin x}{x^2}$ and $q(x) = -\dfrac{\cos x}{x^2}$, so that $x = 0$ is
a singular point. Note that $xp(x) = (\sin x)/x \to 1 = p_0$
as $x \to 0$ and $x^2 q(x) = -\cos x \to -1 = q_0$ as $x \to 0$. In
order to assert that $x = 0$ is a regular singular point we
must demonstrate that $xp(x)$ and $x^2 q(x)$, with $xp(x) = 1$ at
$x = 0$ and $x^2 q(x) = -1$ at $x = 0$, have convergent power
series (are analytic) about $x = 0$. We know that $\cos x$ is
analytic so we need only consider $(\sin x)/x$. Now

$$\sin x = \sum_{n=0}^{\infty} (-1)^n x^{2n+1}/(2n+1)! \quad \text{for } -\infty < x < \infty \quad \text{so}$$

$$(\sin x)/x = \sum_{n=0}^{\infty} (-1)^n x^{2n}/(2n+1)! \quad \text{and hence is analytic.}$$

Thus we may conclude that $x = 0$ is a regular singular
point.

17b. From part a) it follows that the indicial equation is
$r(r-1) + r - 1 = r^2 - 1 = 0$ and the roots are $r_1 = 1$,
$r_2 = -1$.

17c. To find the first few terms of the solution corresponding
to $r_1 = 1$, assume that

$$y(x) = x \sum_{n=0}^{\infty} a_n x^r$$

$$= x(a_0 + a_1 x + a_2 x^2 + \ldots) = a_0 x + a_1 x^2 + a_2 x^3 + \ldots .$$

Substituting this series for y in the D.E. and expanding
$\sin x$ and $\cos x$ about $x = 0$ yields
$x^2(2a_1 + 6a_2 x + 12a_3 x^2 + 20a_4 x^3 + \ldots) +$
$(x - x^3/3! + x^5/5! - \ldots)(a_0 + 2a_1 x + 3a_2 x^2 + 4a_3 x^3 + 5a_4 x^4 +$
$\ldots) - (1 - x^2/2! + x^4/4! - \ldots)(a_0 x + a_1 x^2 + a_2 x^3 + a_3 x^4 +$
$a_4 x^5 + \ldots) = 0$. Collecting terms we have $(a_0 - a_0)x +$
$(2a_1 + 2a_1 - a_1)x^2 + (6a_2 + 3a_2 - a_0/6 - a_2 + a_0/2)x^3 +$
$(12a_3 + 4a_3 - 2a_1/6 - a_3 + a_1/2)x^4 +$
$(20a_4 + 5a_4 - 3a_2/6 + a_0/120 - a_4 + a_2/2 - a_0/24)x^5 + \ldots = 0.$
Simplifying yields, $3a_1 x^2 + (8a_2 + a_0/3)x^3 + (15a_3 + a_1/6)x^4$
$+ (24a_4 - a_0/30)x^5 + \ldots = 0$. Thus, $a_1 = 0$, $a_2 = -a_0/4!$,

$a_3 = 0$, $a_4 = a_0/6!,\ldots$. Hence

$y_1(x) = x - x^3/4! + x^5/6! + \ldots$ where we have set $a_0 = 1$. From Eq. (24) the second solution has the form

$$y_2(x) = ay_1(x)\ln x + x^{-1}(1+\sum_{n=1}^{\infty} c_n x^n)$$

$$= ay_1(x)\ln x + \frac{1}{x} + c_1 + c_2 x + c_3 x^2 + c_4 x^3 + \ldots, \text{ so}$$

$y_2' = ay_1'\ln x + ay_1 x^{-1} - x^{-2} + c_2 + 2c_3 x + 3c_4 x^2 + \ldots,$ and

$y_2'' = ay_1''\ln x + 2ay_1'x^{-1} - ay_1 x^{-2} + 2x^{-3} + 2c_3 + 3c_4 x + \ldots.$

When these are substituted in the given D.E. the terms including $\ln x$ will appear as

$a[x^2 y_1'' + (\sin x)y_1' - (\cos x)y_1]$, which is zero since y_1 is a solution. For the remainder of the terms, use

$y_1 = x - x^3/24 + x^5/720$ and the $\cos x$ and $\sin x$ series as shown earlier to obtain

$-c_1 + (2/3+2a)x + (3c_3+c_1/2)x^2 + (4/45+c_2/3+8c_4)x^3 + \ldots = 0.$

These yield $c_1 = 0$, $a = -1/3$, $c_3 = 0$, and

$c_4 = -c_2/24 - 1/90$. We may take $c_2 = 0$, since this term will simply generate $y_1(x)$ over again. Thus

$y_2(x) = -\frac{1}{3}y_1(x)\ln x + x^{-1} - \frac{1}{90}x^3$. If a computer algebra system is used, then additional terms in each series may be obtained without much additional effort. The next terms, in each case, are shown here:

$$y_1(x) = x - \frac{x^3}{24} + \frac{x^5}{720} - \frac{43x^7}{1451520} + \ldots \text{ and}$$

$$y_2(x) = -\frac{1}{3}y_1(x)\ln x + \frac{1}{x}[1- \frac{x^4}{90} + \frac{41x^6}{120960} - \ldots].$$

18. We first write the D.E. in the standard form as given for Theorem 5.7.1 except that we are expanding in powers of $(x-1)$ rather than powers of x:

$(x-1)^2 y'' + (x-1)[(x-1)/2\ln x]y' + [(x-1)^2/\ln x]y = 0$. Since $\ln 1 = 0$, $x = 1$ is a singular point. To show it is a regular singular point of this D.E. we must show that $(x-1)/\ln x$ is analytic at $x = 1$; it will then follow that $(x-1)^2/\ln x = (x-1)[(x-1)/\ln x]$ is also analytic at $x = 1$. If we expand $\ln x$ in a Taylor series about $x = 1$ we find that $\ln x = (x-1) - \frac{1}{2}(x-1)^2 + \frac{1}{3}(x-1)^3 - \ldots$.

Thus

$(x-1)/\ln x = [1 - \frac{1}{2}(x-1) + \frac{1}{3}(x-1)^2 - \ldots]^{-1} = 1 + \frac{1}{2}(x-1)+\ldots$

has a power series expansion about $x = 1$, and hence is analytic. We can use the above result to obtain the indicial equation at $x = 1$. We have

$(x-1)^2 y'' + (x-1)[\frac{1}{2} + \frac{1}{4}(x-1) + \ldots]y' + [(x-1) +$

$\frac{1}{2}(x-1)^2 + \ldots]y = 0$. Thus $p_0 = 1/2$, $q_0 = 0$ and the

indicial equation is $r(r-1) + r/2 = 0$. Hence $r = 1/2$ and $r = 0$. In order to find the first three non-zero terms in a series solution corresponding to $r = 1/2$, it is better to keep the differential equation in its original form and to substitute the above power series for $\ln x$:

$[(x-1) - \frac{1}{2}(x-1)^2 + \frac{1}{3}(x-1)^3 - \frac{1}{4}(x-1)^4 + \ldots]y'' + \frac{1}{2}y' + y = 0$.

Next we substitute $y = a_0(x-1)^{1/2} + a_1(x-1)^{3/2} + a_2(x-1)^{5/2}$ + ... and collect coefficients of like powers of $(x-1)$ which are then set equal to zero. This requires some algebra before we find that $6a_1/4 + 9a_0/8 = 0$ and $5a_2 + 5a_1/8 - a_0/12 = 0$. These equations yield $a_1 = -3a_0/4$ and $a_2 = 53a_0/480$. With $a_0 = 1$ we obtain the solution

$y_1(x) = (x-1)^{1/2} - \frac{3}{4}(x-1)^{3/2} + \frac{53}{480}(x-1)^{5/2} + \ldots$. Since

the radius of convergence of the Taylor Series for $(x-1)/\ln x$ is 1, we would expect $\rho = 1$.

20a. If we write the D.E. in the standard form as given in Theorem 5.7.1 we obtain $x^2 y'' + x[\alpha/x]y' + [\beta/x]y = 0$ where $xp(x) = \alpha/x$ and $x^2 q(x) = \beta/x$. Neither of these terms are analytic at $x = 0$ so $x = 0$ is an irregular singular point.

20b. Substituting $y = x^r \sum\limits_{n=0}^{\infty} a_n x^n$ in $x^3 y'' + \alpha xy' + \beta y = 0$ gives

$$\sum_{n=0}^{\infty}(n+r)(n+r-1)a_n x^{n+r+1} + \alpha \sum_{n=0}^{\infty}(n+r)a_n x^{n+r} + \beta \sum_{n=0}^{\infty} a_n x^{n+r} = 0.$$

Shifting the index in the first series and collecting coefficients of common powers of x we obtain

$$(\alpha r + \beta) a_0 x^r + \sum_{n=1}^{\infty} \{(n+r-1)(n+r-2)a_{n-1} + [\alpha(n+r) + \beta]a_n\} x^{n+r} = 0.$$

Thus the indicial equation is $\alpha r + \beta = 0$ with the single root $r = -\beta/\alpha$.

20c. From part b, the recurrence relation is

$$a_n = -\frac{(n+r-1)(n+r-2)a_{n-1}}{\alpha(n+r) + \beta}, \quad n = 1,2,\ldots$$

$$= -\frac{(n - \frac{\beta}{\alpha} - 1)(n - \frac{\beta}{\alpha} - 2)a_{n-1}}{\alpha n}, \quad \text{for } r = -\beta/\alpha.$$

For $\frac{\beta}{\alpha} = -1$, $a_n = -\frac{n(n-1)a_{n-1}}{\alpha n}$, so that $a_1 = 0 \cdot a_0 = 0$.
Since all other a_n are multiples of a_1, and hence are

zero, $y(x) = x$ is the solution. Similarly for $\frac{\beta}{\alpha} = 0$,

$$a_n = -\frac{(n-1)(n-2)}{\alpha n} a_{n-1}$$ and again for $n = 1$ $a_1 = 0$ and

$y(x) = 1$ is the solution. Continuing in this fashion, we see that the series solution will terminate for β/α any positive integer as well as 0 and -1. For other values

of β/α, we have $\left| \dfrac{a_n}{a_{n-1}} \right| = \dfrac{(n-\frac{\beta}{\alpha}-1)(n-\frac{\beta}{\alpha}-2)}{\alpha n}$, which

approaches ∞ as $n \to \infty$ and thus the ratio test yields a zero radius of convergence.

21b. Putting the D.E. in standard form and substituting

$$y = \sum_{n=0}^{\infty} a_n x^{n+r} \text{ gives}$$

$$\sum_{n=0}^{\infty}(n+r)(n+r-1)a_n x^{n+r} + \alpha \sum_{n=0}^{\infty}(n+r)a_n x^{n+r+1-s} + \beta \sum_{n=0}^{\infty} a_n x^{n+r+2-t} = 0.$$

If $s = 2$ and $t = 2$ the first term in each of the three series is $r(r-1)a_0 x^r$, $\alpha r a_0 x^{r-1}$, and $\beta a_0 x^r$, respectively. Thus the indicial equation is $F(r) = \alpha r a_0 = 0$, which requires $r = 0$. Hence there is at most one solution of the assumed form.

21d. In order for the indicial equation to be quadratic in r
     it is necessary that the first term in the first series
     contribute to the indicial equation. This means that the
     first term in the second and the third series cannot have
     powers less than x^r. The first terms are $r(r-1)a_0 x^r$,
     $\alpha r a_0 x^{r+1-s}$, and $\beta a_0 x^{r+2-t}$, respectively. Thus if $s \le 1$ and
     $t \le 2$ the quadratic term will appear in the indicial
     equation.

Section 5.8, Page 303

1.   It is clear that $x = 0$ is a singular point. The D.E. is
     in the standard form given in Theorem 5.7.1 with
     $xp(x) = 2$ and $x^2 q(x) = x$. Both are analytic at $x = 0$, so
     $x = 0$ is a regular singular point. Substituting

$$y = \sum_{n=0}^{\infty} a_n x^{n+r} \text{ in the D.E., shifting indices}$$

appropriately, and collecting coefficients of like powers
of x yields

$$[r(r-1) + 2r]a_0 x^r + \sum_{n=1}^{\infty} [(r+n)(r+n+1)a_n + a_{n-1}]x^{r+n} = 0.$$

The indicial equation is $F(r) = r(r+1) = 0$ with roots
$r_1 = 0$, $r_2 = -1$. Treating a_n as a function of r, we see
that $a_n(r) = -a_{n-1}(r)/F(r+n)$, $n = 1,2,\ldots$ if $F(r+n) \ne 0$.
Thus $a_1(r) = -a_0/F(r+1)$, $a_2(r) = a_0/F(r+1)F(r+2),\ldots$, and
$a_n(r) = (-1)^n a_0/F(r+1)F(r+2)\ldots F(r+n)$, provided
$F(r+n) \ne 0$ for $n = 1,2,\ldots$. For the case $r_1 = 0$, we have
$a_n(0) = (-1)^n a_0/F(1)F(2) \ldots F(n) = (-1)^n a_0/n!(n+1)!$ so

one solution is $y_1(x) = \sum_{n=0}^{\infty} (-1)^n x^n/n!(n+1)!$ where we have

set $a_0 = 1$.
     If we try to use the above recurrence relation for
the case $r_2 = -1$ we find that $a_n(-1) = -a_{n-1}/n(n-1)$,
which is undefined for $n = 1$. Thus we must follow the
procedure described at the end of Sect. 5.7 to calculate
a second solution of the form given in Eq.(24).
Specifically, we use Eqs.(19) and (20) of that section to
calculate a and $c_n(r_2)$, where $r_2 = -1$. Since
$r_1 - r_2 = 1 = N$, we have $a_N(r) = a_1(r) = -1/F(r+1)$, with
$a_0 = 1$. Hence

$a = \lim_{r \to -1} [(r+1)(-1)/F(r+1)] = \lim_{r \to -1} [-(r+1)/(r+1)(r+2)] = -1.$

Next

$$c_n(-1) = \frac{d}{dr}[(r+1)a_n(r)]\bigg|_{r=-1} = (-1)^n \frac{d}{dr}[\frac{(r+1)}{F(r+1) \dots F(r+n)}]\bigg|_{r=-1}$$

where we again have set $a_0 = 1$. Observe that

$(r+1)/F(r+1)\dots F(r+n)=1/[(r+2)^2(r+3)^2\dots(r+n)^2(r+n+1)]=1/G_n(r).$

Hence $c_n(-1) = (-1)^{n+1}G_n'(-1)/G_n^2(-1)$. Notice that

$G_n(-1) = 1^2 \cdot 2^2 \cdot 3^2 \dots (n-1)^2 n = (n-1)!n!$ and

$G_n'(-1)/G_n(-1) = 2[1/1 + 1/2 + 1/3 + \dots + 1/(n-1)] + 1/n =$

$H_n + H_{n-1}$. Thus $c_n(-1) = (-1)^{n+1}(H_n + H_{n-1})/(n-1)!n!.$

From Eq.(24) of Section 5.7 we obtain the second solution

$$y_2(x) = - y_1(x)\ln x + x^{-1}[1 - \sum_{n=1}^{\infty}(-1)^n(H_n + H_{n-1})x^n/n!(n-1)!].$$

2. It is clear that $x = 0$ is a singular point. The D.E. is
    in the standard form given in Theorem 5.7.1 with
    $xp(x) = 3$ and $x^2q(x) = 1+x$. Both are analytic at $x = 0$,
    so $x = 0$ is a regular singular point. Substituting

$$y = \sum_{n=0}^{\infty} a_n x^{n+r} \text{ in the D.E., shifting indices}$$

appropriately, and collecting coefficients of like powers
of x yields

$$[r(r-1) + 3r + 1]a_0 x^r + \sum_{n=1}^{\infty}\{[(r+n)(r+n+2) + 1]a_n + a_{n-1}\}x^{n+r} = 0.$$

The indicial equation is $F(r) = r^2 + 2r + 1 = (r+1)^2 = 0$
with the double root $r_1 = r_2 = -1$. Treating a_n as a
function of r, we see that $a_n(r) = -a_{n-1}(r)/F(r+n)$,
$n = 1,2,\dots$. Thus $a_1(r) = -a_0/F(r+1)$,
$a_2(r) = a_0/F(r+1)F(r+2),\dots$, and
$a_n(r) = (-1)^n a_0/F(r+1)F(r+2)\dots F(r+n)$. Setting $r = -1$ we
find that $a_n(-1)=(-1)^n a_0/(n!)^2$, $n = 1,2,\dots$. Hence one

solution is $y_1(x) = x^{-1}\sum_{n=0}^{\infty}(-1)^n x^n/(n!)^2$ where we have set

$a_0 = 1$. To find a second solution we follow the
procedure described in Section 5.7 for the case when the

roots of the indicial equation are equal. Specifically, the second solution will have the form given in Eq.(17) of that section. We must calculate $a_n'(-1)$. If we let

$$G_n(r) = F(r+1)\ldots F(r+n) = (r+2)^2(r+3)^2\ldots(r+n+1)^2 \text{ and}$$

take $a_0 = 1$, then $a_n'(-1) = (-1)^n[1/G_n(r)]'$ evaluated

$r = -1$. Hence $a_n'(-1) = (-1)^{n+1}G_n'(-1)/G_n^2(-1)$. But

$G_n(-1) = (n!)^2$ and $G_n'(-1)/G_n(-1) = 2[1/1 + 1/2 + 1/3 + \ldots + 1/n] = 2H_n$. Thus a second solution is

$$y_2(x) = y_1(x)\ln x - 2x^{-1}\sum_{n=1}^{\infty}(-1)^n H_n x^n/(n!)^2.$$

3. The roots of the indicial equation are r_1 and $r_2 = 0$ and thus the analysis is similar to that for Prob. 2.

4. The roots of the indicial equation are $r_1 = -1$ and $r_2 = -2$ and thus the analysis is similar to that for Prob.1.

5. Since $x = 0$ is a regular singular point, substitute

$$y = \sum_{n=0}^{\infty} a_n x^{n+r}$$ in the D.E., shift indices appropriately,

and collect coefficients of like powers of x to obtain
$[r^2 - 9/4]a_0 x^r + [(r+1)^2 - 9/4]a_1 x^{r+1}$

$$+ \sum_{n=2}^{\infty}\{[(r+n)^2 - 9/4]a_n + a_{n-2}\}\, x^{n+r} = 0.$$

The indicial equation is $F(r) = r^2 - 9/4 = 0$ with roots $r_1 = 3/2$, $r_2 = -3/2$. Treating a_n as a function of r we see that $a_n(r) = -a_{n-2}(r)/F(r+n)$, $n = 2,3,..$ if $F(r+n) \neq 0$. For the case $r_1 = 3/2$, $F(r_1+1)$, which is the coefficient of x^{r_1+1} is $\neq 0$ so we must set $a_1 = 0$. It follows that $a_3 = a_5 = \ldots = 0$. For the even coefficients, set $n = 2m$ so $a_{2m}(3/2) = -a_{2m-2}(3/2)/F(3/2 + 2m) = -a_{2m-2}/2^2 m(m+3/2)$, $m = 1,2\ldots$. Thus $a_2(3/2) = - a_0/2^2\cdot 1(1 + 3/2)$, $a_4(3/2) = a_0/2^4\cdot 2!(1 + 3/2)(2 + 3/2),\ldots$, and $a_{2m}(3/2) = (-1)^m/2^{2m}m!\cdot(1 + 3/2)\ldots(m + 3/2)$. Hence one solution is

$$y_1(x) = x^{3/2}[1 + \sum_{m=1}^{\infty} \frac{(-1)^m}{m!(1 + 3/2)(2 + 3/2)...(m + 3/2)} (\frac{x}{2})^{2m}],$$

where we have set $a_0 = 1$. For this problem, the roots r_1 and r_2 of the indicial equation differ by an integer: $r_1 - r_2 = 3/2 - (-3/2) = 3$. Hence we can anticipate that there may be difficulty in calculating a second solution corresponding to $r = r_2$. This difficulty will occur in calculating $a_3(r) = - a_1(r)/F(r+3)$ since when $r = r_2 = -3/2$ we have $F(r_2+3) = F(r_1) = 0$. However, in this problem we are fortunate because $a_1 = 0$ and it will not be necessary to use the theory described at the end of Section 5.7. Notice for $r = r_2 = -3/2$ that the coefficient of x^{r_2+1} is $[(r_2+1)^2 - 9/4]a_1$, which does not vanish unless $a_1 = 0$. Thus the recurrence relation for the odd coefficients yields $a_5 = -a_3/F(7/2)$, $a_7 = -a_5/F(11/2) = a_3/F(11/2)F(7/2)$ and so forth. Substituting these terms into the assumed form we see that a multiple of $y_1(x)$ has been obtained and thus we may take $a_3 = 0$ without loss of generality. Hence $a_3 = a_5 = a_7 = ... = 0$. The even coefficients are given by $a_{2m}(-3/2) = -a_{2m-2}(-3/2)/F(2m - 3/2)$, $m = 1,2...$. Thus $a_2(-3/2) = -a_0/2^2 \cdot 1 \cdot (1 - 3/2)$, $a_4(-3/2) = a_0/2^4 \cdot 2!(1 - 3/2)(2 - 3/2),...$, and $a_{2m}(-3/2) = (-1)^m a_0/2^{2m} m!(1 - 3/2)(2 - 3/2) ... (m - 3/2)$. Thus a second solution is

$$y_2(x) = x^{-3/2}[1 + \sum_{m=1}^{\infty} \frac{(-1)^m}{m!(1 - 3/2)(2 - 3/2) ... (m - 3/2)} (\frac{x}{2})^{2m}].$$

7. Apply the ratio test:
$$\lim_{m \to \infty} \frac{|(-1)^{m+1} x^{2m+2}/2^{2m+2}[(m+1)!]^2|}{|(-1)^m x^{2m}/2^{2m}(m!)^2|} = |x^2| \lim_{m \to \infty} \frac{1}{2^2(m+1)^2} = 0$$
for every x. Thus the series for $J_0(x)$ converges absolutely for all x.

12. If $\xi = \alpha x^\beta$, then $dy/dx = \frac{1}{2}x^{-1/2}f + x^{1/2}f'\alpha\beta x^{\beta-1}$ where f' denotes $df/d\xi$. Find d^2y/dx^2 in a similar fashion and use algebra to show that f satisfies the D.E.

$\xi^2 f'' + \xi f' + [\xi^2 - \upsilon^2]f = 0$, which is the Bessel Equation of order υ.

13. To compare $y'' - xy = 0$ with the D.E. of Problem 12, we must multiply by x^2 to get $x^2 y'' - x^3 y = 0$. Thus $2\beta = 3$, $\alpha^2\beta^2 = -1$ and $1/4 - \upsilon^2\beta^2 = 0$. Hence $\beta = 3/2$, $\alpha = 2i/3$ and $\upsilon^2 = 1/9$ which yields the desired result.

14. First we verify that $J_0(\lambda_j x)$ satisfies the D.E. We know that $J_0(t)$ is a solution of the Bessel equation of order zero:

$$t^2 J_0''(t) + t J_0'(t) + t^2 J_0(t) = 0 \text{ or}$$

$$J_0''(t) + t^{-1} J_0'(t) + J_0(t) = 0.$$

Let $t = \lambda_j x$. Then

$$\frac{d}{dx} J_0(\lambda_j x) = \frac{d}{dt} J_0(t) \frac{dt}{dx} = \lambda_j J_0'(t)$$

$$\frac{d^2}{dx^2} J_0(\lambda_j x) = \lambda_j \frac{d}{dt}[J_0'(t)]\frac{dt}{dx} = \lambda_j^2 J_0''(t).$$

Substituting $y = J_0(\lambda_j x)$ in the given D.E. and making use of these results, we have

$$\lambda_j^2 J_0''(t) + (\lambda_j/t)\ \lambda_j J_0'(t) + \lambda_j^2 J_0(t) =$$

$$\lambda_j^2 [J_0''(t) + t^{-1} J_0'(t) + J_0(t)] = 0.$$

Thus $y = J_0(\lambda_j x)$ is a solution of the given D.E. For the second part of the problem we follow the hint. First, rewrite the D.E. by multiplying by x to yield
$xy'' + y' + \lambda_j^2 xy = 0$, which can be written as $(xy')' = -\lambda_j^2 xy$.
Now let $y_i(x) = J_0(\lambda_i x)$ and $y_j(x) = J_0(\lambda_j x)$ and we have, respectively:

$$(xy_i')' = -\lambda_i^2 xy_i$$

$$(xy_j')' = -\lambda_j^2 xy_j.$$

Now, multiply the first equation by y_j, the second by y_i, integrate each from 0 to 1, and subtract the second from the first:

$$\int_0^1 [y_j(xy_i')' - y_i(xy_j')']dx = -(\lambda_i^2 - \lambda_j^2) \int_0^1 xy_i y_j dx.$$

If we integrate each term on the left side once by parts and note that $y_i = y_j = 0$ at $x = 0$ and $x = 1$, we find that the left side of this equation is identically zero. Hence the right side is identically zero and for $\lambda_i \neq \lambda_j$ this gives the desired result.

CHAPTER 6

<u>Section 6.1, Page 312</u>

1. The graph of f(t) is shown.
    Since the function is
    continuous on each interval,
    but has a jump discontinuity
    at t = 1, f(t) is piecewise
    continuous.

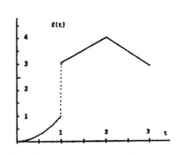

2. Note that $\lim\limits_{t \to 1^+} (t-1)^{-1} = \infty$.

5b. Since t^2 is continuous for $0 \le t \le A$ for any positive A
    and since $t^2 \le e^{at}$ for any a > 0 and for t sufficiently
    large, it follows from Theorem 6.1.2 that $\pounds\{t^2\}$ exists
    for s > 0. $\pounds\{t^2\} = \int_0^\infty e^{-st}t^2dt = \lim\limits_{M \to \infty} \int_0^M e^{-st}t^2dt$

$$= \lim\limits_{M \to \infty} [\frac{-t^2}{s}e^{-st}\Big|_0^M + \frac{2}{s}\int_0^M e^{-st}tdt]$$

$$= \lim\limits_{M \to \infty} \frac{-M^2}{s}e^{-sM} + \frac{2}{s}\lim\limits_{M \to \infty}[-\frac{1}{s}te^{-st}\Big|_0^M + \frac{1}{s}\int_0^M e^{-st}dt]$$

$$= 0 + \lim\limits_{M \to \infty} \frac{-M}{s^2}e^{-sM} + \frac{2}{s^2}\lim\limits_{M \to \infty} -\frac{1}{s}e^{-st}\Big|_0^M = \frac{2}{s^3}.$$

6. That f(t) = cosat satisfies the hypotheses of Theorem
    6.1.2 can be verified by recalling that $|cosat| \le 1$ for
    all t. To determine $\pounds\{cosat\} = \int_0^\infty e^{-st} cosatdt$ we
    must integrate by parts twice to get
    $$\int_0^\infty e^{-st} cosatdt = \lim\limits_{M \to \infty}[(-s^{-1}e^{-st} cosat + as^{-2} e^{-st} sinat)\Big|_0^M$$

    $$- (a^2/s^2)\int_0^M e^{-st} cosatdt].$$ Evaluating the first two
    terms and letting M $\to$ ∞ yields
    $$\int_0^\infty e^{-st} cosatdt = \frac{1}{s} - (a^2/s^2)\int_0^\infty e^{-st} cosatdt \text{ and hence}$$
    $[1+a^2/s^2]\int_0^\infty e^{-st} cosatdt = 1/s, s > 0.$ Division by $[1+a^2/s^2]$
    and simplification yields the desired solution.

9. From the definition for coshbt we have
    $\pounds\{e^{at}coshbt\} = \pounds\{\frac{1}{2}[e^{(a+b)t} + e^{(a-b)t}]\}.$ Using the linearity

property of £, Eq.(5), the right side becomes

$\frac{1}{2}£\{e^{(a+b)t}\} + \frac{1}{2}£\{e^{(a-b)t}\}$ which can be evaluated using the

result of Ex. 5 and thus

$$£\{e^{at}coshbt\} = \frac{1/2}{s-(a+b)} + \frac{1/2}{s-(a-b)}, \text{ for } s-a > |b|$$

$$= \frac{s-a}{(s-a)^2-b^2}.$$

13. We write $sinat = (e^{iat} - e^{-iat})/2i$, then the linearity of
the Laplace transform operator allows us to write
$£\{e^{at}sinbt\} = (1/2i)£\{e^{(a+ib)t}\}-(1/2i)£\{e^{(a-ib)t}\}$. Each of
these two terms can be evaluated by using the result of
Ex. 5, where we now have to require s to be greater than
the real part of the complex numbers a ± ib in order for
the integrals to converge. Complex algebra then gives
the desired result. An alternate method of evaluation
would be to use integration on the integral appearing in
the definition of $£\{e^{at}sinbt\}$, but that method requires
integration by parts twice.

16. As in Prob. 13,
$£\{tsinat\} = (1/2i)£\{te^{iat}\} - (1/2i)£\{te^{-iat}\}$. Using the
result of Prob. 15 we obtain
$£\{tsinat\} = (1/2i)[(s-b)^{-2} - (s+b)^{-2}]$ where b = ia and
s > 0. Combining fractions we obtain
$£\{tsinat\} = 2as/(s^2+a^2)^2$, s > 0.

19. Use the approach shown in Prob. 16 with the result of
Prob. 18, for n = 2. A computer algebra system may also
be used.

21. The integral $\int_0^A (t^2 + 1)^{-1}dt$ can be evaluated in terms of
the arctan function and then Eq. (1) can be used. To
illustrate Theorem 6.1.1, however, consider that

$\frac{1}{t^2+1} < \frac{1}{t^2}$ for t ≥ 1 and, from Ex. 3, $\int_1^\infty t^{-2}dt$ converges

and hence $\int_1^\infty (t^2 + 1)^{-1}dt$ also converges.

$\int_0^1 (t^2 + 1)^{-1}dt$ is finite and hence does not affect the

convergence of $\int_0^\infty (t^2 + 1)^{-1}dt$ at infinity.

25. If we let u = f and dv = $e^{-st}dt$ then

$$F(s) = \int_0^\infty e^{-st}f(t)dt = \lim_{M \to \infty} -\frac{1}{s}e^{-st}f(t)\Big|_0^M + \frac{1}{s}\int_0^\infty e^{-st}f'(t)dt$$

$$= \frac{1}{s}f(0) + \frac{1}{s}\int_0^\infty e^{-st}f'(t)dt.$$ This last integral converges

(and is thus finite) using an argument similar to that given to establish Theoremm 6.1.2. Hence $\lim_{s \to \infty} F(s) = 0$.

27a. Make a transformation of variables with x = st and
     dx = sdt. Then use the definition of $\Gamma(P+1)$ from
     Prob. 26.

27b. From part a, $\mathcal{L}\{t^n\} = \dfrac{1}{s^{n+1}}\int_0^\infty e^{-x}x^n dx = \dfrac{n}{s^{n+1}}\int_0^\infty e^{-x}x^{n-1}dx$

$$= \frac{n!}{s^{n+1}}\int_0^\infty e^{-x}dx,$$ using integration by

parts successively. Evaluation of the last integral yields the desired answer.

27c. From part a, $\mathcal{L}\{t^{-1/2}\} = \dfrac{1}{\sqrt{s}}\int_0^\infty e^{-x}x^{-1/2}dx$. Let $x = y^2$, then

$$2dy = x^{-1/2}dx \text{ and thus } \mathcal{L}\{t^{-1/2}\} = \frac{2}{\sqrt{s}}\int_0^\infty e^{-y^2}dy.$$

27d. Use the definition of $\mathcal{L}\{t^{1/2}\}$ and integrate by parts once
     to get $\mathcal{L}\{t^{1/2}\} = (1/2s)\mathcal{L}\{t^{-1/2}\}$. The result follows from
     part c.

Section 6.2, Page 322

Problems 1 through 10 are solved by using partial fractions
and algebra to manipulate the given function into a form
matching one of the functions appearing in the middle column
of Table 6.2.1.

2. We have $\dfrac{4}{(s-1)^3} = 2\dfrac{2!}{(s-1)^{2+1}}$ and thus the inverse Laplace
    transform is $2t^2e^t$, using line 11.

4. We have $\dfrac{3s}{s^2-s-6} = \dfrac{3s}{(s-3)(s+2)} = \dfrac{9/5}{s-3} + \dfrac{6/5}{s+2}$ using partial
    fractions. Thus $(9/5)e^{3t} + (6/5)e^{-2t}$ is the inverse
    transform, from line 2.

7. We have $\dfrac{2s+1}{s^2-2s+2} = \dfrac{2s+1}{(s-1)^2+1} = \dfrac{2(s-1)}{(s-1)^2+1} + \dfrac{3}{(s-1)^2+1}$, where
    we first used the concept of completing the square (in
    the denominator) and then added and subtracted
    appropriately to put the numerator in the desired form.
    Lines 9 and 10 may now be used to find the desired
    result.

In each of the Problems 11 through 23 it is assumed that the
I.V.P. has a solution y = ϕ(t) which, with its first two
derivatives, satisfies the conditions of the Corollary 6.2.2.

11. Take the Laplace transform of the D.E., using Eq.(1) and
    Eq.(2), to get
    $s^2Y(s) - sy(0) - y'(0) - [sY(s) - y(0)] - 6Y(s) = 0.$
    Using the I.C. and solving for Y(s) we obtain
    $Y(s) = \dfrac{s-2}{s^2-s-6}.$ Following the pattern of Eq.(12) we have
    $\dfrac{s-2}{s^2-s-6} = \dfrac{a}{s+2} + \dfrac{b}{s-3} = \dfrac{a(s-3)+b(s+2)}{(s+2)(s-3)}.$ Equating like
    powers in the numerators we find a+b = 1 and
    -3a + 2b = -2. Thus a = 4/5 and b = 1/5 and
    $Y(s) = \dfrac{4/5}{s+2} + \dfrac{1/5}{s-3},$ which yields the desired solution
    using Table 6.2.1.

14. Taking the Laplace transform we have $s^2Y(s) - sy(0) - y'(0) -$
    $4[sY(s)-y(0)] + 4Y(s) = 0.$ Using the I.C. and solving for
    Y(s) we find $Y(s) = \dfrac{s-3}{s^2-4s+4}.$ Since the denominator is a

    perfect square, the partial fraction form is $\dfrac{s-3}{s^2-4s+4} =$

    $\dfrac{a}{(s-2)^2} + \dfrac{b}{s-2}.$ Solving for a and b, as shown in examples of
    this section or in Prob. 11, we find a = -1 and b = 1. Thus
    $Y(s) = \dfrac{1}{s-2} - \dfrac{1}{(s-2)^2},$ from which we find
    $y(t) = e^{2t} - te^{2t}$ (lines 2 and 11 in Table 6.2.1).

15. Note that $Y(s) = \dfrac{2s-4}{s^2-2s+4} = \dfrac{2s-4}{(s-1)^2+3} = \dfrac{2(s-1)}{(s-1)^2+3} - \dfrac{2}{(s-1)^2+3}.$
    Three formulas in Table 6.2.1 are now needed: F(s-c)

(with c = 1) in line 14 in conjunction with the ones for cosat and sinat (with a = $\sqrt{3}$), lines 5 and 6.

17. The Laplace transform of the D.E. is
$s^4 Y(s) - s^3 y(0) - s^2 y'(0) - sy''(0) - y'''(0) - 4[s^3 Y(s) - s^2 y(0)$
$-sy'(0) - y''(0)] + 6[s^2 Y(s) - sy(0) - y'(0)] - 4[sY(s) - y(0)]$
$+ Y(s) = 0$. Using the I.C. and solving for Y(s) we find
$$Y(s) = \frac{s^2 - 4s + 7}{s^4 - 4s^3 + 6s^2 - 4s + 1}.$$ The correct partial fraction
form for this is $\dfrac{a}{(s-1)^4} + \dfrac{b}{(s-1)^3} + \dfrac{c}{(s-1)^2} + \dfrac{d}{s-1}$.
Setting this equal to Y(s) above and equating the
numerators we have $s^2 - 4s + 7 = a + b(s-1) + c(s-1)^2 +$
$d(s-1)^3$. Solving for a,b,c, and d and use of Table 6.2.1
yields the desired solution.

20. The Laplace transform of the D.E. is
$s^2 Y(s) - sy(0) - y'(0) + \omega^2 Y(s) = s/(s^2+4)$. Applying the
I.C. and solving for Y(s) we get $Y(s) = s/[(s^2+4)(s^2+\omega^2)]$
$+ s/(s^2+\omega^2)$. Decomposing the first term by partial
fractions we have
$$Y(s) = \frac{s}{(\omega^2-4)(s^2+4)} - \frac{s}{(\omega^2-4)(s^2+\omega^2)} + \frac{s}{s^2+\omega^2}$$
$$= (\omega^2-4)^{-1}[\frac{(\omega^2-5)s}{s^2+\omega^2} + \frac{s}{s^2+4}].$$
Then, using Table 6.1.2, we have
$y = (\omega^2-4)^{-1}[(\omega^2-5)\cos\omega t + \cos 2t]$.

22. Solving for Y(s) we find $Y(s) = 1/[(s-1)^2 + 1] +$
$1/(s+1)[(s-1)^2 + 1]$. Using partial fractions on the
second term we obtain
$Y(s) = 1/[(s-1)^2 + 1] + \{1/(s+1) - (s-3)/[(s-1)^2 + 1]\}/5$
$= (1/5)\{(s+1)^{-1} - (s-1)[(s-1)^2 + 1]^{-1} + 7[(s-1)^2 + 1]^{-1}\}$.
Hence, $y = (1/5)(e^{-t} - e^t\cos t + 7e^t\sin t)$.

24. Under the standard assumptions, the Lapace transform of
the left side of the D.E. is $s^2 Y(s) - sy(0) - y'(0) +$
$4Y(s)$. To transform the right side we must revert to the
definition of the Laplace trasnform to determine
$\int_0^\infty e^{-st} f(t)dt$. Since f(t) is piecewise continuous we are
able to calculate $\mathcal{L}\{f(t)\}$ by

$$\int_0^\infty e^{-st}f(t)dt = \int_0^\pi e^{-st} \, dt + \lim_{M \to \infty} \int_\pi^M (e^{-st})(0)dt$$

$$= \int_0^\pi e^{-st}dt = (1 - e^{-\pi s})/s.$$

Hence, the Laplace transform $Y(s)$ of the solution is given by $Y(s) = s/(s^2+4) + (1 - e^{-\pi s})/s(s^2+4)$.

27b. The Taylor series for f about $t = 0$ is

$$f(t) = \sum_{n=0}^\infty (-1)^n t^{2n}/(2n+1)!, \text{ which is obtained from}$$

part(a) by dividing each term of the sine series by t. Also, f is continuous for $t > 0$ since $\lim_{t \to 0+} (\sin t)/t = 1$. Assuming that we can compute the Laplace transform of f

term by term, we obtain $\mathcal{L}\{f(t)\} = \mathcal{L}\{\sum_{n=0}^\infty (-1)^n t^{2n}/(2n+1)!\}$

$$= \sum_{n=0}^\infty [(-1)^n/(2n+1)!\mathcal{L}\{t^{2n}\}$$

$$= \sum_{n=0}^\infty [(-1)^n(2n)!/(2n+1)!]s^{-(2n+1)}$$

$$= \sum_{n=0}^\infty [(-1)^n/(2n+1)]s^{-(2n+1)}, \text{ which converges for } s > 1.$$

The Taylor series for arctan x is given by

$$\sum_{n=0}^\infty (-1)^n x^{2n+1}/(2n+1), \text{ for } |x| < 1. \text{ Comparing } \mathcal{L}\{f(t)\} \text{ with}$$

the Taylor series for arctanx, we conclude that $\mathcal{L}\{f(t)\} = \arctan(1/s)$, $s > 1$.

30. Setting $n = 2$ in Prob. 28b, we have

$$\mathcal{L}\{t^2\sin bt\} = \frac{d^2}{ds^2}[b/(s^2+b^2)] = \frac{d}{ds}[-2bs/(s^2+b^2)^2] =$$

$$-2b/(s^2+b^2)^2 + 8bs^2/(s^2+b^2)^3 = 2b(3s^2-b^2)/(s^2+b^2)^3.$$

32. Using the result of Prob. 28a, repeatedly, we have

$$\mathcal{L}\{te^{at}\} = -\frac{d}{ds}(s-a)^{-1} = (s-a)^{-2},$$

$$\mathcal{L}\{t^2e^{at}\} = -\frac{d}{ds}(s-a)^{-2} = 2(s-a)^{-3}, \text{ and}$$

$\mathcal{L}\{t^3 e^{at}\} = -\dfrac{d}{ds} 2(s-a)^{-3} = 3!(s-a)^{-4}$. Continuing in this fashion, or using induction, we obtain the desired result.

36a. Taking the Laplace transform of the D.E. we obtain
$\mathcal{L}\{y''\} - \mathcal{L}\{ty\} = \mathcal{L}\{y''\} + \mathcal{L}\{-ty\}$
$\qquad\qquad = s^2 Y(s) - sy(0) - y'(0) + Y'(s) = 0$.
Hence, Y satisfies $Y' + s^2 Y = s$.

38a. From Eq(i) we have $A_k = \lim\limits_{s \to r_k} (s-r_k)\,\dfrac{P(r_k)}{Q(r_k)}$, since Q has

distinct zeros. Thus $A_k = P(r_k)\,\lim\limits_{s \to r_k}\dfrac{s-r_k}{Q(r_k)} = \dfrac{P(r_k)}{Q'(r_k)}$, by

L'Hopital's Rule.

38b. Since $\mathcal{L}^{-1}\left\{\dfrac{1}{s-r_k}\right\} = e^{r_k t}$, the result follows.

Section 6.3, Page 329

2.   From the definition of $u_c(t)$
we have:
$g(t) = (t-3)u_2(t) - (t-2)u_3(t)$
$= \begin{cases} 0 - 0 = 0, & 0 \le t < 2 \\ (t-3) - 0 = t-3, & 2 \le t < 3. \\ (t-3) - (t-2) = -1, & 3 \le t \end{cases}$

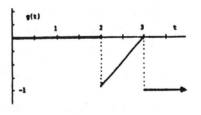

4.   As indicated in the discussion
following Eq.(2), the unit step
function can be used to
translate a given function f,
with domain $t \ge 0$, a distance c
to the right by the
multiplication $u_c(t)f(t-c)$.
Hence the required graph of
$y = u_3(t)f(t-3)$ for $f(t) = \sin t$ is shown.

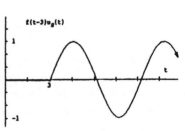

8.   In order to use Theorem 6.3.1 we must write $f(t)$ in terms
of $u_c(t)$. Since $t^2 - 2t + 2 = (t-1)^2 + 1$ (by completing
the square), we can thus write $f(t) = u_1(t)g(t-1)$, where
$g(t) = t^2 + 1$. Now applying Theorem 6.3.1 we have

$\mathcal{L}\{f(t)\} = \mathcal{L}\{u_1(t)g(t-1)\} = e^{-s}\,\mathcal{L}\{g(t)\} = e^{-s}(2/s^3 + 1/s).$

14. Use partial fractions to write

    $F(s) = e^{-2s}\dfrac{1}{3}[\dfrac{1}{s-1} - \dfrac{1}{s+2}]$. For ease in calculations let

    us define $G(s) = (s-1)^{-1}$ and $H(s) = (s+2)^{-1}$. Then

    $F(s) = [e^{-2s}G(s) - e^{-2s}H(s)]/3$. Using the fact that

    $\mathcal{L}\{e^{at}\} = (s-a)^{-1}$ and applying Theorem 6.3.1, we have

    $F(s) = [e^{-2s}\mathcal{L}\{e^t\} - e^{-2s}\mathcal{L}\{e^{-2t}\}]/3$. Thus

    $F(s) = [\mathcal{L}\{u_2(t)e^{(t-2)}\} - \mathcal{L}\{u_2(t)e^{-2(t-2)}\}]/3$. Using the

    linearity of the Laplace transform, we have

    $\mathcal{L}\{f(t)\} = \mathcal{L}\{u_2(t)[e^{t-2} - e^{-2(t-2)}]/3\}$. Hence,

    $f(t) = [u_2(t)(e^{t-2} - e^{-2(t-2)})]/3$. An alternate method is

    to complete the square in the denominator:

    $F(s) = \dfrac{e^{-2s}}{(s+1/2)^2 - 9/4}$. From line 7, Table 6.2.1, this

    gives $f(t) = (2/3)u_2(t)e^{-(t-2)/2}\sinh\dfrac{3}{2}(t-2)$, which can be

    shown to be the same as that found above.

21. By completing the square in the denominator of F we can

    write $F(s) = (2s+1)/[(2s+1)^2 + 4]$. This has the form

    $G(2s+1)$ where $G(u) = u/(u^2+4)$. We must find

    $\mathcal{L}^{-1}\{G(2s+1)\}$. Applying the results of Prob. 19(c), with

    $a = 2$ and $b = 1$, we have $\mathcal{L}^{-1}\{F(s)\} = \dfrac{1}{2}e^{-t/2}\cos(\dfrac{2t}{2})$,

    since $\mathcal{L}^{-1}\{G(s)\} = \cos 2t$.

22. If the approach of Prob. 21 is used we find

    $f(t) = (1/3)e^{2t/3}\sinh(t/3)$, which is equivalent to the

    given answer using the definition of sinh t.

27. Assuming that term-by-term integration of the infinite

    series is permissible and recalling that $\mathcal{L}\{u_c(t)\} = e^{-cs}/s$

    for $s > 0$, we have $\mathcal{L}\{f(t)\} = (1/s) + \displaystyle\sum_{k=1}^{\infty}(-1)^k\,\mathcal{L}\{u_k(t)\}$

    $= (1/s) + \displaystyle\sum_{k=1}^{\infty}(-1)^k\,(e^{-ks}/s) = \dfrac{1}{s}\displaystyle\sum_{k=0}^{\infty}(-e^{-s})^k$. We recognize the last infinite series as the geometric series, $\sum\limits_{k=0}^{\infty} ar^k$,

with $a = 1$ and $r = -e^{-s}$. This series converges to $[1/(1+e^{-s})]$ if $|r| < 1$ (or $s > 0$). Hence, $\mathcal{L}\{f(t)\} = (1/s)[1/(1+e^{-s})]$, $s > 0$.

28. Using the definition of the Laplace transform we have

$F(s) = \mathcal{L}\{f(t)\} = \int_0^{\infty} e^{-st}f(t)dt$. Since f is periodic with period T, we have $f(t+T) = f(t)$. This suggests that we rewrite the improper integral as $\int_0^{\infty} e^{-st}f(t)dt =$

$\sum\limits_{n=0}^{\infty} \int_{nT}^{(n+1)T} e^{-st}f(t)dt$. The periodicity of f also suggests that we make the change of variable $t = r + nT$. Hence,

$F(s) = \sum\limits_{n=0}^{\infty} \int_0^T e^{-s(r+nT)}f(r+nT)dr = \sum\limits_{n=0}^{\infty} (e^{-sT})^n \int_0^T e^{-rs} f(r)dr$,

where we have used the fact that $f(r+nT) = f(r+(n-1)T) = \ldots = f(r+T) = f(r)$, since f is periodic. We recognize this last series as the geometric

series, $\sum\limits_{n=0}^{\infty} au^n$, with $a = \int_0^T e^{-rs}f(r)dr$ and $u = e^{-sT}$. The

geometric series converges to $a/(1-u)$ for $|u| < 1$ and consequently we obtain

$F(s) = (1 - e^{-sT})^{-1} \int_0^T e^{-rs} f(r)dr$, $s > 0$.

30. The function f is periodic with period 2. The result of

Prob. 28 gives us $\mathcal{L}\{f(t)\} = \int_0^2 e^{-st} f(t)dt/(1-e^{-2s})$.

Calculating the integral we have

$\int_0^2 e^{-st}f(t)dt = \int_0^1 e^{-st}dt - \int_1^2 e^{-st}dt$

$= (1-e^{-s})/s + (e^{-2s}-e^{-s})/s$

$= (e^{-2s}-2e^{-s}+1)/s$

$= (1-e^{-s})^2/s$. Since the denominator of

$\mathcal{L}\{f(t)\}$, $1 - e^{-2s}$, may be written as $(1-e^{-s})(1+e^{-s})$ we obtain the desired answer.

1. f(t) can be written in the form $f(t) = 1 - u_{\pi/2}(t)$ and
    thus the Laplace transform of the D.E. is
    $(s^2+1)Y(s) - sy(0) - y'(0) = (1/s) - e^{-\pi s/2}/s$. Using the
    I.C. and solving for Y(s), we obtain
    $Y(s) = (s^2+1)^{-1} + [s(s^2+1)]^{-1} - e^{-\pi s/2}/s(s^2+1)$. Using
    partial fractions on the second and third terms we find
    $Y(s) = (s^2+1)^{-1} + (1/s) - s/(s^2+1) - e^{-\pi s/2}/s + e^{-\pi s/2}s/(s^2+1)$.
    The inverse transform of the first three terms can be
    obtained directly from Table 6.2.1. Using Theorem 6.3.1
    to find the inverse transform of the last two terms we
    have $\mathcal{L}^{-1}\{e^{-\pi s/2}/s\} = u_{\pi/2}(t)g(t - \pi/2)$ where

    $g(t) = \mathcal{L}^{-1}\{1/s\} = 1$ and
    $\mathcal{L}^{-1}\{e^{-\pi s/2}s/(s^2+1)\} = u_{\pi/2}(t)h(t - \pi/2)$ where

    $h(t) = \mathcal{L}^{-1}\{s/(s^2+1)\} = \cos t$. Hence,
    $y = 1 + \sin t - \cos t + u_{\pi/2}(t)[\cos(t - \pi/2) - 1]$

    $= 1 + \sin t - \cos t - u_{\pi/2}(t)[1 - \sin t]$. The graph of the
    forcing function is a unit pulse for $0 \le t < \pi/2$ and 0
    thereafter. The graph of the solution will be composed
    of two segments. The first, for $0 \le t < \pi/2$, is a
    sinusoid oscillating about 1, which represents the system
    response to a unit forcing function and the given initial
    conditions. For $t \ge \pi/2$, the forcing function, f(t), is
    zero and the "initial" conditions are
    $y(\pi/2) = \lim_{t \to \pi/2} (1 + \sin t - \cos t) = 2$ and

    $y'(\pi/2) = \lim_{t \to \pi/2} (\cos t + \sin t) = 1$. For $t \ge \pi/2$ the system

    response is $y(t) = 2\sin t - \cos t$, which is a sinusoid
    oscillating about zero.

3. According to Theorem 6.3.1,
    $\mathcal{L}\{u_{2\pi}(t)\sin(t-2\pi)\} = e^{-2\pi s} \mathcal{L}\{\sin t\} = e^{-2\pi s}/(s^2+1)$.
    Transforming the D.E., we have
    $(s^2+4)Y(s) - sy(0) - y'(0) = 1/(s^2+1) - e^{-2\pi s}/(s^2+1)$.
    Using the I.C. and solving for Y(s), we obtain
    $Y(s) = (1-e^{-2\pi s})/(s^2+1)(s^2+4)$. We apply partial fractions
    to write
    $Y(s) = [s^2+1)^{-1} - (s^2+4)^{-1} - e^{-2\pi s}(s^2+1)^{-1} + e^{-2\pi s}(s^2+4)^{-1}]/3$.
    We compute the inverse transform of the first two terms
    directly from Table 6.2.1 after noting that
    $(s^2+4)^{-1} = (1/2)[2/(s^2+4)]$. We apply Theorem 6.3.1 to the
    last two terms to obtain the solution,

y =(1/3){sint-(1/2)sin2t-$u_{2\pi}$(t)[sin(t-2π)-(1/2)sin2(t-2π)]}.
This may be simplified, using trigonometric identities,
to y = [(2sint - sin2t)(1-$u_{2\pi}$(t))]/6. Note that the
forcing function is sint - sin(t-2π) = 0 for t $\geq$ 2π. The
solution is y(t) = 2sint - sin2t for 0 $\leq$ t < 2π. Thus
y($2\pi^{-}$) = 0 and y'($2\pi^{-}$) = 2cos2π - 2cos4π = 0. Hence the
"initial" value problem for t $\geq$ 2π is y" + 4y = 0,
y(2π) = 0, y'(2π) = 0, which has the trivial solution
y $\equiv$ 0. [Note that 1 - $u_{2\pi}$(t) = 0 for t $\geq$ 2π so this
agrees with the solution y given above].

8.   Taking the Laplace transform, applying the I.C. and using
Theorem 6.3.1 we have (s^2+s+5/4)Y(s) = (1-$e^{-\pi s/2}$)/s^2. Thus

$$Y(s) = \frac{1-e^{-\pi s/2}}{s^2(s^2+s+5/4)}$$

$$= (1-e^{-\pi s/2})\left\{\frac{4/5}{s^2} - \frac{16/25}{s} + \frac{(16/25)s-4/25}{(s+1/2)^2+1}\right\}$$

= (1-$e^{-\pi s/2}$)H(s), where we have used partial
fractions and completed the square in the denominator of
the last term. Since the numerator of the last term of H
can be written as $\frac{16}{25}$[(s+1/2) - 3/4], we see that

$\mathcal{L}^{-1}${H(s)} = (4/25)(5t - 4 + 4$e^{-t/2}$cost - 3$e^{-t/2}$sint),
which yields the desired solution. The graph of the
forcing function is a ramp (f(t) = t) for 0 $\leq$ t < π/2 and
a constant (f(t) = π/2) for t $\geq$ π/2. The solution will
be a damped sinusoid oscillating about the "ramp"
(20t-16)/25 for 0 $\leq$ t < π/2 and oscillating about 2π/5
for t $\geq$ π/2.

10. Note that g(t) = sint - u_{π}(t)sint = sint + u_{π}(t)sin(t-π).
Proceeding as in Prob. 8 we find

$$Y(s) = (1+e^{-\pi s})\frac{1}{(s^2+1)(s^2+s+5/4)}.$$ The correct partial

fraction expansion of the quotient is $\frac{as+b}{s^2+1} + \frac{cs+d}{s^2+s+5/4}$,

where
a+c = 0, a+b+d = 0, (5/4)a+b+c = 0 and (5/4)b+d = 1 by
equating coefficients. Solving for a,b,c,d and following
the steps of Prob. 8 yields the desired solution.

16b. Taking the Laplace transform of the D.E. we obtain
U(s^2 + s/4 + 1) = k($e^{-3s/2}$-$e^{-5s/2}$)/s, since the I.C. are
zero. Solving for U and using partial fractions yields

$$U(s) = k(e^{-3s/2} - e^{-5s/2})(\frac{1}{s} - \frac{s+1/4}{s^2+s/4+1}). \quad \text{Thus, if}$$

$$H(s) = (\frac{1}{s} - \frac{s+1/4}{s^2+s/4+1}), \quad \text{then, since}$$

$$s^2 + s/4 + 1 = (s+1/8)^2 + 63/74,$$

$$h(t) = 1 - e^{-t/8}(\cos\frac{3\sqrt{7}}{8}t + \frac{\sqrt{7}}{21}\sin\frac{3\sqrt{7}}{8}t) \quad \text{and}$$

$$u(t) = ku_{3/2}(t)h(t-3/2) - ku_{5/2}(t)h(t-5/2).$$

16c. In all cases the plot will be zero for $0 \le t < 3/2$. For
$3/2 \le t < 5/2$ the plot will be the system response
(damped sinusoid) to a step input of magnitude k. For
$t \ge 5/2$, the plot will be the system response to the

I.C. $u(5^-/2)$, $u'(5^-/2)$ with
no forcing function. The
graph shown is for $k = 2$.
Varying k will just affect
the amplitude. Note that
the amplitude never
reaches 2, which would be
the steady state response
for the step input $2u_{3/2}(t)$.

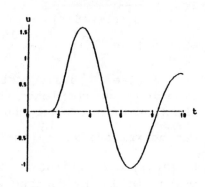

Note also that the solution
and its derivative are
continuous at $t = 5/2$.

19a. The graph on $0 \le t < 6\pi$ will depend on how large n is.
For instance, if $n = 2$ then

$$f(t) = \begin{cases} 1 & 0 \le t < \pi, \ 2\pi \le t < 6\pi \\ -1 & -\pi \le t < 2\pi \end{cases}.$$

For $n \ge 6$, $f(t) = \begin{cases} 1 & 0 \le t < \pi, \ 2\pi \le t < 3\pi, \ 4\pi \le t < 5\pi \\ -1 & \pi \le t < 2\pi, \ 3\pi \le t < 4\pi, \ 5\pi \ t < 6\pi \end{cases}.$

19b. Taking the Laplace transform of the D.E. and using the I.C. we

have $Y(s) = \dfrac{1}{s(s^2+1)}[1 + 2\sum\limits_{k=1}^{n}(-1)^k e^{-\pi ks}]$, since

$\mathcal{L}\{u_{\pi k}(t)\} = \dfrac{e^{-\pi ks}}{s}$ Since $\dfrac{1}{s(s^2+1)} = \dfrac{1}{s} - \dfrac{s}{s^2+1}$, we then obtain

$$y(t) = 1 - \cos t + 2\sum\limits_{k=1}^{n}(-1)^k u_{\pi k}(t)[1 - \cos(t-\pi k)], \quad \text{using}$$

line 13 in Table 6.2.1.

19c. For $0 \le t < \pi$, y(t)=1-cost, which peaks at t=π, just when the forcing function changes from +1 to -1. Thus the forcing function "reinforces" the natural motion, creating a "resonance". This occurs at each π interval until t > 15π, at which time the forcing function no longer changes and the solution continues oscillating about -1. If n = 16, the solution would continue to oscillate about +1 for t > 16π.

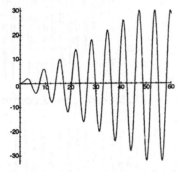

19d. Since $\cos(t-\pi k) = (-1)^k \cos t$, the solution in part b can be written as

$$y(t) = 1 - \cos t + 2\sum_{k=1}^{n}(-1)^k u_{\pi k}(t) - 2\sum_{k=1}^{n}(-1)^{2k}\cos t$$

$$= 1 - \cos t - 2n\cos t + 2\sum_{k=1}^{n}(-1)^k u_{\pi k}(t) \text{ which diverges for } n \to \infty.$$

20. In this case

$$Y(s) = \frac{1}{s(s^2+.1s+1)}[1 + 2\sum_{k=1}^{n}(-1)^k e^{-\pi ks}]. \text{ Using partial}$$

fractions we have

$$H(s) = \frac{1}{s(s^2+.1s+1)} = \frac{1}{s} - \frac{s+.1}{s^2+.1s+1}$$

$$= \frac{1}{s} - \frac{s+.05}{(s+.05)^2+b^2} - \frac{.05}{(s+.05)^2+b^2}, \text{ where}$$

$b^2 = [1-(.05)^2] = .9975$. Now let

$$h(t) = \mathcal{L}^{-1}\{H(s)\} = 1 - e^{-.05t}\cos bt - \frac{.05}{b}e^{-.05t}\sin bt. \text{ Hence,}$$

$$y(t) = h(t) + 2\sum_{k=1}^{n}(-1)^k u_{\pi k}(t) h(t-\pi k), \text{ and thus, for t > n}\pi \text{ the}$$

solution will be approximated by

$$\pm 1 - Ae^{-.05(t-n\pi)}\cos[b(t-n\pi) + \delta], \text{ and therefore}$$

converges as t$\to\infty$.

20a. y(t) for n = 30   y(t) for n = 31

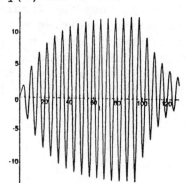   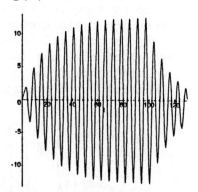

20b. From the graph of part a, A ≅ 12.5 and the frequency is 2π.

20c. From the graph
     (or analytically)
     A = 10 and the
     frequency is 2π.

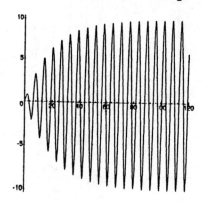

Section 6.5, Page 344

1. Proceeding as in Example 1, we take the Laplace transform
   of the D.E. and apply the I.C.:
   $(s^2 + 2s + 2)Y(s) = s + 2 + e^{-\pi s}$. Thus,
   $Y(s) = (s+2)/[(s+1)^2 + 1] + e^{-\pi s}/[(s+1)^2 + 1]$. We write
   the first term as $(s+1)/[(s+1)^2 + 1] + 1/[(s+1)^2 + 1]$.
   Applying Theorem 6.3.1 and using Table 6.2.1, we obtain
   the solution, $y = e^{-t}\cos t + e^{-t}\sin t + u_\pi(t)e^{-(t-\pi)}\sin(t-\pi)$.

3. Taking the Laplace transform and using the I.C. we have
   $(s^2+3s+2)Y(s) = \dfrac{1}{2} + e^{-5s} + \dfrac{e^{-10s}}{s}$. Thus
   $Y(s) = \dfrac{1/2}{s^2+3s+2} + \dfrac{e^{-5s}}{s^2+3s+2} + e^{-10s}(\dfrac{1/2}{s} + \dfrac{1/2}{s+2} - \dfrac{1}{s+1})$ and
   hence
   $y(t) = \dfrac{1}{2}h(t) + u_5(t)h(t-5) + u_{10}(t)[\dfrac{1}{2}+\dfrac{1}{2}e^{-2(t-10)}-e^{-(t-10)}]$
   where $h(t) = e^{-t} - e^{-2t}$.

5. The Laplace transform of the D.E. is

$$(s^2+2s+3)Y(s) = \frac{1}{s^2+1} + e^{-3\pi s}, \text{ so}$$

$$Y(s) = \frac{1}{(s^2+1)(s^2+2s+3)} + e^{-3\pi s}[\frac{1}{s^2+2s+3}]. \text{ Using partial}$$

fractions or a computer algebra system we obtain

$$y(t) = \frac{1}{4}\sin t - \frac{1}{4}\cos t + \frac{1}{4}e^{-t}\cos\sqrt{2}\,t + \frac{1}{\sqrt{2}}u_{3\pi}(t)h(t-3\pi),$$

where $h(t) = e^{-t}\sin\sqrt{2}\,t$.

7. Taking the Laplace transform of the D.E. yields

$$(s^2+1)Y(s) - y'(0) = \int_0^\infty e^{-st}\delta(t-2\pi)\cos t\,dt. \text{ Since}$$

$\delta(t-2\pi) = 0$ for $t \neq 2\pi$ the integral on the right is equal

to $\int_{-\infty}^\infty e^{-st}\delta(t-2\pi)\cos t\,dt$ which equals $e^{-2\pi s}\cos 2\pi$ from

Eq.(16). Substituting for $y'(0)$ and solving for $Y(s)$

gives $Y(s) = \dfrac{1}{s^2+1} + \dfrac{e^{-2\pi s}}{s^2+1}$ and hence

$$y(t) = \sin t + u_{2\pi}(t)\sin(t-2\pi) = \begin{cases} \sin t & 0 \le t < 2\pi \\ 2\sin t & 2\pi \le t \end{cases}.$$

10. Follow the same steps as in the solution for Prob. 7.

13a. From Eq. (22) y(t) will complete one cycle when
     $\sqrt{15}\,(t-5)/4 = 2\pi$ or $T = t - 5 = 8\pi/\sqrt{15}$, which is
     consistent with the plot in Fig. 6.5.3. Since an impulse
     causes a discontinuity in the first derivative, we need
     to find the value of y' at t = 5 and t = 5 + T. From Eq.
     (22) we have, for t ≥ 5,

$$y' = e^{-(t-5)/4}[\frac{-1}{2\sqrt{15}}\sin\frac{\sqrt{15}}{4}(t-5) + \frac{1}{2}\cos\frac{\sqrt{15}}{4}(t-5)]. \text{ Thus}$$

$y'(5) = \dfrac{1}{2}$ and $y'(5+T) = \dfrac{1}{2}e^{-T/4}$. Since the original

impulse, $\delta(t-5)$, caused a discontinuity in y' of 1/2 at
t = 5, we must choose the impulse at t = 5 + T to be
$-e^{-T/4}$, which is equal and opposite to y' at 5 + T.

13b. Now consider $2y'' + y' + 2y = \delta(t-5) + k\delta(t-5-T)$ with
     $y(0) = 0$, $y'(0) = 0$. Using the results of Ex. 1 we have

$$y(t) = \frac{2}{\sqrt{15}} u_5(t) e^{-(t-5)/4} \sin\frac{\sqrt{15}}{4}(t-5)$$

$$+ \frac{2k}{\sqrt{15}} u_{5+T}(t) e^{-(t-5-T)/4} \sin\frac{\sqrt{15}}{4}(t-5-T)$$

$$= \frac{2}{\sqrt{15}} e^{-(t-5)/4} [u_5(t)\sin\frac{\sqrt{15}}{4}(t-5) + k u_{5+T}(t) e^{T/4} \sin\frac{\sqrt{15}}{4}(t-5-T)]$$

$$= \frac{2}{\sqrt{15}} e^{-(t-5)/4} [u_5(t) + k e^{T/4} u_{5+T}(t)] \sin\frac{\sqrt{15}}{4}(t-5), \text{ since}$$

$T = 8\pi/\sqrt{15}$. If $k = -e^{-T/4}$ then $1 + k e^{-T/4} = 0$ for $t > 5 + T$ and $y(t) \equiv 0$, which is the desired result.

17b. We have $(s^2+1)Y(s) = \sum_{k=1}^{20} e^{-k\pi s}$ so that $Y(s) = \sum_{k=1}^{20} \frac{e^{-ks}}{s^2+1}$ and

hence $y(t) = \sum_{k=1}^{20} u_{k\pi}(t)\sin(t-k\pi)$

$\quad = u_\pi(t)\sin(t-\pi) + u_{2\pi}(t)\sin(t-2\pi) + \ldots + u_{20\pi}\sin(t-20\pi)$.

For $0 \leq t < \pi$, $y(t) \equiv 0$.

For $\pi \leq t < 2\pi$, $y(t) = \sin(t-\pi) = -\sin t$.

For $2\pi \leq t < 3\pi$, $y(t) = \sin(t-\pi)+\sin(t-2\pi) = -\sin t+\sin t \equiv 0$. Due to the periodicity of $\sin t$, the solution will exhibit this behavior in alternate intervals for $0 \leq t < 20\pi$.

17c. After $t = 20\pi$ the solution remains at zero.

21b. Taking the transform and using the I.C. we have

$$(s^2+1)Y(s) = \sum_{k=1}^{15} e^{-(2k-1)\pi} \text{ so that } Y(s) = \sum_{k=1}^{15} \frac{e^{-(2k-1)\pi}}{s^2+1}.$$

Thus $y(t) = \sum_{k=1}^{15} u_{(2k-1)\pi}(t)\sin[t-(2k-1)\pi]$.

21c. For $t > 29\pi$, $y(t) = \sin(t-\pi) + \sin(t-3\pi) +\ldots+ \sin(t-29\pi)$
$$= -\sin t - \sin t -\ldots-\sin t$$
$$= -15\sin t.$$

25b. Substituting for $f(t)$ in the integral of part a, we have

$y = \int_0^t e^{-(t-\tau)}\delta(\tau-\pi)\sin(t-\tau)d\tau$. We know that the integration variable is always less than t (the upper limit) and thus for $t < \pi$ we have $\tau < \pi$ and thus

$\delta(\tau-\pi) = 0$. Hence $y \equiv 0$ for $t < \pi$. For $t > \pi$ utilize Eq.(16).

Section 6.6, Page 351

1c. Using the format of Eqs.(2) and (3) we have

$$f*(g*h) = \int_0^t f(t-\tau)(g*h)(\tau)d\tau$$

$$= \int_0^t f(t-\tau)[\int_0^\tau g(\tau-\eta)h(\eta)d\eta]d\tau$$

$$= \int_0^t [\int_\eta^t f(t-\tau)g(\tau-\eta)d\tau]h(\eta)(d\eta).$$

The last double integral is obtained from the previous line by interchanging the order of the η and τ integrations. Making the change of variable $\omega = \tau - \eta$ on the inside integral yields

$$f*(g*h) = \int_0^t [\int_0^{t-\eta} f(t-\eta-\omega)g(\omega)d\omega]h(\eta)d\eta$$

$$= \int_0^t [(f*g)(t-\eta)]h(\eta)d\eta = (f*g)*h.$$

4. It is possible to determine $f(t)$ explicitly by using integration by parts and then find its transform $F(s)$. However, it is much more convenient to apply Theorem 6.6.1. Let us define $g(t) = t^2$ and $h(t) = \cos 2t$. Then, $f(t) = \int_0^t g(t-\tau)h(\tau)d\tau$. Using Table 6.2.1, we have $G(s) = \mathcal{L}\{g(t)\} = 2/s^3$ and $H(s) = \mathcal{L}\{h(t)\} = s/(s^2+4)$. Hence, by Theorem 6.6.1, $\mathcal{L}\{f(t)\} = F(s) = G(s)H(s) = 2/s^2(s^2+4)$.

8. As was done in Ex. 1 think of $F(s)$ as the product of s^{-4} and $(s^2+1)^{-1}$ which, according to Table 6.2.1, are the transforms of $t^3/6$ and $\sin t$, respectively. Hence, by Theorem 6.6.1, the inverse transform of $F(s)$ is

$$f(t) = (1/6)\int_0^t (t-\tau)^3 \sin\tau d\tau.$$

14. We take the Laplace transform of the D.E. and apply the I.C.: $(s^2 + 2s + 2)Y(s) = \alpha/(s^2 + \alpha^2)$. Solving for $Y(s)$, we have $Y(s) = [\alpha/(s^2+\alpha^2)][(s+1)^2 + 1]^{-1}$, where the second factor has been written in a convenient way by completing the square. Thus $Y(s)$ is seen to be the product of the transforms of $\sin\alpha t$ and $e^{-t}\sin t$ respectively. Hence,

according to Theorem 6.6.1, $y = \int_0^t e^{-(t-\tau)}\sin(t-\tau)\sin\alpha\tau d\tau$.

16. Proceeding as in Prob. 14 we obtain

$$(s^2+s+5/4)Y(s) - s = \frac{1-e^{-\pi s}}{s} \text{ or}$$

$$Y(s) = \frac{s}{s^2+s+5/4} + \frac{1-e^{-\pi s}}{s(s^2+s+5/4)}$$

$$= \frac{(s+1/2) - 1/2}{(s+1/2)^2+1} + \frac{1-e^{-\pi s}}{s} \cdot \frac{1}{(s+1/2)^2+1},$$

where the first term is obtained by completing the square
in the denominator and the second term is written as the
product of two terms whose inverse transforms are known,
so that Theorem 6.6.1 can be used. Note that
$\mathcal{L}^{-1}\{(1-e^{-\pi s})/s\} = 1 - u_\pi(t)$. Also note that a different
form of the same solution would be obtained by writing
the second term as $(1-e^{-\pi s})(\dfrac{a}{s} + \dfrac{bs + c}{(s+1/2)^2+1})$ and solving
for a, b and c. In this case $\mathcal{L}^{-1}\{1-e^{-\pi s}\} = \delta(t) - \delta(t-\pi)$
from Section 6.5.

18. Taking the Laplace transform, using the I.C. and solving,
we have $Y(s) = (s+3)/(s+1)(s+2) + s/[(s^2+\alpha^2)(s+1)(s+2)]$.
As in Prob. 16, there are several correct ways the second
term can be treated in order to use the convolution
integral. In order to obtain the desired answer, write
the second term as
$$\frac{s}{s^2+\alpha^2}(\frac{a}{s+1} + \frac{b}{s+2}) \text{ and solve for a and b.}$$

21. To find $\Phi(s)$ you must recognize the integral that
appears in the equation as a convolution integral.
Taking the transform of both sides then yields

$$\Phi(s) + K(s)\Phi(s) = F(s), \text{ or } \Phi(s) = \frac{F(s)}{1+K(s)}.$$

24a. Again, we recognize $\int_0^t (t-\xi)\phi(\xi)d\xi$ as the convolution of t
and $\phi(t)$. Thus, taking the transform of both sides, we
get $\Phi(s) - \dfrac{1}{s^2}\Phi(s) = \dfrac{1}{s}$. Solving for Φ we get $\Phi(s) =$

$\dfrac{s}{s^2-1}$ and hence $\phi(t) = \cosh t$ from line 8 of Table 6.2.1.

24b. Taking the derivative of the given equation we get

$$\phi'(t) - \frac{d}{dt}\int_0^t (t-\xi)\phi(\xi)d\xi = 0, \text{ or}$$

$$\phi'(t) - \int_0^t \phi(\xi)d\xi - (t-\xi)\phi(\xi)\big|_{\xi=t} = 0, \text{ using Leibnitz's}$$

Rule. Since the last term is zero, we have

$$\phi'(t) - \int_0^t \phi(\xi)d\xi = 0, \text{ Eq.(i), and then } \phi''(t) - \phi(t) = 0,$$

Eq.(ii). The original equation gives $\phi(0) = 1$ and Eq.(i) gives $\phi'(0) = 0$. Eq.(ii) with the IC is the desired IVP.

24c. The solution to $\phi''(t) - \phi(t) = 0$, $\phi(0) = 1$, $\phi'(0) = 0$ is also $\phi(t) = \cosh t$.

26a. Taking the transform, we obtain

$$s\Phi(s) - \phi(0) + \frac{1}{s^2}\Phi(s) = \frac{1}{s^2}, \text{ and thus}$$

$$\Phi(s) = \frac{1}{s^3+1} = \frac{1}{(s+1)(s^2-s+1)} = \frac{1/3}{s+1} - \frac{(s-2)/3}{s^2-s+1}.$$

Completing the square in the denominator of the last term and using lines 9 and 10 of Table 6.2.1 we obtain

$$\phi(t) = e^{-t}/3 - (1/3)e^{t/2}\cos(\sqrt{3}\,t/2) + (1/\sqrt{3})e^{t/2}\sin(\sqrt{3}\,t/2).$$

26b. From the given equation $\phi'(0) = 0$. Differentiating we get

$$\phi''(t) + \int_0^t \phi(\xi)d\xi = 1 \text{ and } \phi''(0) = 1. \text{ Another}$$

differentiation gives $\phi'''(t) + \phi(t) = 0$. Thus the IVP is $\phi'''(t) + \phi(t) = 0$, $\phi(0) = 0$, $\phi'(0) = 0$, and $\phi''(0) = 1$.

26c. The characteristic equation is $r^3 + 1 = 0$, so the roots are the cube roots of -1, which are -1 and $(1 \pm \sqrt{3})/2$.

CHAPTER 7

Section 7.1, Page 360

2. As in Ex. 1, let $x_1 = u$ and $x_2 = u'$. Then $x_1' = x_2$ and
    $x_2' = u'' = 3\sin t - .5u' - 2u = -2x_1 - .5x_2 + 3\sin t$.

4. In this case let $x_1 = u$, $x_2 = u'$, $x_3 = u''$, and $x_4 = u'''$.

6. Let $x_1 = u$ and $x_2 = u'$; then $x_1' = x_2$ is the first of the
    desired pair of equations. The second equation is
    obtained by substituting $u'' = x_2'$, $u' = x_2$, and $u = x_1$ in
    the given D.E. The I.C. become $x_1(0) = u_0$, $x_2(0) = u_0'$.

8. Follow the steps outlined in Prob.7. Solve the first D.E.
    for x_2 to obtain $x_2 = \frac{3}{2}x_1 - \frac{1}{2}x_1'$. Substitute this into
    the second D.E. to obtain $x_1'' - x_1' - 2x_1 = 0$, which has the
    solution $x_1 = c_1e^{2t} + c_2e^{-t}$. Differentiating this and
    substituting into the above equation for x_2 yields $x_2 =$
    $\frac{1}{2}c_1e^{2t} + 2c_2e^{-t}$. The I.C. then give
    $c_1 + c_2 = 3$ and $\frac{1}{2}c_1 + 2c_2 = \frac{1}{2}$, which yield
    $c_1 = \frac{11}{3}$, $c_2 = -\frac{2}{3}$. Thus $x_1 = \frac{11}{3}e^{2t} - \frac{2}{3}e^{-t}$ and
    $x_2 = \frac{11}{6}e^{2t} - \frac{4}{3}e^{-t}$. Note that for large t, the second
    term in each solution vanishes and we have $x_1 \cong \frac{11}{3}e^{2t}$ and
    $x_2 \cong \frac{11}{6}e^{2t}$, so that $x_1 \cong 2x_2$. This says that the graph
    will be asymptotic to the line $x_1 = 2x_2$ for large t.

9. Solving the first D.E. for x_2 gives $x_2 = \frac{4}{3}x_1' - \frac{5}{3}x_1$,
    which substituted into the second D.E. yields
    $x_1'' - 2.5x_1' + x_1 = 0$. Thus $x_1 = c_1e^{t/2} + c_2e^{2t}$ and
    $x_2 = -c_1e^{t/2} + c_2e^{2t}$. Using the I.C. yields $c_1 = -3/2$ and
    $c_2 = -1/2$. For large t, $x_1 \cong (-1/2)e^{2t}$ and $x_2 \cong (-1/2)e^{2t}$
    and thus the graph is asymptotic to $x_1 = x_2$ in the third
    quadrant. The graph is shown on the right.

9. 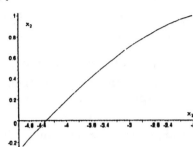   12.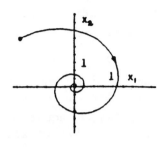

12. Solving the first D.E. for x_2 gives $x_2 = \frac{1}{2}x_1' + \frac{1}{4}x_1$ and

    substitution into the second D.E. gives

    $x_1'' + x_1' + \frac{17}{4}x_1 = 0$. Thus $x_1 = e^{-t/2}(c_1\cos 2t + c_2\sin 2t)$ and

    $x_2 = e^{-t/2}(c_2\cos 2t - c_1\sin 2t)$. The I.C. yields $c_1 = -2$ and

    $c_2 = 2$.

14. If $a_{12} \neq 0$, then solve the first equation for x_2,

    obtaining $x_2 = [x_1' - a_{11}x_1 - g_1(t)]/a_{12}$. Upon substituting

    this expression into the second equation, we have a

    second order linear O.D.E. for x_1. One I.C. is

    $x_1(0) = x_1^0$. The second I.C. is

    $x_2(0) = [x_1'(0) - a_{11}x_1(0) - g_1(0)]/a_{12} = x_2^0$. Solving for

    $x_1'(0)$ gives $x_1'(0) = a_{12}x_2^0 + a_{11}x_1^0 + g_1(0)$. If $a_{12} = 0$, then

    solve the second equation for x_1 and proceed as above.

    These results hold when $a_{11}, \ldots, a_{22}$ are functions of t

    as long as the derivatives exist and $a_{12}(t)$ and

    $a_{21}(t)$ are not both zero on the interval. The I.C.

    will involve $a_{11}(0)$ and $a_{12}(0)$.

20. Number the nodes 1,2, and 3 clockwise beginning with the

    top right node in Fig. 7.1.4. Also let I_1, I_2, I_3, and I_4

    denote the currents through the resistor $R = 1$, the

    inductor $L = 1$, the capacitor $C = \frac{1}{2}$, and the resistor

    $R = 2$, respectively. Let V_1, V_2, V_3, and V_4 be the

    corresponding voltage drops. Kirchhoff's first law

    applied to nodes 1 and 2, respectively, gives

    (i) $I_1 - I_2 = 0$ and (ii) $I_2 - I_3 - I_4 = 0$. Kirchhoff's

    second law applied to each loop gives

(iii) $V_1 + V_2 + V_3 = 0$ and (iv) $V_3 - V_4 = 0$. The current-voltage relation through each circuit element yields four more equations: (v) $V_1 = I_1$, (vi) $I_2' = V_2$,

(vii) $(1/2)V_3' = I_3$ and (viii) $V_4 = 2I_4$. We thus have a system of eight equations in eight unknowns, and we wish to eliminate all of the variables except I_2 and V_3 from this system of equations. For example, we can use Eqs.(i) and (iv) to eliminate I_1 and V_4 in Eqs.(v) and (viii). Then use the new Eqs.(v) and (viii) to eliminate V_1 and I_4 in Eqs.(ii) and (iii). Finally, use the new Eqs. (ii) and (iii) in Eqs.(vi) and (vii) to obtain $I_2' = - I_2 - V_3$, $V_3' = 2I_2 - V_3$. These equations are identical (when subscripts on the remaining variables are dropped) to the equations given in the text.

22a. Note that the amount of water in each tank remains constant. Thus $Q_1(t)/30$ and $Q_2(t)/20$ represent oz./gal of salt in each tank. We assume the mixture in each tank is well stirred. Then, for the first tank we have $$\frac{dQ_1}{dt} = 1.5 - 3\frac{Q_1(t)}{30} + 1.5\frac{Q_2(t)}{20},$$ where the first term on the right represents the amount of salt per minute entering the mixture from an external source, the second term represents the loss of salt per minute going to Tank 2 and the third term represents the gain of salt per minute entering from Tank 2. Similarly, we have $$\frac{dQ_2}{dt} = 3 + 3\frac{Q_1(t)}{30} - 4\frac{Q_2(t)}{20}$$ for Tank 2. We are given that $Q_1(0) = 25$ oz. and $Q_2(0) = 15$ oz.

22b. Solve the second equation for $Q_1(t)$ to obtain $Q_1(t) = 10Q_2' + 2Q_2 - 30$. Substitution into the first equation then yields $10Q_2'' + 3Q_2' + \frac{1}{8}Q_2 = \frac{9}{2}$. Equilibrium is the steady state solution, which is $Q_2^E = 8(9/2) = 36$. Substituting this value into the equation for Q_1 yields $Q_1^E = 72 - 30 = 42$. It can be shown that $Q_1(t)$ satisfies the same second order DE as $Q_2(t)$ (except with the constant 21/4 on the right side) and thus the exponentials in the solutions for each are the same. Hence each tank approaches the equilibrium solution at then same rate.

22c. Substitute $Q_1 = x_1 + 42$ and $Q_2 = x_2 + 36$ into the
     equations found in part a.

Section 7.2, Page 372

1a. $2\mathbf{A} = \begin{pmatrix} 2 & -4 & 0 \\ 6 & 4 & -2 \\ -4 & 2 & 6 \end{pmatrix}$ so that

$$2\mathbf{A} + \mathbf{B} = \begin{pmatrix} 2+4 & -4-2 & 0+3 \\ 6-1 & 4+5 & -2+0 \\ -4+6 & 2+1 & 6+2 \end{pmatrix} = \begin{pmatrix} 6 & -6 & 3 \\ 5 & 9 & -2 \\ 2 & 3 & 8 \end{pmatrix}$$

1c. Using Eq.(9) and following Ex. 1 we have

$$\mathbf{AB} = \begin{pmatrix} 4 + 2 + 0 & -2 - 10 + 0 & 3 + 0 + 0 \\ 12 - 2 - 6 & -6 + 10 - 1 & 9 + 0 - 2 \\ -8 - 1 + 18 & 4 + 5 + 3 & -6 + 0 + 6 \end{pmatrix},$$

which yields the correct answer.

6a. $\mathbf{AB} = \begin{pmatrix} 6 & -5 & -7 \\ 1 & 9 & 1 \\ -1 & -2 & 8 \end{pmatrix}$ and $\mathbf{BC} = \begin{pmatrix} 5 & 3 & 3 \\ -1 & 7 & 3 \\ 2 & 3 & -2 \end{pmatrix}$ so that

$$(\mathbf{AB})\mathbf{C} = \mathbf{A}(\mathbf{BC}) = \begin{pmatrix} 7 & -11 & -3 \\ 11 & 20 & 17 \\ -4 & 3 & -12 \end{pmatrix}.$$

In Problems 10 through 19 the method of row reduction, as
illustrated in Ex. 2, can be used to find the inverse matrix
or else to show that none exists. We start with the original
matrix augmented by the indentity matrix, describe a suitable
sequence of elementary row operations, and show the result of
applying these operations.

10. Start with the given matrix augmented by the identity
     matrix. $\begin{pmatrix} 1 & 4 & . & 1 & 0 \\ & & . & & \\ -2 & 3 & . & 0 & 1 \end{pmatrix}$

     Add 2 times the first row to the second row.

$$\begin{pmatrix} 1 & 4 & . & 1 & 0 \\ & & . & & \\ 0 & 11 & . & 2 & 1 \end{pmatrix}$$

Multiply the second row by (1/11).

$$\begin{pmatrix} 1 & 4 & . & 1 & 0 \\ & & . & & \\ 0 & 1 & . & 2/11 & 1/11 \end{pmatrix}$$

Add (-4) times the second row to the first row.

$$\begin{pmatrix} 1 & 0 & . & 3/11 & -4/11 \\ & & . & & \\ 0 & 1 & . & 2/11 & 1/11 \end{pmatrix}$$

Since we have performed the same operation on the given
matrix and the identity matrix, the 2 x 2 matrix
appearing on the right side of this augmented matrix is
the desired inverse matrix. The answer can be checked by
multiplying it by the given matrix; the result should be
the indentity matrix.

12. The augmented matrix in this case is:

$$\begin{pmatrix} 1 & 2 & 3 & . & 1 & 0 & 0 \\ & & & . & & & \\ 2 & 4 & 5 & . & 0 & 1 & 0 \\ & & & . & & & \\ 3 & 5 & 6 & . & 0 & 0 & 1 \end{pmatrix}$$

Add (-2) times the first row to the second row and (-3)
times the first row to the third row.

$$\begin{pmatrix} 1 & 2 & 3 & . & 1 & 0 & 0 \\ & & & . & & & \\ 0 & 0 & -1 & . & -2 & 1 & 0 \\ & & & . & & & \\ 0 & -1 & -3 & . & -3 & 0 & 1 \end{pmatrix}$$

Multiply the second and third rows by (-1) and
interchange them.

$$\begin{pmatrix} 1 & 2 & 3 & . & 1 & 0 & 0 \\ & & & . & & & \\ 0 & 1 & 3 & . & 3 & 0 & -1 \\ & & & . & & & \\ 0 & 0 & 1 & . & 2 & -1 & 0 \end{pmatrix}$$

Add (-3) times the third row to the first and second

$$\begin{pmatrix} 1 & 2 & 0 & . & -5 & 3 & 0 \\ & & & . & & & \\ 0 & 1 & 0 & . & -3 & 3 & -1 \\ & & & . & & & \\ 0 & 0 & 1 & . & 2 & -1 & 0 \end{pmatrix}$$

rows.

Add (-2) times the second row to the first row.

$$\begin{pmatrix} 1 & 0 & 0 & . & 1 & -3 & 2 \\ & & & . & & & \\ 0 & 1 & 0 & . & -3 & 3 & -1 \\ & & & . & & & \\ 0 & 0 & 1 & . & 2 & -1 & 0 \end{pmatrix}$$

The desired answer appears on the right side of this augmented matrix.

14. Again, start with the given matrix augmented by the

identity matrix:
$$\begin{pmatrix} 1 & 2 & 1 & . & 1 & 0 & 0 \\ & & & . & & & \\ -2 & 1 & 8 & . & 0 & 1 & 0 \\ & & & . & & & \\ 1 & -2 & -7 & . & 0 & 0 & 1 \end{pmatrix}$$

Add (2) times the first row to the second row and add(-1) times the first row to the third row.

$$\begin{pmatrix} 1 & 2 & 1 & . & 1 & 0 & 0 \\ & & & . & & & \\ 0 & 5 & 10 & . & 2 & 1 & 0 \\ & & & . & & & \\ 0 & -4 & -8 & . & -1 & 0 & 1 \end{pmatrix}$$

Add (4/5) times the second row to the third row.

$$\begin{pmatrix} 1 & 2 & 1 & . & 1 & 0 & 0 \\ & & & . & & & \\ 0 & 5 & 10 & . & 2 & 1 & 0 \\ & & & . & & & \\ 0 & 0 & 0 & .3/5 & 4/5 & & 0 \end{pmatrix}$$

Since the third row of the left matrix is all zeros, no further reduction can be performed, and the given matrix is singular.

21c. Differentiate each element of $\mathbf{A}(t)$. For instance, $4e^{2t}$ is the derivative of $a_{33}(t)$ and this will then be the element in the 3rd row 3rd column of $\mathbf{A}'(t)$.

21d. Integrate each element of $\mathbf{A}(t)$ from $t = 0$ to $t = 1$. For instance, $\int_0^1 e^{2t}dt = (1/2)e^{2t}\big|_0^1 = (1/2)(e^2-1) = (e+1)(e-1)/2$ is the integral of $a_{13}(t)$. Thus $(e+1)(e-1)/2$ will be the element in the 1st row 3rd column of $\int_0^1 \mathbf{A}(t)dt$

22. $\mathbf{x}' = \begin{pmatrix} 4 \\ 2 \end{pmatrix} 2e^{2t} = \begin{pmatrix} 8 \\ 4 \end{pmatrix} e^{2t}$; and

$$\begin{pmatrix} 3 & -2 \\ 2 & -2 \end{pmatrix} \mathbf{x} = \begin{pmatrix} 3 & -2 \\ 2 & -2 \end{pmatrix} \begin{pmatrix} 4 \\ 2 \end{pmatrix} e^{2t} = \begin{pmatrix} 12-4 \\ 8-4 \end{pmatrix} e^{2t} = \begin{pmatrix} 8 \\ 4 \end{pmatrix} e^{2t}.$$

25. $\Psi' = \begin{pmatrix} -3e^{-3t} & 2e^{2t} \\ 12e^{-3t} & 2e^{2t} \end{pmatrix} = \begin{pmatrix} 1 & 1 \\ 4 & -2 \end{pmatrix} \begin{pmatrix} e^{-3t} & e^{2t} \\ -4e^{-3t} & e^{2t} \end{pmatrix}.$

Section 7.3, Page 383

1.   As in Ex. 1, form the augmented matrix and use row reduction:

$$\begin{pmatrix} 1 & 0 & -1 & . & 0 \\ & & & . & \\ 3 & 1 & 1 & . & 1 \\ & & & . & \\ -1 & 1 & 2 & . & 2 \end{pmatrix}$$

Add (-3) times the first row to the second and add the first row to the third.

$$\begin{pmatrix} 1 & 0 & -1 & . & 0 \\ & & & . & \\ 0 & 1 & 4 & . & 1 \\ & & & . & \\ 0 & 1 & 1 & . & 2 \end{pmatrix}$$

Add (-1) times the second row to the third.

$$\begin{pmatrix} 1 & 0 & -1 & . & 0 \\ & & & . & \\ 0 & 1 & 4 & . & 1 \\ & & & . & \\ 0 & 0 & -3 & . & 1 \end{pmatrix}$$

The third row is equivalent to $-3x_3 = 1$ or $x_3 = -1/3$. Likewise the second row is equivalent to $x_2 + 4x_3 = 1$, so $x_2 = 7/3$. Finally, from the first row, $x_1 - x_3 = 0$, so $x_1 = -1/3$. The answer can be checked by substituting into the original equations.

2. Forming the augmented matrix and using row reduction will yield an augmented mstrix whose third row is equivalent to: $0x_1 + 0x_2 + 0x_3 = 1$, or $0 = 1$. Thus there is no solution.

3. Form the augmented matrix and use row reduction.

$$\begin{pmatrix} 1 & 2 & -1 & . & 2 \\ & & & . & \\ 2 & 1 & 1 & . & 1 \\ & & & . & \\ 1 & -1 & 2 & . & -1 \end{pmatrix}$$

Add (-2) times the first row to the second and add (-1) times the first row to the third.

$$\begin{pmatrix} 1 & 2 & -1 & . & 2 \\ & & & . & \\ 0 & -3 & 3 & . & -3 \\ & & & . & \\ 0 & -3 & 3 & . & -3 \end{pmatrix}$$

Add (-1) times the second row to the third row and then multiply the second row by (-1/3).

$$\begin{pmatrix} 1 & 2 & -1 & . & 2 \\ & & & . & \\ 0 & 1 & -1 & . & 1 \\ & & & . & \\ 0 & 0 & 0 & . & 0 \end{pmatrix}$$

Since the last row has only zero entries, it may be dropped. The second row corresponds to the equation $x_2 - x_3 = 1$. We can assign an arbitrary value to either x_2 or x_3 and use this equation to solve for the other. For example, let $x_3 = c$, where c is arbitrary. Then $x_2 = 1 + c$. The first row corresponds to the equation $x_1 + 2x_2 - x_3 = 2$, so $x_1 = 2 - 2x_2 + x_3 = 2 - 2(1+c) + c = -c$.

6. To determine whether the given set of vectors is linearly independent we must solve the system $c_1 \mathbf{x}^{(1)} + c_2 \mathbf{x}^{(2)} + c_3 \mathbf{x}^{(3)} = \mathbf{0}$ for c_1, c_2, and c_3. Writing this in scalar form, we have
$$\begin{aligned} c_1 \quad\quad + c_3 &= 0 \\ c_1 + c_2 \quad\quad &= 0, \text{ so the} \\ c_2 + c_3 &= 0 \end{aligned}$$

augmented matrix is $\begin{pmatrix} 1 & 0 & 1 & . & 0 \\ & & & . & \\ 1 & 1 & 0 & . & 0 \\ & & & . & \\ 0 & 1 & 1 & . & 0 \end{pmatrix}$

Row reduction yields $\begin{pmatrix} 1 & 0 & 1 & . & 0 \\ & & & . & \\ 0 & 1 & -1 & . & 0 \\ & & & . & \\ 0 & 0 & 2 & . & 0 \end{pmatrix}$.

From the third row we have $c_3 = 0$. Then from the second row, $c_2 - c_3 = 0$, so $c_2 = 0$. Finally from the first row $c_1 + c_3 = 0$, so $c_1 = 0$. Since $c_1 = c_2 = c_3 = 0$, we conclude that the given vectors are linearly independent.

8. As in Prob. 6 we wish to solve the system
    $c_1\mathbf{x}^{(1)} + c_2\mathbf{x}^{(2)} + c_3\mathbf{x}^{(3)} + c_4\mathbf{x}^{(4)} = \mathbf{0}$ for c_1, c_2, c_3, and
    c_4. Form the augmented matrix and use row reduction:

$$\begin{pmatrix} 1 & -1 & -2 & -3 & . & 0 \\ & & & & . & \\ 2 & 0 & -1 & 0 & . & 0 \\ & & & & . & \\ 2 & 3 & 1 & -1 & . & 0 \\ & & & & . & \\ 3 & 1 & 0 & 3 & . & 0 \end{pmatrix}$$

Add (-2) times the first row to the second, add (-2)
times the first row to the third, and add (-3) times the
first row to the fourth.

$$\begin{pmatrix} 1 & -1 & -2 & -3 & . & 0 \\ & & & & . & \\ 0 & 2 & 3 & 6 & . & 0 \\ & & & & . & \\ 0 & 5 & 5 & 5 & . & 0 \\ & & & & . & \\ 0 & 4 & 6 & 12 & . & 0 \end{pmatrix}$$

Multiply the second row by (1/2) and then add (-5) times
the second row to the third and add (-4) times the second
row to the fourth.

$$\begin{pmatrix} 1 & -1 & -2 & -3 & . & 0 \\ & & & & . & \\ 0 & 1 & 3/2 & 3 & . & 0 \\ & & & & . & \\ 0 & 0 & -5/2 & -10. & & 0 \\ & & & & . & \\ 0 & 0 & 0 & 0 & . & 0 \end{pmatrix}$$

The third row is equivalent to the equation $c_3 + 4c_4 = 0$.
One way to satisfy this equation is by choosing $c_4 = -1$;
then $c_3 = 4$. From the second row we then have
$c_2 = -(3/2)c_3 - 3c_4 = -6 + 3 = -3$. Then, from the first
row, $c_1 = c_2 + 2c_3 + 3c_4 = -3 + 8 - 3 = 2$. Hence the
given vectors are linearly dependent, and satisfy
$2\mathbf{x}^{(1)} - 3\mathbf{x}^{(2)} + 4\mathbf{x}^{(3)} - \mathbf{x}^{(4)} = \mathbf{0}$.

13. Consider $c_1\mathbf{x}^{(1)}(t) + c_2\mathbf{x}^{(2)}(t) = \mathbf{0}$, which has the

augmented matrix $\begin{pmatrix} 2\sin t & \sin t & . & 0 \\ & & . & \\ \sin t & 2\sin t & . & 0 \end{pmatrix}$. Multiplying the

first row by $-1/2$ and adding to the second row yields
$\begin{pmatrix} 2\sin t & \sin t & . & 0 \\ 0 & (3/2)\sin t & . & 0 \end{pmatrix}$. Since $\sin t$ is not identically zero,

we conclude, from the last row, that $c_2 = 0$. Using this
in the first row gives $c_1 = 0$ and thus $\mathbf{x}^{(1)}(t)$ and $\mathbf{x}^{(2)}(t)$
are linearly independent for $-\infty < t < \infty$.

14. Let $t = t_0$ be a fixed value of t in the interval
$0 \le t \le 1$. To determine whether $\mathbf{x}^{(1)}(t_0)$ and $\mathbf{x}^{(2)}(t_0)$ are
linearly dependent we must solve $c_1\mathbf{x}^{(1)}(t_0)+c_2\mathbf{x}^{(2)}(t_0)=\mathbf{0}$.
We have the augmented matrix

$$\begin{pmatrix} e^{t_0} & 1 & . & 0 \\ & & & \\ t_0 e^{t_0} & t_0 & . & 0 \end{pmatrix}.$$

Multiply the first row by $(-t_0)$ and add to the second row

to obtain $\begin{pmatrix} e^{t_0} & 1 & . & 0 \\ & & & \\ 0 & 0 & . & 0 \end{pmatrix}.$

Thus, for example, we can choose $c_1 = 1$ and $c_2 = -e^{t_0}$,
and hence the given vectors are linearly dependent at t_0.
Since t_0 is arbitrary the vectors are linearly dependent
at each point in the interval. However, there is no
linear relation between $\mathbf{x}^{(1)}$ and $\mathbf{x}^{(2)}$ that is valid
throughout the interval $0 \le t \le 1$. For example, if
$t_1 \ne t_0$, and if c_1 and c_2 are chosen as above, then
$c_1\mathbf{x}^{(1)}(t_1) + c_2\mathbf{x}^{(2)}(t_1)$

$$= \begin{pmatrix} e^{t_1} \\ t_1 e^{t_1} \end{pmatrix} + -e^{t_0}\begin{pmatrix} 1 \\ t_1 \end{pmatrix} = \begin{pmatrix} e^{t_1} - e^{t_0} \\ t_1 e^{t_1} - t_1 e^{t_0} \end{pmatrix} \ne \begin{pmatrix} 0 \\ 0 \end{pmatrix}.$$

Hence the given vectors must be linearly independent on
$0 \le t \le 1$. In fact, the same argument applies to any
interval.

15. To find the eigenvalues and eigenvectors of the given

matrix we must solve $\begin{pmatrix} 5-\lambda & -1 \\ 3 & 1-\lambda \end{pmatrix} \begin{pmatrix} x_1 \\ x_2 \end{pmatrix} = \begin{pmatrix} 0 \\ 0 \end{pmatrix}$. The

determinant of coefficients is $(5-\lambda)(1-\lambda) - (-1)(3) = 0$,
or $\lambda^2 - 6\lambda + 8 = 0$. Hence $\lambda_1 = 2$ and $\lambda_2 = 4$ are the
eigenvalues. The eigenvector corresponding to λ_1 must

satisfy $\begin{pmatrix} 3 & -1 \\ 3 & -1 \end{pmatrix} \begin{pmatrix} x_1 \\ x_2 \end{pmatrix} = \begin{pmatrix} 0 \\ 0 \end{pmatrix}$, or $3x_1 - x_2 = 0$. If we let

$x_1 = 1$, then $x_2 = 3$ and the eigenvector is $\mathbf{x}^{(1)} = \begin{pmatrix} 1 \\ 3 \end{pmatrix}$, or

any constant multiple of this vector. Similarly, the
eigenvector corresponding to λ_2 must satisfy

$\begin{pmatrix} 1 & -1 \\ 3 & -3 \end{pmatrix} \begin{pmatrix} x_1 \\ x_2 \end{pmatrix} = \begin{pmatrix} 0 \\ 0 \end{pmatrix}$, or $x_1 - x_2 = 0$. Hence $\mathbf{x}^{(2)} = \begin{pmatrix} 1 \\ 1 \end{pmatrix}$, or

a multiple thereof.

18. Since $a_{12} = \overline{a_{21}}$, the given matrix is Hermitian and we
know in advance that its eigenvalues are real. To find
the eigenvalues and eigenvectors we must solve

$\begin{pmatrix} 1-\lambda & i \\ -i & 1-\lambda \end{pmatrix} \begin{pmatrix} x_1 \\ x_2 \end{pmatrix} = \begin{pmatrix} 0 \\ 0 \end{pmatrix}$. The determinant of coefficients

is $(1-\lambda)^2 - i(-i) = \lambda^2 - 2\lambda$, so the eigenvalues are
$\lambda_1 = 0$ and $\lambda_2 = 2$; observe that they are indeed real even
though the given matrix has imaginary entries. The
eigenvector corresponding to λ_1 must satisfy

$\begin{pmatrix} 1 & i \\ -i & 1 \end{pmatrix} \begin{pmatrix} x_1 \\ x_2 \end{pmatrix} = \begin{pmatrix} 0 \\ 0 \end{pmatrix}$, or $x_1 + ix_2 = 0$. Note that the second

equation $-ix_1 + x_2 = 0$ is a multiple of the first.
If $x_1 = 1$, then $x_2 = i$, and the eigenvector is

$\mathbf{x}^{(1)} = \begin{pmatrix} 1 \\ i \end{pmatrix}$. In a similar way we find that the

eigenvector associated with λ_2 is $\mathbf{x}^{(2)} = \begin{pmatrix} 1 \\ -i \end{pmatrix}$.

21. The eigenvalues and eigenvectors satisfy

$$\begin{pmatrix} 1-\lambda & 0 & 0 \\ 2 & 1-\lambda & -2 \\ 3 & 2 & 1-\lambda \end{pmatrix} \begin{pmatrix} x_1 \\ x_2 \\ x_3 \end{pmatrix} = \begin{pmatrix} 0 \\ 0 \\ 0 \end{pmatrix}.$$ The determinant of coefficients is

$(1-\lambda)[(1-\lambda)^2 + 4] = 0$, which has roots $\lambda = 1$, $1 \pm 2i$. For
$\lambda = 1$, we then have $2x_1 - 2x_3 = 0$ and $3x_1 + 2x_2 = 0$. Choosing

$x_1 = 2$ then yields $\begin{pmatrix} 2 \\ -3 \\ 2 \end{pmatrix}$ as the eigenvector corresponding to

$\lambda = 1$. For $\lambda = 1 + 2i$ we have
$-2ix_1 = 0$, $2x_1 - 2ix_2 - 2x_3 = 0$ and $3x_1 + 2x_2 - 2ix_3 = 0$,

yielding $x_1 = 0$ and $x_3 = -ix_2$. Thus $\begin{pmatrix} 0 \\ 1 \\ -i \end{pmatrix}$ is the eigenvector

corresponding to $\lambda = 1 + 2i$. A similar calculation shows that
$\begin{pmatrix} 0 \\ 1 \\ i \end{pmatrix}$ is the eigenvector corresponding to $\lambda = 1 - 2i$.

24. Since the given matrix is real and symmetric, we know
    that the eigenvalues are real. Further, even if there
    are repeated eigenvalues, there will be a full set of
    three linearly independent eigenvectors. To find the
    eigenvalues and eigenvectors we must solve

$$\begin{pmatrix} 3-\lambda & 2 & 4 \\ 2 & -\lambda & 2 \\ 4 & 2 & 3-\lambda \end{pmatrix} \begin{pmatrix} x_1 \\ x_2 \\ x_3 \end{pmatrix} = \begin{pmatrix} 0 \\ 0 \\ 0 \end{pmatrix}.$$ The determinant of

coefficients is $(3-\lambda)[-\lambda(3-\lambda)-4] - 2[2(3-\lambda) -8] + 4[4+4\lambda]$
$= -\lambda^3 + 6\lambda^2 + 15\lambda + 8$. Setting this equal to zero and
solving we find $\lambda_1 = \lambda_2 = -1$, $\lambda_3 = 8$. The eigenvectors
corresponding to λ_1 and λ_2 must satisfy

$$\begin{pmatrix} 4 & 2 & 4 \\ 2 & 1 & 2 \\ 4 & 2 & 4 \end{pmatrix} \begin{pmatrix} x_1 \\ x_2 \\ x_3 \end{pmatrix} = \begin{pmatrix} 0 \\ 0 \\ 0 \end{pmatrix};$$ hence there is only the single

relation $2x_1 + x_2 + 2x_3 = 0$ to be satisfied.

Consequently, two of the variables can be selected
arbitrarily and the third is then determined by this
equation. For example, if $x_1 = 1$ and $x_3 = 1$, then $x_2 = -4$,

and we obtain the eigenvector $\mathbf{x}^{(1)} = \begin{pmatrix} 1 \\ -4 \\ 1 \end{pmatrix}$. Similarly, if

$x_1 = 1$ and $x_2 = 0$, then $x_3 = -1$, and we have the

eigenvector $\mathbf{x}^{(2)} = \begin{pmatrix} 1 \\ 0 \\ -1 \end{pmatrix}$, which is linearly independent of

$\mathbf{x}^{(1)}$. There are many other choices that could have been
made; however, by Eq.(38) there can be no
more than two linearly independent eigenvectors
corresponding to the eigenvalue -1. To find the
eigenvector corresponding to λ_3 we must solve

$$\begin{pmatrix} -5 & 2 & 4 \\ 2 & -8 & 2 \\ 4 & 2 & -5 \end{pmatrix} \begin{pmatrix} x_1 \\ x_2 \\ x_3 \end{pmatrix} = \begin{pmatrix} 0 \\ 0 \\ 0 \end{pmatrix}.$$ Interchange the first and

second rows and use row reduction to obtain the
equivalent system $x_1 - 4x_2 + x_3 = 0$, $2x_2 - x_3 = 0$. Since
there are two equations to satisfy only one variable can
be assigned an arbitrary value. If we let $x_2 = 1$, then

$x_3 = 2$ and $x_1 = 2$, so we find that $\mathbf{x}^{(3)} = \begin{pmatrix} 2 \\ 1 \\ 2 \end{pmatrix}$.

27. We are given that $\mathbf{Ax} = \mathbf{b}$ has solutions and thus we have
    $(\mathbf{Ax}, \mathbf{y}) = (\mathbf{b}, \mathbf{y})$. Using $\mathbf{A}^*\mathbf{y} = \mathbf{0}$ and the result of Prob.26,
    we have $(\mathbf{Ax}, \mathbf{y}) = (\mathbf{x}, \mathbf{A}^*\mathbf{y}) = 0$. Thus $(\mathbf{b}, \mathbf{y}) = 0$.
    For Ex. 2, since $\mathbf{A}$ is real,

    $\mathbf{A}^* = \mathbf{A}^T = \begin{pmatrix} 1 & -1 & 2 \\ -2 & 1 & -1 \\ 3 & -2 & 3 \end{pmatrix}$ and, using row reduction, the augmented

    matrix for $\mathbf{A}^*\mathbf{y} = \mathbf{0}$ becomes $\begin{pmatrix} 1 & -1 & 2 & 0 \\ 0 & 1 & -3 & 0 \\ 0 & 0 & 0 & 0 \end{pmatrix}$. Thus $\mathbf{y} = c\begin{pmatrix} 1 \\ 3 \\ 1 \end{pmatrix}$ and

    hence $(\mathbf{b}, \mathbf{y}) = b_1 + 3b_2 + b_3 = 0$.

Section 7.4, Page 389

1. Use Mathematical Induction. It has already been proven that if $x^{(1)}$ and $x^{(2)}$ are solutions, then so is $c_1 x^{(1)} + c_2 x^{(2)}$. Assume that if $x^{(1)}$, $x^{(2)}$, ..., $x^{(k)}$ are solutions, then $x = c_1 x^{(1)} + \cdots + c_k x^{(k)}$ is a solution. Then use Theorem 7.4.1 to conclude that $x + c_{k+1} x^{(k+1)}$ is also a solution and thus $c_1 x^{(1)} + \cdots + c_{k+1} x^{(k+1)}$ is a solution if $x^{(1)}$, ..., $x^{(k+1)}$ are solutions.

2a. From Eq.(10) we have
$$W = \begin{vmatrix} x_1^{(1)} & x_1^{(2)} \\ x_2^{(1)} & x_2^{(2)} \end{vmatrix} = x_1^{(1)} x_2^{(2)} - x_2^{(1)} x_1^{(2)}.$$
Taking the derivative of these two products yields four terms which may be written as
$$\frac{dW}{dt} = [\frac{dx_1^{(1)}}{dt} x_2^{(2)} - x_2^{(1)} \frac{dx_1^{(2)}}{dt}] + [x_1^{(1)} \frac{dx_2^{(2)}}{dt} - \frac{dx_2^{(1)}}{dt} x_1^{(2)}].$$
The terms in the square brackets can now be recognized as the respective determinants appearing in the desired result. A similar result was mentioned in Prob. 20 of Sect. 4.1.

2b. If $x^{(1)}$ is substituted into Eq.(3) we have
$$\frac{dx_1^{(1)}}{dt} = p_{11} x_1^{(1)} + p_{12} x_2^{(1)}$$
$$\frac{dx_2^{(1)}}{dt} = p_{21} x_1^{(1)} + p_{22} x_2^{(1)}.$$
Substituting the first equation above and its counterpart for $x^{(2)}$ into the first determinant appearing in dW/dt and evaluating the result yields $p_{11} \begin{vmatrix} x_1^{(1)} & x_1^{(2)} \\ x_2^{(1)} & x_2^{(2)} \end{vmatrix} = p_{11}W$.

Similarly, the second determinant in dW/dt is evaluated as $p_{22}W$, yielding the desired result.

2c. From part (b) we have $\frac{dW}{W} = [p_{11}(t) + p_{22}(t)]dt$ which gives $W(t) = c \exp\int[p_{11}(t) + p_{22}(t)]dt$.

6a. $W = \begin{vmatrix} t & t^2 \\ 1 & 2t \end{vmatrix} = 2t^2 - t^2 = t^2$.

6b. Pick $t = t_0$, then $c_1 x^{(1)}(t_0) + c_2 x^{(2)}(t_0) = 0$ implies

$c_1 \begin{pmatrix} t_0 \\ 1 \end{pmatrix} + c_2 \begin{pmatrix} t_0^2 \\ 2t_0 \end{pmatrix} = \begin{pmatrix} 0 \\ 0 \end{pmatrix}$, which has a non-zero solution

for c_1 and c_2 if and only if $\begin{vmatrix} t_0 & t_0^2 \\ 1 & 2t_0 \end{vmatrix} = 2t_0^2 - t_0^2 = t_0^2 = 0.$

Thus $\mathbf{x}^{(1)}(t)$ and $\mathbf{x}^{(2)}(t)$ are linearly independent at each point except $t = 0$. Thus they are linearly independent on every interval.

6c. From part (a) we see that the Wronskian vanishes at $t = 0$, but not at any other point. By Theorem 7.4.3, if $\mathbf{P}(t)$, from Eq.(3), is continuous, then the Wronskian is either identically zero or else never vanishes. Hence, we conclude that the D.E. satisfied by $\mathbf{x}^{(1)}(t)$ and $\mathbf{x}^{(2)}(t)$ must have at least one discontinuous coefficient at $t = 0$.

6d. To obtain the system satisfied by $\mathbf{x}^{(1)}$ and $\mathbf{x}^{(2)}$ we consider

$\mathbf{x} = c_1 \mathbf{x}^{(1)} + c_2 \mathbf{x}^{(2)}$, or $\begin{pmatrix} x_1 \\ x_2 \end{pmatrix} = c_1 \begin{pmatrix} t \\ 1 \end{pmatrix} + c_2 \begin{pmatrix} t^2 \\ 2t \end{pmatrix}.$

Taking the derivative we obtain $\begin{pmatrix} x_1' \\ x_2' \end{pmatrix} = c_1 \begin{pmatrix} 1 \\ 0 \end{pmatrix} + c_2 \begin{pmatrix} 2t \\ 2 \end{pmatrix}.$

Solving this last system for c_1 and c_2 we find $c_1 = x_1' - t x_2'$ and $c_2 = x_2'/2$. Thus

$\begin{pmatrix} x_1 \\ x_2 \end{pmatrix} = (x_1' - t x_2') \begin{pmatrix} t \\ 1 \end{pmatrix} + \frac{x_2'}{2} \begin{pmatrix} t^2 \\ 2t \end{pmatrix}$, which yields

$x_1 = t x_1' - \frac{t^2}{2} x_2'$ and $x_2 = x_1'$. Writing this system in

matrix form we have $\mathbf{x} = \begin{pmatrix} t & -t^2/2 \\ 1 & 0 \end{pmatrix} \mathbf{x}'$. Finding the

inverse of the matrix multiplying $\mathbf{x}'$ yields the desired solution. Note that at $t = 0$ two of the elements in $\mathbf{P}(t)$ are discontinuous.

Section 7.5, Page 398

1. Assuming that there are solutions of the form $\mathbf{x} = \xi e^{rt}$, we substitute into the D.E. to find

$r \xi e^{rt} = \begin{pmatrix} 3 & -2 \\ 2 & -2 \end{pmatrix} \xi e^{rt}.$ Since $\xi = I \xi = \begin{pmatrix} 1 & 0 \\ 0 & 1 \end{pmatrix} \xi$, we can write this equation as $\begin{pmatrix} 3 & -2 \\ 2 & -2 \end{pmatrix} \xi - r \begin{pmatrix} 1 & 0 \\ 0 & 1 \end{pmatrix} \xi = \mathbf{0}$ and

thus we must solve $\begin{pmatrix} 3-r & -2 \\ 2 & -2-r \end{pmatrix} \begin{pmatrix} \xi_1 \\ \xi_2 \end{pmatrix} = \begin{pmatrix} 0 \\ 0 \end{pmatrix}$ for r, ξ_1, ξ_2.

The determinant of the coefficients is
$(3-r)(-2-r) + 4 = r^2 - r - 2$, so the eigenvalues are
$r = -1, 2$. The eigenvector corresponding to $r = -1$

satisfies $\begin{pmatrix} 4 & -2 \\ 2 & -1 \end{pmatrix} \begin{pmatrix} \xi_1 \\ \xi_2 \end{pmatrix} = \begin{pmatrix} 0 \\ 0 \end{pmatrix}$, which yields $2\xi_1 - \xi_2 = 0$.

Thus $\mathbf{x}^{(1)}(t) = \xi^{(1)}e^{-t} = \begin{pmatrix} 1 \\ 2 \end{pmatrix} e^{-t}$, where we have set $\xi_1 = 1$.

(Any other non zero choice would also work). In a

similar fashion, for $r = 2$, we have $\begin{pmatrix} 1 & -2 \\ 2 & -4 \end{pmatrix} \begin{pmatrix} \xi_1 \\ \xi_2 \end{pmatrix} = \begin{pmatrix} 0 \\ 0 \end{pmatrix}$,

or $\xi_1 - 2\xi_2 = 0$. Hence $\mathbf{x}^{(2)}(t) = \xi^{(2)}e^{2t} = \begin{pmatrix} 2 \\ 1 \end{pmatrix} e^{2t}$ by

setting $\xi_2 = 1$. The general solution is then
$\mathbf{x} = c_1\mathbf{x}^{(1)}(t) + c_2\mathbf{x}^{(2)}(t)$. To sketch the trajectories we
follow the steps illustrated in Exs. 1 and 2. Setting

$c_2 = 0$ we have $\mathbf{x} = \begin{pmatrix} x_1 \\ x_2 \end{pmatrix} = c_1 \begin{pmatrix} 1 \\ 2 \end{pmatrix} e^{-t}$ or $x_1 = c_1e^{-t}$ and

$x_2 = 2c_1e^{-t}$ and thus one asymptote is given by
$x_2 = 2x_1$. In a similar
fashion $c_1 = 0$ gives
$x_2 = (1/2)x_1$ as a second
asymptote. Since the
roots differ in sign,
the trajectories for
this problem are similar
in nature to those in
Ex. 1. For $c_2 \neq 0$, all
solutions will be

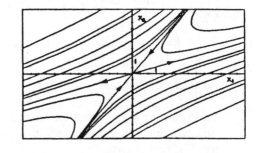

asymptotic to $x_2 = (1/2)x_1$ as $t \to \infty$. For $c_2 = 0$, the
solution approaches the origin along the line $x_2 = 2x_1$.

5. Proceeding as in Prob. 1 we assume a solution of the form
$\mathbf{x} = \xi e^{rt}$, where r, ξ_1, ξ_2 must now satisfy
$\begin{pmatrix} -2-r & 1 \\ 1 & -2-r \end{pmatrix} \begin{pmatrix} \xi_1 \\ \xi_2 \end{pmatrix} = \begin{pmatrix} 0 \\ 0 \end{pmatrix}$. Evaluating the determinant of the
coefficients set equal to zero yields $r = -1, -3$ as the

eigenvalues. For $r = -1$ we find $\xi_1 = \xi_2$ and thus

$\xi^{(1)} = \begin{pmatrix} 1 \\ 1 \end{pmatrix}$ and for $r = -3$ we find $\xi_2 = -\xi_1$ and hence

$\xi^{(2)} = \begin{pmatrix} 1 \\ -1 \end{pmatrix}$. The general solution is then

$\mathbf{x} = c_1 \begin{pmatrix} 1 \\ 1 \end{pmatrix} e^{-t} + c_2 \begin{pmatrix} 1 \\ -1 \end{pmatrix} e^{-3t}$. Since there are two negative

eigenvalues, we would expect the trajectories to be
similar to those of Ex. 2. Setting $c_2 = 0$ and eliminating
t (as in Prob. 1) we find

that $\begin{pmatrix} 1 \\ 1 \end{pmatrix} e^{-t}$ approaches the

origin along the line

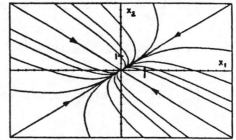

$x_2 = x_1$. Similarly $\begin{pmatrix} 1 \\ -1 \end{pmatrix} e^{-3t}$

approaches the origin along
the line $x_2 = -x_1$. As long
as $c_1 \neq 0$ (since e^{-t} is the dominant term as $t \to 0$), all
trajectories approach the origin asymptotic to $x_2 = x_1$.
For $c_1 = 0$, the trajectory approaches the origin along x_2
$= -x_1$, as shown in the graph.

6. The characteristic equation is $(5/4 - r)^2 - 9/16 = 0$, so
    $r = 2, 1/2$. Since the roots are of the same sign, the
    behavior of the solutions is similar to Prob. 5, except
    the trajectories are reversed, since the roots are
    positive.

7. Again assuming $\mathbf{x} = \xi e^{rt}$ we find that r, ξ_1, ξ_2 must

    satisfy $\begin{pmatrix} 4-r & -3 \\ 8 & -6-r \end{pmatrix} \begin{pmatrix} \xi_1 \\ \xi_2 \end{pmatrix} = \begin{pmatrix} 0 \\ 0 \end{pmatrix}$. The determinant of the

    coefficients set equal to zero yields $r = 0, -2$. For
    $r = 0$ we find $4\xi_1 = 3\xi_2$. Choosing $\xi_2 = 4$ we find $\xi_1 = 3$

    and thus $\xi^{(1)} = \begin{pmatrix} 3 \\ 4 \end{pmatrix}$. Similarly for $r = -2$ we have

    $\xi^{(2)} = \begin{pmatrix} 1 \\ 2 \end{pmatrix}$ and thus $\mathbf{x} = c_1 \begin{pmatrix} 3 \\ 4 \end{pmatrix} + c_2 \begin{pmatrix} 1 \\ 2 \end{pmatrix} e^{-2t}$. To sketch the

    trajectories, note that the general solution is

equivalent to the simultaneous equations $x_1 = 3c_1 + c_2e^{-2t}$
and $x_2 = 4c_1 + 2c_2e^{-2t}$. Solving the first equation for
c_2e^{-2t} (assuming $c_2 \neq 0$) and substituting into the second
yields $x_2 = 2x_1 - 2c_1$ and thus the trajectories are
parallel straight lines. If $c_2 = 0$, the solution is fixed
at a point.

9. The eigvalues are given by $\begin{vmatrix} 1-r & i \\ -i & 1-r \end{vmatrix} = (1-r)^2 + i^2 =$

$r(r-2) = 0$. For $r= 0$ we have $\begin{pmatrix} 1 & i \\ -i & 1 \end{pmatrix}\begin{pmatrix} \xi_1 \\ \xi_2 \end{pmatrix} = 0$ or

$-i\xi_1 + \xi_2 = 0$ and thus $\begin{pmatrix} 1 \\ i \end{pmatrix}$ is one eigenvector. Similarly

$\begin{pmatrix} 1 \\ -i \end{pmatrix}$ is the eigenvector for $r = 2$.

14. The eigenvalues and eigenvectors of the coefficient

matrix satisfy $\begin{pmatrix} 1-r & -1 & 4 \\ 3 & 2-r & -1 \\ 2 & 1 & -1-r \end{pmatrix}\begin{pmatrix} \xi_1 \\ \xi_2 \\ \xi_3 \end{pmatrix} = \begin{pmatrix} 0 \\ 0 \\ 0 \end{pmatrix}$. The determinant

of coefficients set equal to zero reduces to
$r^3 - 2r^2 - 5r + 6 = 0$, so the eigenvalues are
$r_1 = 1$, $r_2 = -2$, and $r_3 = 3$. The eigenvector

corresponding to r_1 must satisfy $\begin{pmatrix} 0 & -1 & 4 \\ 3 & 1 & -1 \\ 2 & 1 & -2 \end{pmatrix}\begin{pmatrix} \xi_1 \\ \xi_2 \\ \xi_3 \end{pmatrix} = \begin{pmatrix} 0 \\ 0 \\ 0 \end{pmatrix}$.

Using row reduction we obtain the equivalent system
$\xi_1 + \xi_3 = 0$, $\xi_2 - 4\xi_3 = 0$. Letting $\xi_1 = 1$, it follows
that

$\xi_3 = -1$ and $\xi_2 = -4$, so $\xi^{(1)} = \begin{pmatrix} 1 \\ -4 \\ -1 \end{pmatrix}$. In a similar way the

eigenvectors corresponding to r_2 and r_3 are found to be

$\xi^{(2)} = \begin{pmatrix} 1 \\ -1 \\ -1 \end{pmatrix}$ and $\xi^{(3)} = \begin{pmatrix} 1 \\ 2 \\ 1 \end{pmatrix}$, respectively. Thus the

general solution of the given D.E. is

$$\mathbf{x} = c_1 \begin{pmatrix} 1 \\ -4 \\ -1 \end{pmatrix} e^t + c_2 \begin{pmatrix} 1 \\ -1 \\ -1 \end{pmatrix} e^{-2t} + c_3 \begin{pmatrix} 1 \\ 2 \\ 1 \end{pmatrix} e^{3t}.$$ Notice that the

"trajectories" of this solution would lie in the $x_1\ x_2\ x_3$ three dimensional space.

16. The eigenvalues and eigenvectors of the coefficient

matrix are found to be $r_1 = -1$, $\xi^{(1)} = \begin{pmatrix} 1 \\ 1 \end{pmatrix}$ and $r_2 = 3$,

$\xi^{(2)} = \begin{pmatrix} 1 \\ 5 \end{pmatrix}$. Thus the general solution of the given D.E.

is $\mathbf{x} = c_1 \begin{pmatrix} 1 \\ 1 \end{pmatrix} e^{-t} + c_2 \begin{pmatrix} 1 \\ 5 \end{pmatrix} e^{3t}$. The I.C. yields the

system of equations $c_1 \begin{pmatrix} 1 \\ 1 \end{pmatrix} + c_2 \begin{pmatrix} 1 \\ 5 \end{pmatrix} = \begin{pmatrix} 1 \\ 3 \end{pmatrix}$. The augmented

matrix of this system is $\begin{pmatrix} 1 & 1 & . & 1 \\ & & . & \\ 1 & 5 & . & 3 \end{pmatrix}$ and by row reduction

we obtain $\begin{pmatrix} 1 & 1 & . & 1 \\ & & . & \\ 0 & 1 & . & 1/2 \end{pmatrix}$. Thus $c_2 = 1/2$ and $c_1 = 1/2$.

Substituting these values in the general solution gives the solution of the I.V.P. As $t \to \infty$, the solution

becomes asymptotic to $\mathbf{x} = \dfrac{1}{2}\begin{pmatrix} 1 \\ 5 \end{pmatrix} e^{3t}$, or $x_2 = 5x_1$.

20. Substituting $\mathbf{x} = \xi t^r$ into the D.E. we obtain

$r\xi t^r = \begin{pmatrix} 2 & -1 \\ 3 & -2 \end{pmatrix} \xi t^r$. For $t \neq 0$ this equation can be

written as $\begin{pmatrix} 2-r & -1 \\ 3 & -2-r \end{pmatrix} \begin{pmatrix} \xi_1 \\ \xi_2 \end{pmatrix} = \begin{pmatrix} 0 \\ 0 \end{pmatrix}$. The eigenvalues and

eigenvectors are $r_1 = 1$, $\xi^{(1)} = \begin{pmatrix} 1 \\ 1 \end{pmatrix}$ and $r_2 = -1$,

$\xi^{(2)} = \begin{pmatrix} 1 \\ 3 \end{pmatrix}$. Substituting these in the assumed form we

obtain the general solution $\mathbf{x} = c_1 \begin{pmatrix} 1 \\ 1 \end{pmatrix} t + c_2 \begin{pmatrix} 1 \\ 3 \end{pmatrix} t^{-1}$.

25.

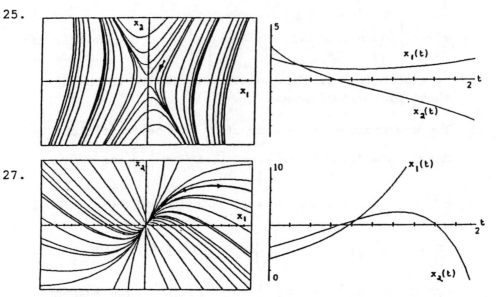

27.

31c. The eigevalues are given by

$$\begin{vmatrix} -1-r & -1 \\ -\alpha & -1-r \end{vmatrix} = r^2 + 2r + 1 - \alpha = 0. \quad \text{Thus } r_{1,2} = -1\pm\sqrt{\alpha}\,.$$

Note that in Part (a) the eigenvalues are both negative while in Part (b) they differ in sign. Thus, in this part, if we choose $\alpha = 1$, then one eigenvalue is zero, which is the transition of the one root from negative to positive. This is the desired bifurcation point.

Section 7.6, Page 410

1. We assume a solution of the form $\mathbf{x} = \xi e^{rt}$ thus r and ξ are solutions of $\begin{pmatrix} 3-r & -2 \\ 4 & -1-r \end{pmatrix} \begin{pmatrix} \xi_1 \\ \xi_2 \end{pmatrix} = \begin{pmatrix} 0 \\ 0 \end{pmatrix}$. The determinant of coefficients is $(r^2-2r-3) + 8 = r^2 - 2r + 5$, so the eigenvalues are $r = 1 \pm 2i$. The eigenvector corresponding to $1 + 2i$ satisfies $\begin{pmatrix} 2-2i & -2 \\ 4 & -2-2i \end{pmatrix} \begin{pmatrix} \xi_1 \\ \xi_2 \end{pmatrix} = \begin{pmatrix} 0 \\ 0 \end{pmatrix}$, or $(2-2i)\xi_1 - 2\xi_2 = 0$. If $\xi_1 = 1$, then $\xi_2 = 1-i$ and $\xi^{(1)} = \begin{pmatrix} 1 \\ 1-i \end{pmatrix}$ and thus one complex-valued solution of the D.E. is $\mathbf{x}^{(1)}(t) = \begin{pmatrix} 1 \\ 1-i \end{pmatrix} e^{(1+2i)t}$. To find real-valued solutions (see Eqs.8 and 9) we take the real and

imaginary parts, respectively of $\mathbf{x}^{(1)}(t)$. Thus $\mathbf{x}^{(1)}(t) =$

$$\begin{pmatrix} 1 \\ 1-i \end{pmatrix} e^t (\cos 2t + i \sin 2t)$$

$$= e^t \begin{pmatrix} \cos 2t + i \sin 2t \\ \cos 2t + \sin 2t + i(\sin 2t - \cos 2t) \end{pmatrix}$$

$$= e^t \begin{pmatrix} \cos 2t \\ \cos 2t + \sin 2t \end{pmatrix} + i e^t \begin{pmatrix} \sin 2t \\ \sin 2t - \cos 2t \end{pmatrix}$$

Hence the general solution of the D.E. is

$$\mathbf{x} = c_1 e^t \begin{pmatrix} \cos 2t \\ \cos 2t + \sin 2t \end{pmatrix} + c_2 e^t \begin{pmatrix} \sin 2t \\ \sin 2t - \cos 2t \end{pmatrix}.$$ The

solutions spiral to ∞ as $t \to \infty$ due to the e^t terms.

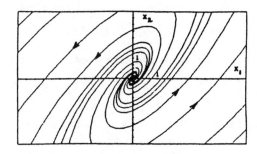

7.   The eigenvalues and eigenvectors of the coefficient

matrix satisfy $\begin{pmatrix} 1-r & 0 & 0 \\ 2 & 1-r & -2 \\ 3 & 2 & 1-r \end{pmatrix} \begin{pmatrix} \xi_1 \\ \xi_2 \\ \xi_3 \end{pmatrix} = \begin{pmatrix} 0 \\ 0 \\ 0 \end{pmatrix}$.   The

determinant of coefficients reduces to $(1-r)(r^2 - 2r + 5)$
so the eigenvalues are $r_1 = 1$, $r_2 = 1 + 2i$, and
$r_3 = 1 - 2i$. The eigenvector corresponding to r_1
satisfies

$\begin{pmatrix} 0 & 0 & 0 \\ 2 & 0 & -2 \\ 3 & 2 & 0 \end{pmatrix} \begin{pmatrix} \xi_1 \\ \xi_2 \\ \xi_3 \end{pmatrix} = \begin{pmatrix} 0 \\ 0 \\ 0 \end{pmatrix}$; hence $\xi_1 - \xi_3 = 0$ and

$3\xi_1 + 2\xi_2 = 0$.   If we let $\xi_2 = -3$ then $\xi_1 = 2$ and $\xi_3 = 2$,

so one solution of the D.E. is $\begin{pmatrix} 2 \\ -3 \\ 2 \end{pmatrix} e^t$.   The eigenvector

corresponding to r_2 satisfies $\begin{pmatrix} -2i & 0 & 0 \\ 2 & -2i & -2 \\ 3 & 2 & -2i \end{pmatrix} \begin{pmatrix} \xi_1 \\ \xi_2 \\ \xi_3 \end{pmatrix} = \begin{pmatrix} 0 \\ 0 \\ 0 \end{pmatrix}$.

Hence $\xi_1 = 0$ and $i\xi_2 + \xi_3 = 0$. If we let $\xi_2 = 1$, then $\xi_3 = -i$. Thus a complex-valued solution is

$$\begin{pmatrix} 0 \\ 1 \\ -i \end{pmatrix} e^t(\cos 2t + i \sin 2t).$$ Taking the real and imaginary

parts, see Prob. 1, we obtain $\begin{pmatrix} 0 \\ \cos 2t \\ \sin 2t \end{pmatrix} e^t$ and $\begin{pmatrix} 0 \\ \sin 2t \\ -\cos 2t \end{pmatrix} e^t$,

respectively. Thus the general solution is

$$\mathbf{x} = c_1 \begin{pmatrix} 2 \\ -3 \\ 2 \end{pmatrix} e^t + c_2 e^t \begin{pmatrix} 0 \\ \cos 2t \\ \sin 2t \end{pmatrix} + c_3 e^t \begin{pmatrix} 0 \\ \sin 2t \\ -\cos 2t \end{pmatrix},$$ which spirals

to ∞ about the x_1 axis in the $x_1 x_2 x_3$ space as $t \to \infty$.

9.   The eigenvalues and eigenvectors of the coefficient

matrix satisfy $\begin{pmatrix} 1-r & -5 \\ 1 & -3-r \end{pmatrix}\begin{pmatrix} \xi_1 \\ \xi_2 \end{pmatrix} = \begin{pmatrix} 0 \\ 0 \end{pmatrix}$. The determinant of

coefficients is $r^2 + 2r + 2$ so that the eigenvalues are $r = -1 \pm i$. The eigenvector corresponding to $r = -1 + i$

is given by $\begin{pmatrix} 2-i & -5 \\ 1 & -2-i \end{pmatrix}\begin{pmatrix} \xi_1 \\ \xi_2 \end{pmatrix} = \mathbf{0}$ so that $\xi_1 = (2+i)\xi_2$ and

thus one complex-valued solution is

$$\mathbf{x}^{(1)}(t) = \begin{pmatrix} 2+i \\ 1 \end{pmatrix} e^{(-1+i)t}.$$ Finding the real and complex

parts of $\mathbf{x}^{(1)}$, as in Prob.1, leads to the general

solution $\mathbf{x} = c_1 e^{-t}\begin{pmatrix} 2\cos t - \sin t \\ \cos t \end{pmatrix} + c_2 e^{-t}\begin{pmatrix} 2\sin t + \cos t \\ \sin t \end{pmatrix}.$

Setting $t = 0$ we find $\mathbf{x}(0) = \begin{pmatrix} 1 \\ 1 \end{pmatrix} = c_1 \begin{pmatrix} 2 \\ 1 \end{pmatrix} + c_2 \begin{pmatrix} 1 \\ 0 \end{pmatrix}$, which

is equivalent to the system $\begin{array}{l} 2c_1 + c_2 = 1 \\ c_1 + 0 = 1 \end{array}$. Thus $c_1 = 1$,

$c_2 = -1$ and $\mathbf{x}(t) = e^{-t}\begin{pmatrix} 2\cos t - \sin t \\ \cos t \end{pmatrix} - e^{-t}\begin{pmatrix} 2\sin t + \cos t \\ \sin t \end{pmatrix}$

$$= e^{-t}\begin{pmatrix} \cos t - 3\sin t \\ \cos t - \sin t \end{pmatrix},$$ which spirals to

zero as $t \to \infty$, due to the e^{-t} term.

11a. The eigenvalues are given by

$$\begin{vmatrix} 3/4-r & -2 \\ 1 & -5/4-r \end{vmatrix} = r^2 + r/2 + 17/16 = 0, \text{ so } r = -1/4 \pm i.$$

11d. Choose $\mathbf{x}(0) = \begin{pmatrix} 5 \\ 5 \end{pmatrix}$, then

the trajectory starts at
(5,5) in the x_1x_2 plane
and spirals around the
t-axis and converges
to the t axis as $t \to \infty$.

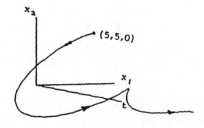

15a. The eigenvalues satisfy $\begin{vmatrix} 2-r & -5 \\ \alpha & -2-r \end{vmatrix} = r^2 - 4 + 5\alpha = 0$, so

$r_1, r_2 = \pm\sqrt{4-5\alpha}$.

15b. The critical value of α yields $r_1 = r_2 = 0$, or $\alpha = 4/5$.

15c.

16a. $\begin{vmatrix} 5/4-r & 3/4 \\ \alpha & 5/4-r \end{vmatrix} = r^2 - 5r/2 + (25/16 - 3\alpha/4) = 0$, so

$r_{1,2} = 5/4 \pm \sqrt{3\alpha}/2$.

16b. There are two critical values of α. For $\alpha < 0$ the
eigenvalues are complex, while for $\alpha > 0$ they are real.
There will be a second critical value of α when $r_2 = 0$,
or $\alpha = 25/12$. In this case the second real eigenvalue
goes from positive to negative.

16c.

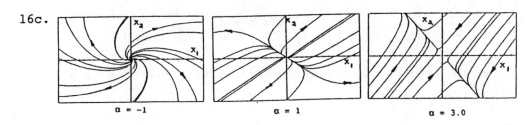

18a. We have $\begin{vmatrix} 3-r & \alpha \\ -6 & -4-r \end{vmatrix} = r^2 + r - 12 + 6\alpha = 0$, so

$r_1, r_2 = -1/2 \pm \sqrt{49-24\alpha}\,/2$.

18b. The critical values occur when $49 - 24\alpha = 1$ (in which case $r_2 = 0$) and when $49 - 24\alpha = 0$, in which case $r_1 = r_2 = -1/2$. Thus $\alpha = 2$ and $\alpha = 49/24 \approx 2.04$.

18c.

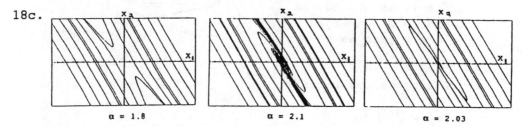

$\alpha = 1.8$ $\alpha = 2.1$ $\alpha = 2.03$

21. If we seek solutions of the form $\mathbf{x} = \xi t^r$, then r must be an eigenvalue and ξ a corresponding eigenvector of the coefficient matrix. Thus r and ξ satisfy

$\begin{pmatrix} -1-r & -1 \\ 2 & -1-r \end{pmatrix}\begin{pmatrix} \xi_1 \\ \xi_2 \end{pmatrix} = \begin{pmatrix} 0 \\ 0 \end{pmatrix}$. The determinant of coefficients

is $(-1-r)^2 + 2 = r^2 + 2r + 3$, so the eigenvalues are $r = -1 \pm \sqrt{2}\,i$. The eigenvector corresponding to

$-1 + \sqrt{2}\,i$ satisfies $\begin{pmatrix} -\sqrt{2}\,i & -1 \\ 2 & -\sqrt{2}\,i \end{pmatrix}\begin{pmatrix} \xi_1 \\ \xi_2 \end{pmatrix} = \begin{pmatrix} 0 \\ 0 \end{pmatrix}$ or.

$\sqrt{2}\,i\xi_1 + \xi_2 = 0$. If we let $\xi_1 = 1$, then $\xi_2 = -\sqrt{2}\,i$, and

$\xi^{(1)} = \begin{pmatrix} 1 \\ -\sqrt{2}\,i \end{pmatrix}$. Thus a complex-valued solution of the

given D.E. is $\begin{pmatrix} 1 \\ -\sqrt{2}\,i \end{pmatrix} t^{-1+\sqrt{2}\,i}$. From Eq. (15) of

Sect. 5.5 we have (since $t^{\sqrt{2}\,i} = e^{\ln t^{\sqrt{2}\,i}} = e^{\sqrt{2}\,i\ln t}$)
$t^{-1+\sqrt{2}\,i} = t^{-1}[\cos(\sqrt{2}\,\ln t) + i\sin(\sqrt{2}\,\ln t)]$ for $t > 0$.
Separating the complex valued solution into real and imaginary parts, we obtain the two real-valued solutions

$\mathbf{u} = t^{-1}\begin{pmatrix} \cos(\sqrt{2}\,\ln t) \\ \sqrt{2}\,\sin(\sqrt{2}\,\ln t) \end{pmatrix}$ and $\mathbf{v} = t^{-1}\begin{pmatrix} \sin(\sqrt{2}\,\ln t) \\ -\sqrt{2}\,\cos(\sqrt{2}\,\ln t) \end{pmatrix}$.

23a. The eigenvalues are given by $(r+1/4)[(r+1/4)^2 + 1] = 0$.

23b.

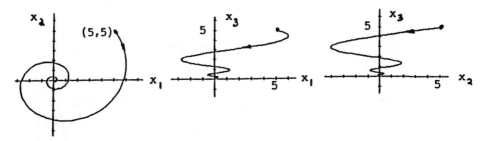

23c. Graph starts in the
first octant and
spirals around the
x_3 axis, converging
to zero.

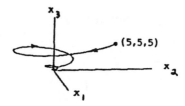

30a. Following the steps leading to Eq.(24), and using the
given values for the m's and k's, we obtain $y_1' = y_3$,
$y_2' = y_4$, $y_3' = -4y_1 + 3y_2$, and $y_4' = (9/4)y_1 - (13/4)y_2$.

Thus $\mathbf{Y}' = \mathbf{AY}$, where $\mathbf{A} = \begin{pmatrix} 0 & 0 & 1 & 0 \\ 0 & 0 & 0 & 1 \\ -4 & 3 & 0 & 0 \\ 9/4 & -13/4 & 0 & 0 \end{pmatrix}$.

30b. The eigenvalues of $\mathbf{A}$ are given by
$\det(\mathbf{A} - r\mathbf{I}) = r^4 + (29/4)r^2 + 25/4 = 0$, and thus
$r_{1,2} = \pm i$ and $r_{3,4} = \pm(5/2)i$. The eigenvector
corresponding ro $r_1 = i$ satisfies

$\begin{pmatrix} -i & 0 & 1 & 0 \\ 0 & -i & 0 & 1 \\ -4 & 3 & -i & 0 \\ 9/4 & -13/4 & 0 & -i \end{pmatrix} \begin{pmatrix} \xi_1 \\ \xi_2 \\ \xi_3 \\ \xi_4 \end{pmatrix} = \begin{pmatrix} 0 \\ 0 \\ 0 \\ 0 \end{pmatrix}$, or $\xi_3 = i\xi_1$, $\xi_4 = i\xi_2$,

$-4\xi_1 + 3\xi_2 - i\xi_3 = 0$, and $(9/4)\xi_1 - (13/4)\xi_2 = i\xi_4$. Setting

$\xi_1 = 1$, the first three equations yield $\xi^{(1)} = \begin{pmatrix} 1 \\ 1 \\ i \\ i \end{pmatrix}$ and thus $\xi^{(2)} = \begin{pmatrix} 1 \\ 1 \\ -i \\ -i \end{pmatrix}$ by Eq.(5). It should be noted that the

fourth equation is also satisfied by this choice.

Similarly, $\xi^{(3)}$ satisfies $\begin{pmatrix} -5i/2 & 0 & 1 & 0 \\ 0 & -5i/2 & 0 & 1 \\ -4 & 3 & -5i/2 & 0 \\ 9/4 & -13/4 & 0 & -5i/2 \end{pmatrix} \begin{pmatrix} \xi_1 \\ \xi_2 \\ \xi_3 \\ \xi_4 \end{pmatrix} = \begin{pmatrix} 0 \\ 0 \\ 0 \\ 0 \end{pmatrix}$,

which yields $\xi^{(3)} = \begin{pmatrix} 4 \\ -3 \\ 10i \\ -15i/2 \end{pmatrix}$ and $\xi^{(4)} = \begin{pmatrix} 4 \\ -3 \\ -10i \\ 15i/2 \end{pmatrix}$.

30c. Taking the real and imaginary parts of $e^{it} \begin{pmatrix} 1 \\ 1 \\ i \\ i \end{pmatrix}$ yields

$w_1(t) = (\cos t, \cos t, -\sin t, -\sin t)^T$ and

$w_2(t) = (\sin t, \sin t, \cos t, \cos t)^T$ as the corresponding two

real valued solutions. Similarly, $e^{(5/2)it} \begin{pmatrix} 4 \\ -3 \\ 10i \\ -15i/2 \end{pmatrix}$ yields

the other two real valued solutions, denoted as $w_3(t)$ and
$w_4(t)$. The general solution of the system is then
$y(t) = c_1 w_1(t) + c_2 w_2(t) + c_3 w_3(t) + c_4 w_4(t)$.

30d. There are two fundamental modes, one represented by
$\cos(t - \delta_1)$, of frequency 1, and the other represented by
$\cos(5t/2 - \delta_2)$, of frequency 5/2 (see Sect. 3.8).

30e. From part(c),

$y(0) = c_1 \begin{pmatrix} 1 \\ 1 \\ 0 \\ 0 \end{pmatrix} + c_2 \begin{pmatrix} 0 \\ 0 \\ 1 \\ 1 \end{pmatrix} + c_3 \begin{pmatrix} 4 \\ -3 \\ 0 \\ 0 \end{pmatrix} + c_4 \begin{pmatrix} 0 \\ 0 \\ 10 \\ -15/2 \end{pmatrix} = \begin{pmatrix} 2 \\ 1 \\ 0 \\ 0 \end{pmatrix}$, which

yields $c_2 = c_4 = 0$, $c_1 = 10/7$, and $c_3 = 1/7$.

Section 7.7, Page 420

Each of the Problems 1 through 10, except 2 and 8, has been
solved in one of the previous sections. Thus a fundamental
matrix for the given systems can be readily written down.
The fundamental matrix $\Phi(t)$ satisfying $\Phi(0) = I$ can then be
found, as shown in the following problems.

2. The characteristic equation is given by $\begin{vmatrix} -3/4-r & 1/2 \\ 1/8 & -3/4-r \end{vmatrix} =$

$r^2 + 3r/2 + 1/2 = 0$, so $r = -1, -1/2$. For $r = -1$ we have

$\begin{pmatrix} 1/4 & 1/2 \\ 1/8 & 1/4 \end{pmatrix} \begin{pmatrix} \xi_1 \\ \xi_2 \end{pmatrix} = \begin{pmatrix} 0 \\ 0 \end{pmatrix}$, and thus $\xi^{(1)} = \begin{pmatrix} -2 \\ 1 \end{pmatrix}$. Likewise

$\xi^{(2)} = \begin{pmatrix} 2 \\ 1 \end{pmatrix}$ and thus $x^{(1)}(t) = \begin{pmatrix} -2 \\ 1 \end{pmatrix} e^{-t}$ and $x^{(2)}(t) = \begin{pmatrix} 2 \\ 1 \end{pmatrix} e^{-t/2}$. To

find the first column of Φ we choose c_1 and c_2 so that

$c_1 x^{(1)}(0) + c_2 x^{(2)}(0) = \begin{pmatrix} 1 \\ 0 \end{pmatrix}$, which yields $-2c_1 + 2c_2 = 1$ and

$c_1 + c_2 = 0$. Thus $c_1 = -1/4$ and $c_2 = 1/4$ and the first column

of Φ is $\begin{pmatrix} e^{-t/2}/2 & e^{-t}/2 \\ e^{-t/2}/4 & -e^{-t}/4 \end{pmatrix}$. The second colunm of Φ is determined

by $d_1 x^{(1)}(0) + d_2 x^{(2)}(0) = \begin{pmatrix} 0 \\ 1 \end{pmatrix}$ which yields $d_1 = d_2 = 1/2$ and

thus the second column of Φ is $\begin{pmatrix} e^{-t/2} & -e^{-t} \\ e^{-t/2}/2 & e^{-t}/2 \end{pmatrix}$.

4. From Prob. 4 of Sect. 7.5 we have the two linearly

independent solutions $x^{(1)}(t) = \begin{pmatrix} 1 \\ -4 \end{pmatrix} e^{-3t}$ and

$x^{(2)}(t) = \begin{pmatrix} 1 \\ 1 \end{pmatrix} e^{2t}$. Hence a fundamental matrix Ψ is given

by $\Psi(t) = \begin{pmatrix} e^{-3t} & e^{2t} \\ -4e^{-3t} & e^{2t} \end{pmatrix}$. To find the fundamental matrix

$\Phi(t)$ satisfying the I.C. $\Phi(0) = I$ we can proceed in
either of two ways. One way is to find $\Psi(0)$, invert it
to obtain $\Psi^{-1}(0)$, and then to form the product
$\Psi(t)\Psi^{-1}(0)$, which is $\Phi(t)$. Alternatively, we can find
the first column of Φ by determining the linear
combination

$c_1 \mathbf{x}^{(1)}(t) + c_2 \mathbf{x}^{(2)}(t)$ that satisfies the I.C. $\begin{pmatrix} 1 \\ 0 \end{pmatrix}$. This

requires that $c_1 + c_2 = 1$, $-4c_1 + c_2 = 0$, so we obtain $c_1 = 1/5$ and $c_2 = 4/5$. Thus the first column of $\Phi(t)$ is

$\begin{pmatrix} (1/5)e^{-3t} + (4/5)e^{2t} \\ -(4/5)e^{-3t} + (4/5)e^{2t} \end{pmatrix}$. Similarly, the second column of

Φ is that linear combination of $\mathbf{x}^{(1)}(t)$ and $\mathbf{x}^{(2)}(t)$ that

satisfies the I.C. $\begin{pmatrix} 0 \\ 1 \end{pmatrix}$. Thus we must have

$c_1 + c_2 = 0$, $-4c_1 + c_2 = 1$; therefore $c_1 = -1/5$ and $c_2 = 1/5$. Hence the second column of $\Phi(t)$ is

$\begin{pmatrix} -(1/5)e^{-3t} + (1/5)e^{2t} \\ (4/5)e^{-3t} + (1/5)e^{2t} \end{pmatrix}$.

6. Two linearly independent real-valued solutions of the given D.E. were found in Prob. 2 of Sect. 7.6. Using the result of that problem, we have

$\Psi(t) = \begin{pmatrix} -2e^{-t}\sin 2t & 2e^{-t}\cos 2t \\ e^{-t}\cos 2t & e^{-t}\sin 2t \end{pmatrix}$. To find $\Phi(t)$

we determine the linear combinations of the columns of

$\Psi(t)$ that satisfy the I.C. $\begin{pmatrix} 1 \\ 0 \end{pmatrix}$ and $\begin{pmatrix} 0 \\ 1 \end{pmatrix}$, respectively.

In the first case c_1 and c_2 satisfy $0c_1 + 2c_2 = 1$ and $c_1 + 0c_2 = 0$. Thus $c_1 = 0$ and $c_2 = 1/2$. In the second case we have $0c_1 + 2c_2 = 0$ and $c_1 + 0c_2 = 1$, so $c_1 = 1$ and $c_2 = 0$. Using these values of c_1 and c_2 to form the first and second columns of $\Phi(t)$ respectively, we obtain

$\Phi(t) = \begin{pmatrix} e^{-t}\cos 2t & -2e^{-t}\sin 2t \\ e^{-t}\sin 2t/2 & e^{-t}\cos 2t \end{pmatrix}$.

10. From Prob. 14 Sect. 7.5 we have $\mathbf{x}^{(1)} = \begin{pmatrix} 1 \\ -4 \\ -1 \end{pmatrix} e^{t}$,

$\mathbf{x}^{(2)} = \begin{pmatrix} 1 \\ -1 \\ -1 \end{pmatrix} e^{-2t}$ and $\mathbf{x}^{(3)} = \begin{pmatrix} 1 \\ 2 \\ 1 \end{pmatrix} e^{3t}$. For the first column

of Φ we want to choose c_1, c_2, c_3 such that

$$c_1 \mathbf{x}^{(1)}(0) + c_2 \mathbf{x}^{(2)}(0) + c_3 \mathbf{x}^{(3)}(0) = \begin{pmatrix} 1 \\ 0 \\ 0 \end{pmatrix}. \quad \text{Thus}$$

$c_1 + c_2 + c_3 = 1$, $-4c_1 - c_2 + 2c_3 = 0$ and $-c_1 - c_2 + c_3 = 0$, which yield $c_1 = 1/6$, $c_2 = 1/3$ and $c_3 = 1/2$. The first column of Φ is then
$(e^t/6 + e^{-2t}/3 + e^{3t}/2, -2e^t/3 - e^{-2t}/3 + e^{3t},$
$$-e^t/6 - e^{-2t}/3 + e^{3t}/2)^T.$$
Likewise, for the second column we have

$$d_1 \mathbf{x}^{(1)}(0) + d_2 \mathbf{x}^{(2)}(0) + d_3 \mathbf{x}^{(3)}(0) = \begin{pmatrix} 0 \\ 1 \\ 0 \end{pmatrix}, \quad \text{which yields}$$

$d_1 = -1/3$, $d_2 = 1/3$ and $d_3 = 0$ and thus
$(-e^t/3 + e^{-2t}/2, 4e^t/3 - e^{-2t}/3, e^t/3 - e^{-2t}/3)^T$ is the second column of $\Phi(t)$. Finally, for the third column we

$$\text{have } e_1 \mathbf{x}^{(1)}(0) + e_2 \mathbf{x}^{(2)}(0) + e_3 \mathbf{x}^{(3)}(0) = \begin{pmatrix} 0 \\ 0 \\ 1 \end{pmatrix}, \quad \text{which gives}$$

$e_1 = 1/2$, $e_2 = -1$ and $e_3 = 1/2$ and hence
$(e^t/2 - e^{-2t} + e^{3t}/2, -2e^t + e^{-2t} + e^{3t}, -e^t/2 + e^{-2t} + e^{3t}/2)^T$
is the third column of $\Phi(t)$.

11. From Eq. (14) the solution is given by $\Phi(t)\mathbf{x}^0$. Thus

$$\mathbf{x} = \begin{pmatrix} 3e^t/2 - e^{-t}/2 & -e^t/2 + e^{-t}/2 \\ 3e^t/2 - 3e^{-t}/2 & -e^t/2 + 3e^{-t}/2 \end{pmatrix} \begin{pmatrix} 2 \\ -1 \end{pmatrix}$$

$$= \begin{pmatrix} 7e^t/2 - 3e^{-t}/2 \\ 7e^t/2 - 9e^{-t}/2 \end{pmatrix} = \frac{7}{2}\begin{pmatrix} 1 \\ 1 \end{pmatrix}e^t - \frac{3}{2}\begin{pmatrix} 1 \\ 3 \end{pmatrix}e^{-t}.$$

Section 7.8, Page 428

1. The eigenvalues and eigenvectors of the given coefficient matrix satisfy $\begin{pmatrix} 3-r & -4 \\ 1 & -1-r \end{pmatrix}\begin{pmatrix} \xi_1 \\ \xi_2 \end{pmatrix} = \begin{pmatrix} 0 \\ 0 \end{pmatrix}$. The determinant of coefficients is $(3-r)(-1-r) + 4 = r^2 - 2r + 1 = (r-1)^2$ so $r_1 = 1$ and $r_2 = 1$. The eigenvectors corresponding to this double eigenvalue satisfy $\begin{pmatrix} 2 & -4 \\ 1 & -2 \end{pmatrix}\begin{pmatrix} \xi_1 \\ \xi_2 \end{pmatrix} = \begin{pmatrix} 0 \\ 0 \end{pmatrix}$, or $\xi_1 - 2\xi_2 = 0$. Thus the only eigenvectors are multiples

of $\xi^{(1)} = \begin{pmatrix} 2 \\ 1 \end{pmatrix}$. One solution of the given D.E. is

$\mathbf{x}^{(1)}(t) = \begin{pmatrix} 2 \\ 1 \end{pmatrix} e^t$, but there is no second solution of this

form. To find a second solution we assume, as in Eq. (13), that $\mathbf{x} = \xi t e^t + \eta e^t$ and substitute this expression into the D.E. As in Ex. 2 we find that ξ is

an eigenvector, so we choose $\xi = \begin{pmatrix} 2 \\ 1 \end{pmatrix}$. Then η must

satisfy Eq.(24): $(\mathbf{A} - r\mathbf{I})\eta = \xi$, or $\begin{pmatrix} 2 & -4 \\ 1 & -2 \end{pmatrix}\begin{pmatrix} \eta_1 \\ \eta_2 \end{pmatrix} = \begin{pmatrix} 2 \\ 1 \end{pmatrix}$ for

this problem. Solving these equations yields $\eta_1 - 2\eta_2 = 1$. If $\eta_2 = k$, where k is an arbitrary constant, then $\eta_1 = 1 + 2k$. Hence the second solution that we obtain is

$\mathbf{x}^{(2)}(t) = \begin{pmatrix} 2 \\ 1 \end{pmatrix} t e^t + \begin{pmatrix} 1 + 2k \\ k \end{pmatrix} e^t = \begin{pmatrix} 2 \\ 1 \end{pmatrix} t e^t + \begin{pmatrix} 1 \\ 0 \end{pmatrix} e^t + k\begin{pmatrix} 2 \\ 1 \end{pmatrix} e^t.$

The last term is a multiple of the first solution $\mathbf{x}^{(1)}(t)$ and may be neglected, that is, we may set k = 0. Thus

$\mathbf{x}^{(2)}(t) = \begin{pmatrix} 2 \\ 1 \end{pmatrix} t e^t + \begin{pmatrix} 1 \\ 0 \end{pmatrix} e^t$ and

the general solution is
$\mathbf{x} = c_1 \mathbf{x}^{(1)}(t) + c_2 \mathbf{x}^{(2)}(t)$.
All solutions diverge
to infinity along lines
of slope 1/2 as $t \to \infty$.

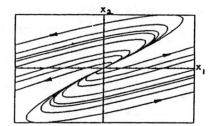

3. The origin is attracting.
That is, as $t \to \infty$
the solution approaches
the origin tangent to
the line y = 3x/2.

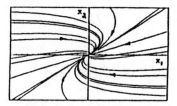

5. Substituting $\mathbf{x} = \xi e^{rt}$ into the given system, we find that the eigenvalues and eigenvectors satisfy

$\begin{pmatrix} 1-r & 1 & 1 \\ 2 & 1-r & -1 \\ 0 & -1 & 1-r \end{pmatrix}\begin{pmatrix} \xi_1 \\ \xi_2 \\ \xi_3 \end{pmatrix} = \begin{pmatrix} 0 \\ 0 \\ 0 \end{pmatrix}$. The determinant of coefficients

is $-r^3 + 3r^2 - 4$ and thus $r_1 = -1$, $r_2 = 2$ and $r_3 = 2$.
The eigenvector corresponding to r_1 satisfies

$$\begin{pmatrix} 2 & 1 & 1 \\ 2 & 2 & -1 \\ 0 & -1 & 2 \end{pmatrix} \begin{pmatrix} \xi_1 \\ \xi_2 \\ \xi_3 \end{pmatrix} = \begin{pmatrix} 0 \\ 0 \\ 0 \end{pmatrix} \text{ which yields } \xi^{(1)} = \begin{pmatrix} -3 \\ 4 \\ 2 \end{pmatrix} \text{ and}$$

$\mathbf{x}^{(1)} = \begin{pmatrix} -3 \\ 4 \\ 2 \end{pmatrix} e^{-t}$. The eigenvectors corresponding to the

double eigenvalue must satsify $\begin{pmatrix} -1 & 1 & 1 \\ 2 & -1 & -1 \\ 0 & -1 & -1 \end{pmatrix} \begin{pmatrix} \xi_1 \\ \xi_2 \\ \xi_3 \end{pmatrix} = \begin{pmatrix} 0 \\ 0 \\ 0 \end{pmatrix}$,

which yields the single eigenvector $\xi^{(2)} = \begin{pmatrix} 0 \\ 1 \\ -1 \end{pmatrix}$ and hence

$\mathbf{x}^{(2)}(t) = \begin{pmatrix} 0 \\ 1 \\ -1 \end{pmatrix} e^{2t}$. The second solution corresponding to

the double eigenvalue will have the form specified by

Eq.(13), which yields $\mathbf{x}^{(3)} = \begin{pmatrix} 0 \\ 1 \\ -1 \end{pmatrix} t e^{2t} + \eta e^{2t}$.

Substituting this into the given system, or using

Eq.(24), we find that η satisfies $\begin{pmatrix} -1 & 1 & 1 \\ 2 & -1 & -1 \\ 0 & -1 & -1 \end{pmatrix} \begin{pmatrix} \eta_1 \\ \eta_2 \\ \eta_3 \end{pmatrix} = \begin{pmatrix} 0 \\ 1 \\ -1 \end{pmatrix}$.

Using row reduction we find that $\eta_1 = 1$ and $\eta_2 + \eta_3 = 1$,
where either η_2 or η_3 is arbitrary. If we choose $\eta_2 = 0$,

then $\eta = \begin{pmatrix} 1 \\ 0 \\ 1 \end{pmatrix}$ and thus $\mathbf{x}^{(3)} = \begin{pmatrix} 0 \\ 1 \\ -1 \end{pmatrix} t e^{2t} + \begin{pmatrix} 1 \\ 0 \\ 1 \end{pmatrix} e^{2t}$. The

general solution is then $\mathbf{x} = c_1 \mathbf{x}^{(1)} + c_2 \mathbf{x}^{(2)} + c_3 \mathbf{x}^{(3)}$.

9. We have $\begin{vmatrix} 2-r & 3/2 \\ -3/2 & -1-r \end{vmatrix} = (r-1/2)^2 = 0$. For $r = 1/2$, the

eigenvector is given by $\begin{pmatrix} 3/2 & 3/2 \\ -3/2 & -3/2 \end{pmatrix} \begin{pmatrix} \xi_1 \\ \xi_2 \end{pmatrix} = 0$, so $\xi = \begin{pmatrix} 1 \\ -1 \end{pmatrix}$

and $\begin{pmatrix} 1 \\ -1 \end{pmatrix} e^{t/2}$ is one solution. For the second solution we

have $\mathbf{x} = \xi t e^{t/2} + \eta e^{t/2}$, where $(\mathbf{A} - \dfrac{1}{2}\mathbf{I})\eta = \xi$, $\mathbf{A}$ being

the coefficient matrix for this problem. This last
equation reduces to $3\eta_1/2 + 3\eta_2/2 = 1$ and
$-3\eta_1/2 - 3\eta_2/2 = -1$. Choosing $\eta_2 = 0$ yields $\eta_1 = 2/3$
and hence the general solution is

$$\mathbf{x} = c_1 \begin{pmatrix} 1 \\ -1 \end{pmatrix} e^{t/2} + c_2 \begin{pmatrix} 2/3 \\ 0 \end{pmatrix} e^{t/2} + c_2 \begin{pmatrix} 1 \\ -1 \end{pmatrix} t e^{t/2}. \quad \mathbf{x}(0) = \begin{pmatrix} 3 \\ -2 \end{pmatrix}$$

gives $c_1 + 2c_2/3 = 3$ and $-c_1 = -2$, and hence $c_1 = 2$,
$c_2 = 3/2$. The graphs are shown for $-10 \le t \le 1$.

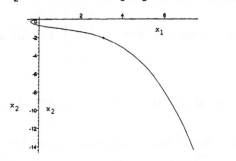 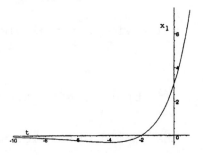

11. Since the coefficient matrix is lower triangular, the
    eigenvalues are easily found to be $r = 1, 1, 2$. For $r = 2$,

    we have $\begin{pmatrix} -1 & 0 & 0 \\ -4 & -1 & 0 \\ 3 & 6 & 0 \end{pmatrix} \begin{pmatrix} \xi_1 \\ \xi_2 \\ \xi_3 \end{pmatrix} = \begin{pmatrix} 0 \\ 0 \\ 0 \end{pmatrix}$, which yields $\xi = \begin{pmatrix} 0 \\ 0 \\ 1 \end{pmatrix}$, so one

    solution is $\mathbf{x}^{(1)} = \begin{pmatrix} 0 \\ 0 \\ 1 \end{pmatrix} e^{2t}$. For $r = 1$, we have

$$\begin{pmatrix} 0 & 0 & 0 \\ -4 & 0 & 0 \\ 3 & 6 & 1 \end{pmatrix} \begin{pmatrix} \xi_1 \\ \xi_2 \\ \xi_3 \end{pmatrix} = \begin{pmatrix} 0 \\ 0 \\ 0 \end{pmatrix}, \text{ which yields the second solution}$$

$\mathbf{x}^{(2)} = \begin{pmatrix} 0 \\ 1 \\ -6 \end{pmatrix} e^t$. The third solution is of the form

$\mathbf{x}^{(3)} = \begin{pmatrix} 0 \\ 1 \\ -6 \end{pmatrix} t e^t + \eta e^t$, where $\begin{pmatrix} 0 & 0 & 0 \\ -4 & 0 & 0 \\ 3 & 6 & 1 \end{pmatrix} \eta = \begin{pmatrix} 0 \\ 1 \\ -6 \end{pmatrix}$ and thus

$\eta_1 = -1/4$ and $6\eta_2 + \eta_3 = -21/4$. Choosing $\eta_2 = 0$ gives

$\eta_3 = -21/4$ and hence

$$\mathbf{x}(t) = c_1 \begin{pmatrix} 0 \\ 1 \\ -6 \end{pmatrix} e^t + c_2 \left[\begin{pmatrix} -1/4 \\ 0 \\ -21/4 \end{pmatrix} e^t + \begin{pmatrix} 0 \\ 1 \\ -6 \end{pmatrix} t e^t \right] + c_3 \begin{pmatrix} 0 \\ 0 \\ 1 \end{pmatrix} e^{2t}. \quad \text{The}$$

I.C. then yield $c_1 = 2$, $c_2 = 4$ and $c_3 = 3$ and hence

$$\mathbf{x} = \begin{pmatrix} -1 \\ 2 \\ -33 \end{pmatrix} e^t + 4 \begin{pmatrix} 0 \\ 1 \\ -6 \end{pmatrix} t e^t + 3 \begin{pmatrix} 0 \\ 0 \\ 1 \end{pmatrix} e^{2t}, \quad \text{which become unbounded}$$

as $t \to \infty$.

12.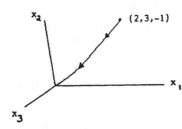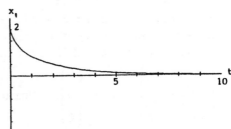

14. Assuming $\mathbf{x} = \xi t^r$ and substituting into the given system, we find r and ξ must satisfy $\begin{pmatrix} 1-r & -4 \\ 4 & -7-r \end{pmatrix} \begin{pmatrix} \xi_1 \\ \xi_2 \end{pmatrix} = \begin{pmatrix} 0 \\ 0 \end{pmatrix}$, which has the double eigenvalue $r = -3$ and single eigenvector $\begin{pmatrix} 1 \\ 1 \end{pmatrix}$. Hence one solution of the given D.E. is

$\mathbf{x}^{(1)}(t) = \begin{pmatrix} 1 \\ 1 \end{pmatrix} t^{-3}$. By analogy with the scalar case considered in Sect. 5.5 and Ex. 2 of this section, we seek a second solution of the form $\mathbf{x} = \eta t^{-3}\ln t + \zeta t^{-3}$. Substituting this expression into the D.E. we find that η and ζ satisfy the equations $(\mathbf{A} + 3\mathbf{I})\eta = \mathbf{0}$ and

$(\mathbf{A} + 3\mathbf{I})\zeta = \eta$, where $\mathbf{A} = \begin{pmatrix} 1 & -4 \\ 4 & -7 \end{pmatrix}$ and $\mathbf{I}$ is the identity

matrix. Thus $\eta = \begin{pmatrix} 1 \\ 1 \end{pmatrix}$, from above, and ζ then satisfies

$\begin{pmatrix} 4 & -4 \\ 4 & -4 \end{pmatrix} \begin{pmatrix} \zeta_1 \\ \zeta_2 \end{pmatrix} = \begin{pmatrix} 1 \\ 1 \end{pmatrix}$. Choosing $\zeta_1 = 0$ we obtain $\zeta_2 = -1/4$ and

hence a second solution is $\mathbf{x}^{(2)}(t) = \begin{pmatrix} 1 \\ 1 \end{pmatrix} t^{-3}\ln t + \begin{pmatrix} 0 \\ -1/4 \end{pmatrix} t^{-3}$.

15. The eigenvalues are given by $r^2 - (a+d)r + ad-bc = 0$. Use
     the quadratic formula to find the roots. Then show that
     the roots are either real and negative or else are
     complex with negative real part when $a+d < 0$ and $ad-bc>0$.
     In both these cases the solution approaches zero as
     $t \to \infty$.

17a. The eigenvalues and eigenvectors of the coefficient

     matrix satisfy $\begin{pmatrix} 1-r & 1 & 1 \\ 2 & 1-r & -1 \\ -3 & 2 & 4-r \end{pmatrix} \begin{pmatrix} \xi_1 \\ \xi_2 \\ \xi_3 \end{pmatrix} = \begin{pmatrix} 0 \\ 0 \\ 0 \end{pmatrix}$. The determinant

     of coefficients is $8 - 12r + 6r^2 - r^3 = (2-r)^3$, so the
     eigenvalues are $r_1 = r_2 = r_3 = 2$. The eigenvectors
     corresponding to this triple eigenvalue satisfy
     $\begin{pmatrix} -1 & 1 & 1 \\ 2 & -1 & -1 \\ -3 & 2 & 2 \end{pmatrix} \begin{pmatrix} \xi_1 \\ \xi_2 \\ \xi_3 \end{pmatrix} = \begin{pmatrix} 0 \\ 0 \\ 0 \end{pmatrix}$. Using row reduction we can reduce

     this to the equivalent system $\xi_1 - \xi_2 - \xi_3 = 0$, and
     $\xi_2 + \xi_3 = 0$. If we let $\xi_2 = 1$, then $\xi_1 = 0$ and $\xi_3 = -1$,

     so the only eigenvectors are multiples of $\xi = \begin{pmatrix} 0 \\ 1 \\ -1 \end{pmatrix}$.

17b. From part (a), one solution of the given D.E. is

     $\mathbf{x}^{(1)}(t) = \begin{pmatrix} 0 \\ 1 \\ -1 \end{pmatrix} e^{2t}$, but there are no other linearly

     independent solutions of this form.

17c. We now seek a second solution of the form
     $\mathbf{x} = \xi t e^{2t} + \eta e^{2t}$. Thus $\mathbf{Ax} = \mathbf{A}\xi t e^{2t} + \mathbf{A}\eta e^{2t}$ and
     $\mathbf{x}' = 2\xi t e^{2t} + \xi e^{2t} + 2\eta e^{2t}$. Equating like terms, we then
     have $(\mathbf{A}-2\mathbf{I})\xi = 0$ and $(\mathbf{A}-2\mathbf{I})\eta = \xi$. Thus ξ is the same as
     in part (a) and the second equation yields
     $\begin{pmatrix} -1 & 1 & 1 \\ 2 & -1 & -1 \\ -3 & 2 & 2 \end{pmatrix} \begin{pmatrix} \eta_1 \\ \eta_2 \\ \eta_3 \end{pmatrix} = \begin{pmatrix} 0 \\ 1 \\ -1 \end{pmatrix}$. By row reduction this is

     equivalent to the system $\begin{pmatrix} 1 & -1 & -1 \\ 0 & 1 & 1 \\ 0 & 0 & 0 \end{pmatrix} \begin{pmatrix} \eta_1 \\ \eta_2 \\ \eta_3 \end{pmatrix} = \begin{pmatrix} 0 \\ 1 \\ 0 \end{pmatrix}$. If we choose $\eta_3 = 0$, then $\eta_2 = 1$ and $\eta_1 = 1$, so $\eta = \begin{pmatrix} 1 \\ 1 \\ 0 \end{pmatrix}$. Hence

a second solution of the D.E. is

$$\mathbf{x}^{(2)}(t) = \begin{pmatrix} 0 \\ 1 \\ -1 \end{pmatrix} te^{2t} + \begin{pmatrix} 1 \\ 1 \\ 0 \end{pmatrix} e^{2t}.$$

17d. Assuming $\mathbf{x} = \xi(t^2/2)e^{2t} + \eta te^{2t} + \zeta e^{2t}$, we have
$\mathbf{Ax} = \mathbf{A}\xi(t^2/2)e^{2t} + \mathbf{A}\eta te^{2t} + \mathbf{A}\xi e^{2t}$ and
$\mathbf{x}' = \xi te^{2t} + 2\xi(t^2/2)e^{2t} + \eta e^{2t} + 2\eta te^{2t} + 2\zeta e^{2t}$ and thus
$(\mathbf{A}-2\mathbf{I})\xi = \mathbf{0}$, $(\mathbf{A}-2\mathbf{I})\eta = \xi$ and $(\mathbf{A}-2\mathbf{I})\zeta = \eta$. Again, ξ and η
are as found previously and the last equation is
equivalent to
$$\begin{pmatrix} -1 & 1 & 1 \\ 2 & -1 & -1 \\ -3 & 2 & 2 \end{pmatrix} \begin{pmatrix} \zeta_1 \\ \zeta_2 \\ \zeta_3 \end{pmatrix} = \begin{pmatrix} 1 \\ 1 \\ 0 \end{pmatrix}.$$ By row reduction we find the

equivalent system $\begin{pmatrix} 1 & -1 & -1 \\ 0 & 1 & 1 \\ 0 & 0 & 0 \end{pmatrix} \begin{pmatrix} \zeta_1 \\ \zeta_2 \\ \zeta_3 \end{pmatrix} = \begin{pmatrix} -1 \\ 3 \\ 0 \end{pmatrix}.$ If we let

$\zeta_2 = 0$, then $\zeta_3 = 3$ and $\zeta_1 = 2$, so $\zeta = \begin{pmatrix} 2 \\ 0 \\ 3 \end{pmatrix}$ and

$$\mathbf{x}^{(3)}(t) = \begin{pmatrix} 0 \\ 1 \\ -1 \end{pmatrix} (t^2/2)e^{2t} + \begin{pmatrix} 1 \\ 1 \\ 0 \end{pmatrix} te^{2t} + \begin{pmatrix} 2 \\ 0 \\ 3 \end{pmatrix} e^{2t}.$$

17e. Ψ is the matrix with $\mathbf{x}^{(1)}$ as the first column, $\mathbf{x}^{(2)}$ as
the second column and $\mathbf{x}^{(3)}$ as the third column.

17f. $\mathbf{T} = \begin{pmatrix} 0 & 1 & 2 \\ 1 & 1 & 0 \\ -1 & 0 & 3 \end{pmatrix}$ and using row operations on $\mathbf{T}$ and $\mathbf{I}$, or a

computer algebra system, $\mathbf{T}^{-1} = \begin{pmatrix} -3 & 3 & 2 \\ 3 & -2 & -2 \\ -1 & 1 & 1 \end{pmatrix}$ and thus

$$\mathbf{T^{-1}AT} = \begin{pmatrix} 2 & 1 & 0 \\ 0 & 2 & 1 \\ 0 & 0 & 2 \end{pmatrix} = \mathbf{J}, \text{ which is equivalent to Eq.(29) for}$$

this problem.

19a. $\mathbf{J^2} = \mathbf{JJ} = \begin{pmatrix} \lambda & 1 \\ 0 & \lambda \end{pmatrix} \begin{pmatrix} \lambda & 1 \\ 0 & \lambda \end{pmatrix} = \begin{pmatrix} \lambda^2 & 2\lambda \\ 0 & \lambda^2 \end{pmatrix}$

$\mathbf{J^3} = \mathbf{JJ^2} = \begin{pmatrix} \lambda & 1 \\ 0 & \lambda \end{pmatrix} \begin{pmatrix} \lambda^2 & 2\lambda \\ 0 & \lambda^2 \end{pmatrix} = \begin{pmatrix} \lambda^3 & 3\lambda^2 \\ 0 & \lambda^3 \end{pmatrix}$

19b. Based upon the results of part a, assume

$\mathbf{J^n} = \begin{pmatrix} \lambda^n & n\lambda^{n-1} \\ 0 & \lambda^n \end{pmatrix}$, then

$\mathbf{J^{n+1}} = \mathbf{JJ^n} = \begin{pmatrix} \lambda & 1 \\ 0 & \lambda \end{pmatrix} \begin{pmatrix} \lambda^n & n\lambda^{n-1} \\ 0 & \lambda^n \end{pmatrix}$

$= \begin{pmatrix} \lambda^{n+1} & (n+1)\lambda^n \\ 0 & \lambda^{n+1} \end{pmatrix}$, which is the same as $\mathbf{J^n}$ with n

replaced by n+1. Thus, by mathematical induction, $\mathbf{J^n}$ has the desired form.

19c. From Eq.(23), Sect. 7.7, we have

$$\exp(\mathbf{J}t) = \mathbf{I} + \sum_{n=1}^{\infty} \frac{\mathbf{J^n}t^n}{n!}$$

$$= \mathbf{I} + \sum_{n=1}^{\infty} \begin{pmatrix} \dfrac{\lambda^n t^n}{n!} & \dfrac{n\lambda^{n-1}t^n}{n!} \\ 0 & \dfrac{\lambda^n t^n}{n!} \end{pmatrix}$$

$$= \begin{pmatrix} 1 + \sum_{n=1}^{\infty} \dfrac{\lambda^n t^n}{n!} & \sum_{n=1}^{\infty} \dfrac{\lambda^{n-1}t^n}{(n-1)!} \\ 0 & 1 + \sum_{n=1}^{\infty} \dfrac{\lambda^n t^n}{n!} \end{pmatrix}$$

$$= \begin{pmatrix} e^{\lambda t} & te^{\lambda t} \\ 0 & e^{\lambda t} \end{pmatrix}, \text{ since}$$

$$\sum_{n=1}^{\infty} \frac{\lambda^{n-1}t^n}{(n-1)!} = t(1 + \sum_{n=1}^{\infty} \frac{\lambda^n t^n}{n!}) = te^{\lambda t}.$$

19d. From Eq.(28), Sect. 7.7, we have

$$\mathbf{x} = \exp(\mathbf{J}t)\mathbf{x}^0 = \begin{pmatrix} e^{\lambda t} & te^{\lambda t} \\ 0 & e^{\lambda t} \end{pmatrix}\begin{pmatrix} x_1^0 \\ x_2^0 \end{pmatrix} = \begin{pmatrix} x_1^0 e^{\lambda t} + x_2^0 te^{\lambda t} \\ x_2^0 e^{\lambda t} \end{pmatrix}$$

$$= \begin{pmatrix} x_1^0 \\ x_2^0 \end{pmatrix} e^{\lambda t} + \begin{pmatrix} x_2^0 \\ 0 \end{pmatrix} te^{\lambda t}.$$

Section 7.9, Page 439

1. From Sect. 7.5 Prob. 3 we have

$$\mathbf{x}^{(c)} = c_1 \begin{pmatrix} 1 \\ 1 \end{pmatrix} e^t + c_2 \begin{pmatrix} 1 \\ 3 \end{pmatrix} e^{-t}. \text{ Note that}$$

$$\mathbf{g}(t) = \begin{pmatrix} 1 \\ 0 \end{pmatrix} e^t + \begin{pmatrix} 0 \\ 1 \end{pmatrix} t \text{ and that } r = 1 \text{ is an eigenvalue of}$$

the coefficient matrix. Thus if the method of undetermined coefficients is used, the assumed form is given by Eq.(18).

2. Using methods of previous sections, we find that the eigenvalues are $r_1 = 2$ and $r_2 = -2$, with corresponding

eigenvectors $\begin{pmatrix} \sqrt{3} \\ 1 \end{pmatrix}$ and $\begin{pmatrix} 1 \\ -\sqrt{3} \end{pmatrix}$. Thus

$$\mathbf{x}^{(c)} = c_1 \begin{pmatrix} \sqrt{3} \\ 1 \end{pmatrix} e^{2t} + c_2 \begin{pmatrix} 1 \\ -\sqrt{3} \end{pmatrix} e^{-2t}. \text{ Writing the}$$

nonhomogeneous term as $\begin{pmatrix} 1 \\ 0 \end{pmatrix} e^t + \begin{pmatrix} 0 \\ \sqrt{3} \end{pmatrix} e^{-t}$ we see that we

can assume $\mathbf{v}(t) = \mathbf{a}e^t + \mathbf{b}e^{-t}$ as the particlar solution. Substituting this in the D.E., we obtain

$$\mathbf{a}e^t - \mathbf{b}e^{-t} = \mathbf{A}\mathbf{a}e^t + \mathbf{A}\mathbf{b}e^{-t} + \begin{pmatrix} 1 \\ 0 \end{pmatrix} e^t + \begin{pmatrix} 0 \\ \sqrt{3} \end{pmatrix} e^{-t}, \text{ where } \mathbf{A}$$

is the given coefficient matrix. All the terms involving e^t must add to zero and thus we have $\mathbf{A}\mathbf{a} - \mathbf{a} + \begin{pmatrix} 1 \\ 0 \end{pmatrix} = \begin{pmatrix} 0 \\ 0 \end{pmatrix}$.

This is equivalent to the system $\sqrt{3} a_2 = -1$ and $\sqrt{3} a_1 - 2a_2 = 0$, or $a_1 = -2/3$ and $a_2 = -1/\sqrt{3}$. Likewise the terms involving e^{-t} must add

to zero, which yields $\mathbf{A}\mathbf{b} + \mathbf{b} + \begin{pmatrix} 0 \\ \sqrt{3} \end{pmatrix} = \begin{pmatrix} 0 \\ 0 \end{pmatrix}$. This is equilvalent to $2b_1 + \sqrt{3}\,b_2 = 0$ and $\sqrt{3}\,b_1 = -\sqrt{3}$ and thus $b_1 = -1$ and $b_2 = 2/\sqrt{3}$. Substituting these values for **a** and **b** into $\mathbf{v}(t)$ and adding $\mathbf{v}(t)$ to $\mathbf{x}^{(c)}$ yields the desired solution.

3.   The method of undetermined coefficients is not straight forward since the assumed form of $\mathbf{v}(t) = \mathbf{a}\cos t + \mathbf{b}\sin t$ leads to singular equations for **a** and **b**. From Prob. 3 of Sect. 7.6 we find that a fundamental matrix is

$$\Psi(t) = \begin{pmatrix} 5\cos t & 5\sin t \\ 2\cos t + \sin t & -\cos t + 2\sin t \end{pmatrix}.$$ The inverse

matrix is

$$\Psi^{-1}(t) = \begin{pmatrix} \dfrac{\cos t - 2\sin t}{5} & \sin t \\ \dfrac{2\cos t + \sin t}{5} & -\cos t \end{pmatrix},$$ which may be found as

in Sect. 7.2 or, more efficiently, by using a computer algebra system. Thus we may use the method of variation of parameters where $\mathbf{x} = \Psi(t)\mathbf{u}(t)$ and $\mathbf{u}(t)$ is given by $\mathbf{u}'(t) = \Psi^{-1}(t)\mathbf{g}(t)$ from Eq.(27). For this problem

$$\mathbf{g}(t) = \begin{pmatrix} -\cos t \\ \sin t \end{pmatrix} \text{ and thus}$$

$$\mathbf{u}'(t) = \begin{pmatrix} \dfrac{\cos t - 2\sin t}{5} & \sin t \\ \dfrac{2\cos t + \sin t}{5} & -\cos t \end{pmatrix} \begin{pmatrix} -\cos t \\ \sin t \end{pmatrix}$$

$$= \frac{1}{5} \begin{pmatrix} 2 - 3\cos 2t + \sin 2t \\ -1 - \cos 2t - 3\sin 2t \end{pmatrix},$$

after multiplying and using appropriate trigonometric identities. Integration and multiplication by Ψ yields the desired solution, using trigonometric identities. This problem may also be solved using undetermined coefficients, where it should be noted that the assumed form requires $t\cos t$ and $t\sin t$ be used, as illustrated in Eq.(18).

4.   In this problem we use the method illustrated in Ex.1. From Prob 4 of Sect.7.5 we have the transformation matrix $\mathbf{T} = \begin{pmatrix} 1 & 1 \\ -4 & 1 \end{pmatrix}$. Inverting **T** we find that $\mathbf{T}^{-1} = \dfrac{1}{5}\begin{pmatrix} 1 & -1 \\ 4 & 1 \end{pmatrix}$.

If we let $\mathbf{x} = \mathbf{T}\mathbf{y}$ and substitute into the D.E., we obtain

$$\mathbf{y}' = \frac{1}{5}\begin{pmatrix} 1 & -1 \\ 4 & 1 \end{pmatrix}\begin{pmatrix} 1 & 1 \\ 4 & -2 \end{pmatrix}\begin{pmatrix} 1 & 1 \\ -4 & 1 \end{pmatrix}\mathbf{y} + \frac{1}{5}\begin{pmatrix} 1 & -1 \\ 4 & 1 \end{pmatrix}\begin{pmatrix} e^{-2t} \\ -2e^{t} \end{pmatrix}$$

$$= \begin{pmatrix} -3 & 0 \\ 0 & 2 \end{pmatrix}\mathbf{y} + \frac{1}{5}\begin{pmatrix} e^{-2t} + 2e^{t} \\ 4e^{-2t} - 2e^{t} \end{pmatrix}. \quad \text{This corresponds to}$$

the two scalar equations

$$y_1' + 3y_1 = (1/5)e^{-2t} + (2/5)e^{t},$$
$$y_2' - 2y_2 = (4/5)e^{-2t} - (2/5)e^{t},$$

which may be solved by the methods of Sect. 2.1. For the first equation the integrating factor is e^{3t} and we obtain $(e^{3t}y_1)' = (1/5)e^{t} + (2/5)e^{4t}$, so $e^{3t}y_1 = (1/5)e^{t} + (1/10)e^{4t} + c_1$. For the second equation the integrating factor is e^{-2t}, so $(e^{-2t}y_2)' = (4/5)e^{-4t} - (2/5)e^{-t}$. Hence $e^{-2t}y_2 = -(1/5)e^{-4t} + (2/5)e^{-t} + c_2$. Thus

$$\mathbf{y} = \begin{pmatrix} 1/5 \\ -1/5 \end{pmatrix}e^{-2t} + \begin{pmatrix} 1/10 \\ 2/5 \end{pmatrix}e^{t} + \begin{pmatrix} c_1 e^{-3t} \\ c_2 e^{2t} \end{pmatrix}. \quad \text{Finally,}$$

multiplying by $\mathbf{T}$, we obtain

$$\mathbf{x} = \mathbf{T}\mathbf{y} = \begin{pmatrix} 0 \\ -1 \end{pmatrix}e^{-2t} + \begin{pmatrix} 1/2 \\ 0 \end{pmatrix}e^{t} + c_1\begin{pmatrix} 1 \\ -4 \end{pmatrix}e^{-3t} + c_2\begin{pmatrix} 1 \\ 1 \end{pmatrix}e^{2t}.$$

The last two terms are the general solution of the corresponding homogeneous system, while the first two terms constitute a particular solution of the nonhomogeneous system.

8. For this problem we illustrate the use of Laplace

Transforms. As in Eq.(43), $(s\mathbf{I} - \mathbf{A})\mathbf{X} = \begin{pmatrix} \dfrac{1}{s-1} \\ \dfrac{-1}{s-1} \end{pmatrix}$, where

$\mathbf{A} = \begin{pmatrix} 2 & -1 \\ 3 & -2 \end{pmatrix}$ (and we have assumed zero I.C. in order to

find a particular solution), thus $\mathbf{X} = (s\mathbf{I} - \mathbf{A})^{-1}\begin{pmatrix} \dfrac{1}{s-1} \\ \dfrac{-1}{s-1} \end{pmatrix}$.

$(s\mathbf{I} - \mathbf{A})^{-1}$ is found to be $\dfrac{1}{s^2-1}\begin{pmatrix} s+2 & -1 \\ 3 & s-2 \end{pmatrix}$ and hence

$$X(s) = \frac{1}{(s^2-1)(s-1)}\binom{s+3}{5-s} = \begin{pmatrix} \dfrac{2}{(s-1)^2} - \dfrac{1/2}{s-1} + \dfrac{1/2}{s+1} \\[3mm] \dfrac{2}{(s-1)^2} - \dfrac{3/2}{s-1} + \dfrac{3/2}{s+1} \end{pmatrix}, \text{ using}$$

partial fractions. The inverse transform gives

$$x(t) = 2\binom{1}{1}te^t - \frac{1}{2}\binom{1}{3}e^t + \frac{1}{2}\binom{1}{3}e^{-t}. \text{ Note that this}$$

particular solution differs from the one shown in the text by a multiple of the homogeneous solution.

12. Since the coefficient matrix is the same as that of Prob.3, use the same procedure as done in that problem, including the Ψ^{-1} found there. In the interval $\pi/2 < t < \pi$ $\sin t > 0$ and $\cos t < 0$; hence $|\sin t| = \sin t$, but $|\cos t| = -\cos t$.

14. To verify that $x^{(c)}$ is the general solution of the corresponding homogeneous system it is sufficient to substitute $x_1(t) = \binom{1}{1}t$ and $x_2(t) = \binom{1}{3}t^{-1}$ individually

into $tx' = \begin{pmatrix} 2 & -1 \\ 3 & -2 \end{pmatrix}x$, since x_1 and x_2 are linearly

independent. For the nonhomogeous solution, substitute

$x = \Psi(t)u(t)$, where $\Psi(t) = \begin{pmatrix} t & 1/t \\ t & 3/t \end{pmatrix}$, into the given

D.E. to obtain $t\Psi'u + t\Psi u' = A\Psi u + g(t)$. Here A is the

coefficeint matrix and $g(t) = \begin{pmatrix} 1-t^2 \\ 2t \end{pmatrix}$. Since $t\Psi' = A\Psi$,

we then have $u' = (1/t)\Psi^{-1}(t)g(t)$. Using a computer algebra system or row operations on Ψ and I, we find

that $\Psi^{-1} = \begin{pmatrix} 3/2t & -1/2t \\ -t/2 & t/2 \end{pmatrix}$ and hence

$u_1' = \dfrac{3}{2t^2} - \dfrac{3}{2} - \dfrac{1}{t}$ and $u_2' = \dfrac{-1}{2} + \dfrac{t^2}{2} + t$, which yields

$u_1 = \dfrac{-3}{2t} - \dfrac{3t}{2} - \ln t + c_1$ and $u_2 = -\dfrac{1}{2}t + \dfrac{t^3}{6} + \dfrac{t^2}{2} + c_2.$

Multiplication of u by $\Psi(t)$ yields the desired solution.

CHAPTER 8

<u>Section 8.1, Page 449</u>

In the following problems that ask for a large number of
numerical calculations the first few steps are shown. It is
then necessary to use these samples as a model to format a
computer program or calculator to find the remaining values.

1a. The Euler formulas is $y_{n+1} = y_n + h(3 + t_n - y_n)$ for
     $n = 0,1,2,3...$ and with $t_0 = 0$ and $y_0 = 1$. Thus
     $y_1 = 1 + .05(3 + 0 - 1) = 1.1$
     $y_2 = 1.1 + .05(3 + .05 - 1.1) = 1.1975 \cong y(.1)$
     $y_3 = 1.1975 + .05(3 + .1 - 1.1975) = 1.29263$
     $y_4 = 1.29263 + .05(3 + .15 - 1.29263) = 1.38549 \cong y(.2)$.

1c. The backward Euler formula is $y_{n+1} = y_n + h(3 + t_{n+1} - y_{n+1})$.
     Solving this for y_{n+1} we find $y_{n+1} = [y_n + h(3 + t_{n+1})]/(1+h)$.
     Thus $y_1 = \dfrac{1 + .05(3.05)}{1.05} = 1.097619$,

     $y_2 = \dfrac{1.097619 + .05(3.1)}{1.05} = 1.192971 \cong y(.1)$,

     $y_3 = \dfrac{1.192971 + .05(3.15)}{1.05} = 1.286162$, and

     $y_4 = \dfrac{1.286162 + .05(3.2)}{1.05} = 1.377298 \cong y(.2)$.

5a. $y_1 = y_0 + h\dfrac{y_0^2 + 2t_0 y_0}{3 + t_0^2} = .5 + .05\dfrac{(.5)^2 + 0}{3 + 0} = .504167$

     $y_2 = .504167 + .05\dfrac{(.504167)^2 + 2(.05)(.504167)}{3 + (.05)^2} = .509239$

5c. $y_1 = .5 + .05\dfrac{y_1^2 + 2(.05)y_1}{3 + (.05)^2}$, which is a quadratic
     equation in y_1. Using the quadratic formula, or an
     equation solver, we obtain $y_1 = .5050895$. Thus
     $y_2 = .5050895 + .05\dfrac{y_2^2 + 2(.1)y_2}{3 + (.1)^2}$ which is again quadratic
     in y_2, yielding $y_2 = .5111273$.

7a. For part a eighty steps must be taken, that is,
     $n = 0,1,...79$ and for part b 160 steps must taken with
     $n = 0,1,...159$. Thus use of a programmable calculator or
     a computer is required.

7c. We have $y_{n+1} = y_n + h(.5 - t_{n+1} + 2y_{n+1})$, which is linear
     in y_{n+1} and thus we have $y_{n+1} = \dfrac{y_n + .5h - ht_{n+1}}{1 - 2h}$. Again,
     80 steps are needed here and 160 steps in part d. In
     This case a spreadsheet is very useful. The first few
     lines, the middle three lines and the last two lines are
     shown for $h = .025$:

n	y_n	t_n	y_{n+1}
0	1	0	1.06513
1	1.06513	.025	1.13303
2	1.13303	.050	1.20381
⋮			
38	7.49768	.950	7.87980
39	7.87980	.975	8.28137
40	8.28137	1.000	8.70341
⋮			
78	55.62105	1.950	58.50966
79	58.50966	1.975	61.54964
80	61.54964	2.000	

At least eight decimal places were used in all
calculations.

9c. The backward Euler formula gives
     $Y_{n+1} = y_n + h\sqrt{t_{n+1} + y_{n+1}}$. Subtracting y_n from both
     sides, squaring both sides, and solving for y_{n+1} yields
     $Y_{n+1} = y_n + \dfrac{h^2}{2} + h\sqrt{y_n + t_{n+1} + h^2/4}$, where the
     positive root is chosen since y' is always positive and
     thus y is increasing. Alternately, an equation solver can
     be used to solve $y_{n+1} = y_n + h\sqrt{t_{n+1} + y_{n+1}}$ for y_{n+1}. The
     first few values, for $h = 0.25$, are $y_1 = 3.043795$,
     $y_2 = 2.088082$, $y_3 = 3.132858$ and $y_4 = 3.178122 \cong y(.1)$.

15. If $y' = 1 - t + 4y$ then differentiation of both sides
     yields $y'' = -1 + 4y' = -1 + 4(1-t+4y) = 3 - 4t + 16y$. In
     Eq.(12) we let y_n, y'_n and y''_n denote the approximate
     values of $\phi(t_n)$, $\phi'(t_n)$, and $\phi''(t_n)$, respectively.
     Keeping the first three terms in the Taylor series we
     have
     $Y_{n+1} = y_n + y'_n h + y''_n h^2/2$
       $= y_n + (1 - t_n + 4y_n)h + (3 - 4t_n + 16y_n)h^2/2$. For $n = 0$,
     $t_0 = 0$ and $y_0 = 1$ we have

$$y_1 = 1 + (1 - 0 + 4)(.1) + (3 - 0 + 16)\frac{(.1)^2}{2} = 1.595 \approx y(.1) \text{ and}$$

$$y_2 = 1.595 + [1 - .1 + 4(1.595)](.1) + [3 - .4 + 16(1.595)]\frac{(.1)^2}{2} = 2.4636 \approx y(.2).$$

16. If $y = \phi(t)$ is the exact solution of the I.V.P., then
    $\phi'(t) = 2\phi(t) - 1$ and $\phi''(t) = 2\phi'(t) = 4\phi(t) - 2$. From
    Eq.(21), $e_{n+1} = [2\phi(\bar{t}_n) - 1]h^2$, $t_n < \bar{t}_n < t_n + h$. Thus a
    bound for e_{n+1} is $|e_{n+1}| \le [1 + 2\max_{0 \le t \le 1}|\phi(t)|]h^2$. Since the
    exact solution is $y = \phi(t) = [1 + \exp(2t)]/2$,
    $e_{n+1} = h^2 \exp(2\bar{t}_n)$. Therefore
    $|e_1| \le (0.1)^2 \exp(0.2) = 0.012$ and
    $|e_4| \le (0.1)^2 \exp(0.8) = 0.022$, since the maximum value of
    $\exp(2\bar{t}_n)$ occurs at $t = .1$ and $t = .4$ respectively. From
    Prob. 2 of Sect. 2.7, the actual error in the first step
    is .0107.

19. The local truncation error is $e_{n+1} = \phi''(t_n)h^2/2$. For this
    problem $\phi'(t) = 5t - 3\phi^{1/2}(t)$ and thus
    $\phi''(t) = 5 - (3/2)\phi^{-1/2}\phi' = 19/2 - (15/2)t\phi^{-1/2}$.
    Substituting this last expression into e_{n+1} yields the
    desired answer.

22d. Since $y'' = -5\pi\sin 5\pi t$, Eq.(21) gives $e_{n+1} = -(5\pi/2)\sin(5\pi\bar{t}_n)h^2$.
    Thus $|e_{n+1}| < \frac{5\pi}{2}h^2$, since $|\sin 5\pi\bar{t}_n| < 1$. For $|e_{n+1}| < .05$ we
    must then have $h < \dfrac{1}{\sqrt{50\pi}} \approx .08$.

23a. From Eq.(14) we have $E_n = \phi(t_n) - y_n$. Using this in
    Eq.(20) we obtain
    $E_{n+1} = E_n + h\{f[t_n,\phi(t_n)] - f(t_n,y_n)\} + \phi''(\bar{t}_n)h^2/2$. Using
    the given inequality involving L we have
    $|f[t_n,\phi(t_n)] - f(t_n,y_n)| \le L|\phi(t_n) - y_n| = L|E_n|$ and thus
    $|E_{n+1}| \le |E_n| + hL|E_n| + \max_{t_0 \le t \le t_n}|\phi''(t)|h^2/2 = \alpha|E_n| + \beta h^2$.

23b. Since $\alpha = 1 + hL$, $\alpha - 1 = hL$. Hence $\beta h^2(\alpha^n - 1)/(\alpha - 1) =$
    $\beta h^2[(1+hL)^n - 1]/hL = \beta h[(1+hL)^n - 1]/L$.

23c. $(1+hL)^n \le \exp(nhL)$ follows from the observation that
    $\exp(nhL) = [\exp(nL)]^n = (1 + hL + h^2L^2/2! + \ldots)^n$ and
    that all the neglected terms are positive. Noting that
    $nh = t_n - t_0$, the rest follows from Eq.(ii).

24. The Taylor series for $\phi(t)$ about $t = t_{n+1}$ is

$$\phi(t) = \phi(t_{n+1}) + \phi'(t_{n+1})(t-t_{n+1}) + \phi''(t_{n+1})\frac{(t-t_{n+1})^2}{2} + \ldots.$$

Letting $\phi'(t) = f(t,\phi(t))$, $t = t_n$ and $h = t_{n+1} - t_n$ we

have $\phi(t_n) = \phi(t_{n+1}) - f(t_{n+1},\phi(t_{n+1}))h + \phi''(\bar{t}_n)h^2/2$, where

$t_n < \bar{t}_n < t_{n+1}$. Thus

$\phi(t_{n+1}) = \phi(t_n) + f(t_{n+1},\phi(t_{n+1}))h - \phi''(\bar{t}_n)h^2/2$. Comparing

this to Eq. 13 we then have $e_{n+1} = -\phi''(\bar{t}_n)h^2/2$.

25b. From Prob. 1 we have $y_{n+1} = y_n + h(3 + t_n - y_n)$, so
$y_1 = 1 + .05(3 + 0 - 1) = 1.1$
$y_2 = 1.1 + .05(3 + .05 - 1.1) = 1.20 \cong y(.1)$
$y_3 = 1.20 + .05(3 + .1 - 1.20) = 1.30$
$y_4 = 1.30 + .05(3 + .15 - 1.30) = 1.39 \cong y(.2)$.

Section 8.2, Page 456

1a. The improved Euler formula is
$y_{n+1} = y_n + [y_n' + f(t_n + h, y_n + hy_n')]h/2$ where
$y' = f(t,y) = 3 + t - y$. Hence $y_n' = 3 + t_n - y_n$ and
$f(t_n + h, y_n + hy_n') = 3 + t_{n+1} - (y_n + hy_n')$. Thus we obtain

$$y_{n+1} = y_n + (3 + t_n - y_n)h + \frac{h^2}{2}(1 - y_n')$$

$$= y_n + (3 + t_n - y_n)h + \frac{h^2}{2}(-2 - t_n + y_n).$$ Thus

$$y_1 = 1 + (3-1)(.05) + \frac{(.05)^2}{2}(-2+1) = 1.098750 \text{ and}$$

$$y_2 = y_1 + (3 +.05 - y_1)(.05) + \frac{(.05)^2}{2}(-2 -.05 + y_1) = 1.19512$$

are the first two steps. In this case, the equation
specifying y_{n+1} is somewhat more complicated when
$y_n' = 3 + t_n - y_n$ was substituted. When designing the steps to
calculate y_{n+1} on a computer, y_n' can be calculated first and
thus the simpler formula for y_{n+1} can be used. The exact
solution is $y(t) = 2 + t - e^{-t}$, so $y(.1) = 1.19516$,
$y(.2) = 1.38127$, $y(.3) = 1.55918$ and $y(.04) = 1.72968$, so the
approximations using $h = .0125$ are quite accurate to five
decimal places.

4. In this case $y_n' = 2t_n + e^{-t_n y_n}$ and thus the improved Euler
formula is

$$y_{n+1} = y_n + \frac{[(2t_n + e^{-t_ny_n}) + 2t_{n+1} + e^{-t_{n+1}(y_n + hy_n')}]h}{2}. \quad \text{For}$$

n = 0, 1, 2 we get $y_1 = 1.05122$, $y_2 = 1.10483$ and $y_3 = 1.16072$
for h = .05.

10. See Problem 4.

11. The improved Euler formula is
$$y_{n+1} = y_n + \frac{f(t_n,y_n) + f(t_{n+1},y_n+hf(t_n,y_n))}{2} h. \quad \text{As suggested in}$$
the text, it's best to perform the following steps when
implementing this formula: let $k_1 = (4 - t_ny_n)/(1 + y_n^2)$,
$k_2 = y_n + hk_1$ and $k_3 = (4 - t_{n+1}k_2)/(1 + k_2^2)$. Then
$y_{n+1} = y_n + (k_1 + k_3)h/2$.

14a. Since $\phi(t_n + h) = \phi(t_{n+1})$ we have, using the first part of
Eq.(5) and the given equation,
$e_{n+1} = \phi(t_{n+1}) - y_{n+1} = [\phi(t_n)-y_n] + [\phi'(t_n) -$
$\frac{y_n' + f(t_n+h, \ y_n+hy_n')}{2}]h + \phi''(t_n)h^2/2! + \phi'''(\bar{t}_n)h^3/3!$.
Since $y_n = \phi(t_n)$ and $y_n' = \phi'(t_n) = f(t_n,y_n)$ this reduces
to
$e_{n+1} = \phi''(t_n)h^2/2! - \{f[t_n+h, \ y_n + hf(t_n,y_n)]$
$$- f(t_n,y_n)\}h/2! + \phi'''(\bar{t}_n)h^3/3!,$$
which can be written in the form of Eq.(i).

14b. First observe that $y' = f(t,y)$ and $y'' = f_t(t,y) +$
$f_y(t,y)y'$. Hence $\phi''(t_n) = f_t(t_n,y_n) + f_y(t_n,y_n)f(t_n,y_n)$.
Using the given Taylor series, with $a = t_n$, $h = h$, $b = y_n$
and $k = hf(t_n,y_n)$ we have

$f[t_n+h,y_n+hf(t_n,y_n)] = f(t_n,y_n)+f_t(t_n,y_n)h+f_y(t_n,y_n)hf(t_n,y_n)$
$+ [f_{tt}(\xi,\eta)h^2+2f_{ty}(\xi,\eta)h^2f(t_n,y_n)+f_{yy}(\xi,\eta)h^2f^2(t_n,y_n)]/2!$

where $t_n < \xi < t_n + h$ and $|\eta-y_n| < h|f(t_n,y_n)|$.
Substituting this in Eq.(i) and using the earlier
expression for $\phi''(t_n)$ we find that the first term on the
right side of Eq.(i) reduces to
$-[f_{tt}(\xi,\eta) + 2f_{ty}(\xi,\eta)f(t_n,y_n) + f_{yy}(\xi,\eta)f^2(t_n,y_n)]h^3/4$,
which is proportional to h^3 plus, possibly, higher order
terms. The reason that there may be higher order terms
is because ξ and η will, in general, depend upon h.

14c. If $f(t,y)$ is linear in t and y, then $f_{tt} = f_{ty} = f_{yy} = 0$
and the terms appearing in the last formula of part (b)
are all zero.

15. Since $\phi(t) = [4t - 3 + 19\exp(4t)]/16$ we have
$\phi'''(t) = 76\exp(4t)$ and thus from Prob. 14c, since f is
linear in t and y, we find $e_{n+1} = 38[\exp(4\bar{t}_n)]h^3/3$.
Thus, since e^{4t} is increasing on $[0,2]$,
$|e_{n+1}| \leq (38h^3/3)\exp(8) = 37,758.8h^3$ on $0 \leq t \leq 2$.
For n = 1, we have
$|e_1| = |\phi(t_1) - y_1| \leq (38/3)\exp(0.2)(.05)^3 = .001934$,
which is approximately 1/15 of the error indicated in
Eq.(27) of the previous section.

19. The Euler method gives
$y_1 = y_0 + h(5t_0 - 3\sqrt{y_0}) = 2 + .1(-3\sqrt{2}) = 1.57574$ and the
improved Euler method gives

$$y_1 = y_0 + \frac{f(t_0,y_0) + f(t_1,y_1)}{2} h$$
$$= 2 + [-3\sqrt{2} + (.5 - 3\sqrt{1.57574})].05 = 1.62458.$$

Thus, the estimated error in using the Euler method is
$1.62458 - 1.57574 = .04884$. Since we want our error tolerance
to be no greater than .0025 we need to adjust the step size
(see the text discussion on the variation of the step size)
downward by a factor of $\sqrt{.0025/.04884} \approx .226$. Thus a step
size of $h = (.1)(.23) = .023$ would be needed for the required
local truncation error bound of .0025.

24. The modified Euler formula is
$y_{n+1} = y_n + hf[t_n + h/2, y_n + (h/2)f(t_n,y_n)]$ where
$f(t,y) = 5t - 3\sqrt{y}$. Thus
$y_1 = 2 + .05[5(t_0 + .025) - 3\mathrm{sqrt}(2 + .025(5t_0 - 3\sqrt{2}))]$
   $= 1.79982$ for $t_0 = 0$. Likewise
$y_2 = y_1 + .05[5(.075) - 3\mathrm{sqrt}\{y_1 + .025(5(.05) - 3\sqrt{2})\}]$
   $= 1.62268$, which is between the values for $h = .05$ and
for $h = .025$ using the improved Euler method in Prob. 2.

Section 8.3, Page 461

4. The Runge-Kutta formula is
$y_{n+1} = y_n + h(k_{n1} + 2k_{n2} + 2k_{n3} + k_{n4})/6$ where k_{n1}, k_{n2}
etc. are given by Eqs.(3). Thus for

$f(t,y) = 2t + e^{-ty}$, $(t_0, y_0) = (0,1)$ and $h = .1$ we have

$k_{01} = 0 + e^0 = 1$

$k_{02} = 2(0 + .05) + e^{-(0+.05)(1+.05k_{01})} = 1.048854$

$k_{03} = 2(.05) + e^{-(.05)(1+.05k_{02})} = 1.048738$

$k_{04} = 2(.1) + e^{-(.1)(1+.1k_{03})} = 1.095398$ and hence

$y(.1) \cong y_1 = 1 + .1(k_{01} + 2k_{02} + 2k_{03} + k_{04})/6 = 1.104843$.
Note that for $h = .05$ we get the same result for $y(.1)$,
indicating the high accuracy of the Runge-Kutta formula.
For comparison to the Euler and improved Euler results
see Prob. 4 in Sects, 8.1 and 8.2.

11. We have $f(t_n, y_n) = (4 - t_n y_n)/(1 + y_n^2)$. Thus for $t_0 = 0$,
    $y_0 = -2$ and $h = .1$ we have
    $k_{01} = f(0,-2) = .8$
    $k_{02} = f(.05, -2+.05(.8)) = f(.05, -1.96) = .846414$,
    $k_{03} = f(.05, -2+.05k_{02}) = f(.05, -1.957679) = .847983$,
    $k_{04} = f(.1, -2+.1k_{03}) = f(.1, -1.915202) = .897927$, and
    $y_1 = -2 + .1(k_{01} + 2k_{02} + 2k_{03} + k_{04})/6 = -1.915221$.   For
    comparison, see Prob. 11 in Sects. 8.1 and 8.2.

14a.

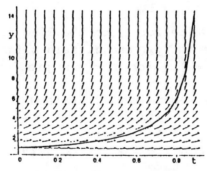

14b. We have $f(t_n, y_n) = t_n^2 + y_n^2$, $t_0 = 0$, $y_0 = 1$ and $h = .1$ so
     $k_{01} = 0^2 + 1^2 = 1$
     $k_{02} = (.05)^2 + [1 + .05(1)]^2 = 1.105$
     $k_{03} = (.05)^2 + [1 + .05(1.105)]^2 = 1.11605$
     $k_{04} = (.1)^2 + [1 + .1(1.11605)]^2 = 1.245666$ and thus
     $y_1 = 1 + .1(k_{01} + 2k_{02} + 2k_{03} + k_{04})/6 = 1.111463$.   Using these
     steps in a computer program, we obtain the following values
     for y:

t	h =.1	h = .05	h = .025	h = 0.0125
.8	5.842	5.8481	5.8483	5.8486
.9	14.0218	14.2712	14.3021	14.3046
.95		46.578	49.757	50.3935

14c. No accurate solution can be obtained for $y(1)$, as the values
     at $t = .975$ for $h = .025$ and $h = .0125$ are 1218 and 23,279
     respectively. These are caused by the slope field becoming
     vertical as $t \to 1$.

Section 8.4, Page 467

4a. The predictor formula, Eq.(6), is
$y_{n+1} = y_n + h(55f_n - 59f_{n-1} + 37f_{n-2} - 9f_{n-3})/24$
and the corrector formula, Eq.(10), is
$y_{n+1} = y_n + h(9f_{n+1} + 19f_n - 5f_{n-1} + f_{n-2})/24$, where
$f_n = 2t_n + \exp(-t_n y_n)$. Using the Runge-Kutta method, from
Sect. 8.3, Prob. 4a, we have for $t_0 = 0$ and $y_0 = 1$,
$y_1 = 1.1048431$, $y_2 = 1.2188411$ and $y_3 = 1.3414680$. Thus the
predictor formula gives $y_4 = 1.4725974$, so $f_4 = 1.3548603$ and
the corrector formula then gives $y_4 = 1.4726173$, which is the
desired value. These results, and the next step, are
summarized in the following table:

n	y_n	f_n	y_{n+1} Predicted	f_{n+1}	y_{n+1} Corrected
0	1	1	Predicted		Corrected
1	1.1048431	1.0954004			
2	1.2188411	1.1836692			
3	1.3414680	1.2686862	1.4725974	1.3548603	1.4726173
4	1.4726173	1.3548559	1.6126246	1.4465016	1.6126215
5	1.6126215				

Note that the value for f_4 on the line for $n = 4$ uses the
corrected value for y_4, and differs slightly from f_4 on
the line for $n = 3$, which uses the predicted value for y_4.

4b. The fourth order Adams-Moulton method is given by
Eq.(10): $y_{n+1} = y_n + (h/24)(9f_{n+1} + 19f_n - 5f_{n-1} + f_{n-2})$.
Substituting $h = .1$ we obtain
$y_{n+1} = y_n + (.1)(19f_n - 5f_{n-1} + f_{n-2})/24 + .0375f_{n+1}$.
For $n = 2$ we then have

$y_3 = y_2 + (.1)(19f_2 - 5f_1 + f_0)/24 + .0375f_3$
$= 1.293894103 + .0375(.6 + e^{-.3y_3})$, using values for
y_2, f_0, f_1, f_2 from part (a). An equation solver then yields
$y_3 = 1.341469821$. Likewise

$y_4 = y_3 + (.1)(19f_3 - 5f_2 + f_1)/24 + .0375f_4$
$= 1.421811841 + .0375(.8 + e^{-.4y_4})$, where f_3 is calculated
using the y_3 found above. This last equation yields
$y_4 = 1.472618922$. Finally

$y_5 = y_4 + (.1)(19f_4 - 5f_3 + f_2)/24 + .0375f_5$
$= 1.558379316 + .0375(1.0 + e^{-.5y_5})$, which gives
$y_5 = 1.612623138$.

4c. We use Eq.(16):

$y_{n+1} = (1/25)(48y_n - 36y_{n-1} + 16y_{n-2} - 3y_{n-3} + 12hf_{n+1})$.

Thus $y_4 = .04(48y_3 - 36y_2 + 16y_1 - 3y_0) + .048f_4$

$\qquad = 1.40758686 + .048(.8 + e^{-.4y_4})$, using values

for y_0, y_1. y_2, y_3 from part (a). An equation solver
then yields $y_4 = 1.472619913$. Likewise

$y_5 = .04(48y_4 - 36y_3 + 16y_2 - 3y_1) + .048f_5$

$\qquad = 1.54319349 + .048(1 + e^{-.5y_5})$, which gives

$y_5 = 1.612625556$.

7a. Using the predictor and corrector formulas (Eqs.6 and 10)
with $f_n = .5 - t_n + 2y_n$ and using the Runge–Kutta method
to calculate y_1, y_2 and y_3, we obtain the following table
for $h = .05$, $t_0 = 0$, $y_0 = 1$:

n	y_n	f_n	y_{n+1} Predicted	f_{n+1}	y_{n+1} Corrected
0	1	2.5			
1	1.130171	2.710342			
2	1.271403	2.9420805			
3	1.424858	3.199717	1.591820	3.483640	1.591825
4	1.591825	3.483649	1.773716	3.797433	1.773721
5	1.773721	3.797443	1.972114	4.144227	1.972119
6	1.972119	4.144238	2.188747	4.527495	2.188753
7	2.188753	4.527507	2.425535	4.951070	2.425542
8	2.425542	4.951084	2.684597	5.419194	2.684604
9	2.684604	5.419209	2.968276	5.936551	2.968284
10	2.968284				

7b. From Eq.(10) we have

$y_{n+1} = y_n + \dfrac{h}{24}(9f_{n+1} + 19f_n - 5f_{n-1} + f_{n-2})$

$\qquad = y_n + \dfrac{h}{24}[9(.5 - t_{n+1} + 2y_{n+1}) + 19f_n - 5f_{n-1} + f_{n-2}]$.

Solving for y_{n+1} we obtain

$y_{n+1} = [y_n + \dfrac{h}{24}(19f_n - 5f_{n-1} + f_{n-2} + 4.5 - 9t_{n+1})]/(1-.75h)$.

For $h = .05$, $t_0 = 0$, $y_0 = 1$ and using y_1 and y_2 as calculated
using the Runge–Kutta formula, we obtain the following table:

n	y_n	f_n	y_{n+1}
0	1	2.5	
1	1.130171	2.710342	
2	1.271403	2.942805	1.424859
3	1.424859	3.199718	1.591825
4	1.591825	3.483650	1.773722
5	1.773722	3.797444	1.972120
6	1.972120	4.144241	2.188755
7	2.188755	4.527510	2.425544
8	2.425544	4.951088	2.684607
9	2.684607	5.419214	2.968287
10	2.968287		

7c. From Eq.(16) we have

$y_{n+1} = (48y_n - 36y_{n-1} + 16y_{n-2} - 3y_{n-3} + 12hf_{n+1})/25$

$= [48y_n - 36y_{n-1} + 16y_{n-2} - 3y_{n-3} + 12h(.5-t_{n+1})]/25+(24/25)hy_{n+1}.$

Solving for y_{n+1} we have

$y_{n+1} = [48y_n - 36y_{n-1} + 16y_{n-2} - 3y_{n-3} + 12h(.5-t_{n+1})]/(25-24h).$

Again, using Runge-Kutta to find y_1 and y_2, we then obtain
the following table:

n	y_n	y_{n+1}
0	1	
1	1.130170833	
2	1.271402571	
3	1.424858497	1.591825573
4	1.591825573	1.773724801
5	1.773724801	1.972125968
6	1.972125968	2.188764173
7	2.188764173	2.425557376
8	2.425557376	2.684625416
9	2.684625416	2.968311063
10	2.968311063	

The exact solution is $y(t) = e^{2t} + t/2$ so $y(.5) = 2.9682818$
and $y(2) = 55.59815$, so we see that the predictor-corrector
method in part (a) is accurate at $t = .5$ through four decimal
places and at $t = 2$ through two decimal places.

16. Let $P_2(t) = At^2 + Bt + C$. As in Eqs. (12) and (13) let
$P_2(t_{n-1}) = y_{n-1}$, $P_2(t_n) = y_n$, $P_2(t_{n+1}) = y_{n+1}$ and
$P_2'(t_{n+1}) = y_{n+1}' = f(t_{n+1},y_{n+1}) = f_{n+1}$. Recall that
$t_{n-1} = t_n - h$ and $t_{n+1} = t_n + h$ and thus we have the four
equations:

$$A(t_n-h)^2 + B(t_n-h) + C = y_{n-1} \qquad (i)$$
$$At_n^2 \qquad + Bt_n \qquad + C = y_n \qquad (ii)$$
$$A(t_n+h)^2 + B(t_n+h) + C = y_{n+1} \qquad (iii)$$
$$2A(t_n+h) + B = f_{n+1} \qquad (iv)$$

Subtracting Eq.(i) from Eq.(ii) to get Eq.(v) (not shown) and subtracting Eq.(ii) from Eq.(iii) to get Eq.(vi) (not shown), then subtracting Eq.(v) from Eq.(vi) yields $y_{n+1} - 2y_n + y_{n-1} = 2Ah^2$, which can be solved for A. Now $B = f_{n+1} - 2A(t_n+h)$ [from Eq.(iv)] and $C = y_n - t_n f_{n+1} + At_n^2 + 2At_n h$ [from Eq.(ii)]. Using these values for A, B and C in Eq.(iii) yields $y_{n+1} = (1/3)(4y_n - y_{n-1} + 2hf_{n+1})$, which is Eq.(15).

Section 8.5, Page 477

2a. If $0 \le t \le 1$ then we know $0 \le t^2 \le 1$ and hence $e^y \le t^2 + e^y \le 1 + e^y$. Since each of these terms represents a slope, we may conclude that the solution of Eq.(i) is bounded above by the solution of Eq.(ii) and is bounded below by the solution of Eq.(iii).

2b. $\phi_1(t)$ and $\phi_2(t)$ can each be found by separation of variables. For $\phi_1(t)$ we have $\dfrac{1}{1+e^y}dy = dt$, or $\dfrac{e^{-y}}{e^{-y}+1}dy = dt$. Integrating both sides yields $-\ln(e^{-y}+1) = t + c$. Solving for y we find $y = \ln[1/(c_1 e^{-t}-1)]$. Setting $t = 0$ and $y = 0$, we obtain $c_1 = 2$ and thus $\phi_1(t) = \ln[e^t/(2-e^t)]$. As $t \to \ln 2$, we see that $\phi_1(t) \to \infty$. A similar analysis shows that $\phi_2(t) = \ln[1/(c_2-t)]$, where $c_2 = 1$ when the I.C. are used. Thus $\phi_2(t) \to \infty$ as $t \to 1$ and thus we conclude that $\phi(t) \to \infty$ for some t such that $\ln 2 \le t \le 1$.

2c. From part (b): $\phi_1(.9) = \ln[1/(c_1 e^{-.9}-1)] = 3.4298$ yields $c_1 = 2.5393$ and thus $\phi_1(t) \to \infty$ when $t \approx .9319$. Similarly for $\phi_2(t)$ we have $c_2 = .9324$ and thus $\phi_2(t) \to \infty$ when $t \approx .932$.

4a. The D.E. is $y' + 10y = 2.5t^2 + .5t$. So $y_h = ce^{-10t}$ is the solution of the related homogeneous equation and the particular solution, using undetermined coefficients, is

$y_p = At^2 + Bt + C.$
Substituting this
into the D.E. yields
$A = 1/4$, $B = C = 0$.
To satisfy the I.C.,
$c = 4$, so
$y(t) = 4e^{-10t} + (t^2/4)$,
which is shown in the
graph.

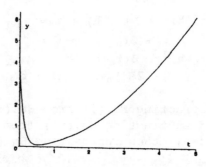

4b. From the discussion following Eq.(15), we see that h must

be less than $\dfrac{2}{|r|}$ for the Euler method to be stable.

Thus, for r = 10, h < .2. For h = .2 we obtain the
following values:

        t = 4   4.2   4.4   4.6   4.8   5.0
        y = 8   .4   8.84   1.28   9.76   2.24

and for h = .18 we obtain:

        t = 4.14   4.32   4.50   4.68   4.86   5.04
        y = 4.26   4.68   5.04   5.48   5.89   6.35.

Clearly the second set of values is stable, although not
very accurate.

4c. For a step size of .25 we find

        t = 4   4.25   4.75   5.00
        y = 4.018 4.533   5.656   6.205,

for a step size of .28 we find

        t = 4.2   4.48   4.76   5.00
        y = 10.14 10.89   11.68   12.51,

and for a step size of .3 we find

        t = 4.2   4.5   4.8   5.1
        y = 353   484   664   912.

Thus instability appears to occur about h = .28 and
certainly by h = .3. Note that the exact solution for
t = 5 is y = 6.2500, so for h = .25 we do obtain a good
approximation.

4d. For h = .5 the error at t = 5 is .013, while for h = .385, the error at t = 5.005 is .01.

5a. The general solution of the D.E. is $y(t) = t + ce^{\lambda t}$, where $y(0) = 0 \Rightarrow c = 0$ and thus $y(t) = t$, which is independent of λ.

5c. Your result in part (b) will depend upon the particular computer system and software that you use. If there is sufficient accuracy, you will obtain the solution $y = t$ for t on $0 \le t \le 1$ for each value of λ that is given, since there is no discretization error. If there is not sufficient accuracy, then round-off error will affect your calculations. For the larger values of λ, the numerical solution will quickly diverge from the exact solution, $y = t$, to the general solution $y = t + ce^{\lambda t}$, where the value of c depends upon the round-off error. If the latter case does not occur, you may simulate it by computing the numerical solution to the I.V.P. $y' - \lambda y = 1 - \lambda t$, $y(.1) = .10000001$. Here we have assumed that the numerical solution, with a step size of .01, is exact up to the point $t = .09$ [i.e. $y(.09) = .09$] and that at $t = .1$ round-off error has occurred as indicated by the slight error in the I.C. It has also been found that a larger step size (h = .05 or h = .1) may also lead to round-off error.

Section 8.6, Page 480

2a. The Euler formula is $\mathbf{x}_{n+1} = \mathbf{x}_n + h\mathbf{f}_n$, where

$$\mathbf{f}_n = \begin{pmatrix} 2x_n+t_ny_n \\ x_ny_n \end{pmatrix}, \quad x_0 = 1 \text{ and } y_0 = 1. \text{ Thus}$$

$$\mathbf{f}_0 = \begin{pmatrix} 2-0 \\ (1)(1) \end{pmatrix} = \begin{pmatrix} 2 \\ 1 \end{pmatrix}, \quad \mathbf{x}_1 = \begin{pmatrix} 1+.1(2) \\ 1+.1(1) \end{pmatrix} = \begin{pmatrix} 1.2 \\ 1.1 \end{pmatrix},$$

$$\mathbf{f}_1 = \begin{pmatrix} 2.4+.1(1.1) \\ (1.2)(1.1) \end{pmatrix} = \begin{pmatrix} 2.51 \\ 1.32 \end{pmatrix}$$

$$\text{and } \mathbf{x}_2 = \begin{pmatrix} 1.2+.1(2.51) \\ 1.1+.1(1.32) \end{pmatrix} = \begin{pmatrix} 1.451 \\ 1.232 \end{pmatrix} \cong \begin{pmatrix} \phi(.2) \\ \psi(.2) \end{pmatrix}$$

2b. Eqs. (7) give:

$$\mathbf{k}_{01} = \begin{pmatrix} f(0,1,1) \\ g(0,1,1) \end{pmatrix} = \begin{pmatrix} 2+0 \\ (1)(1) \end{pmatrix} = \begin{pmatrix} 2 \\ 1 \end{pmatrix}$$

$$\mathbf{k}_{02} = \begin{pmatrix} 2.4+.1(1.1) \\ (1.2)(1.1) \end{pmatrix} = \begin{pmatrix} 2.51 \\ 1.32 \end{pmatrix}$$

$$\mathbf{k}_{03} = \begin{pmatrix} 2.502+.1(1.132) \\ (1.251)(1.132) \end{pmatrix} = \begin{pmatrix} 2.6152 \\ 1.41613 \end{pmatrix}$$

$$\mathbf{k}_{04} = \begin{pmatrix} 3.04608+.2(1.28323) \\ (1.52304)(1.28323) \end{pmatrix} = \begin{pmatrix} 3.30273 \\ 1.95441 \end{pmatrix}$$

Using Eq. (6) in scalar form, we then have
$x_1 = 1+(.2/6)[2+2(2.51)+2(2.6152)+3.30273] = 1.51844$
$y_1 = 1+(.2/6)[1+2(1.32)+2(1.41613)+1.95441] = 1.28089$,
which are approximations to $\phi(.2)$ and $\psi(.2)$
respectively.

7. Write a computer program to do this problem as there are twenty steps or more for $h \le .05$.

8. If we let $y = x'$, then $y' = x''$ and thus we obtain the system $x' = y$ and $y' = t-3x-t^2 y$, with $x(0) = 1$ and $y(0) = x'(0) = 2$. Thus $f(t,x,y) = y$, $g(t,x,y) = t - 3x - t^2 y$, $t_0 = 0$, $x_0 = 1$ and $y_0 = 2$. If a program has been written for an earlier problem, then its best to use that. Otherwise, the first two steps are as follows:

$$\mathbf{k}_{01} = \begin{pmatrix} 2 \\ -3 \end{pmatrix}$$

$$\mathbf{k}_{02} = \begin{pmatrix} 2+(-.15) \\ .05-3(1.1)-(.05)^2(1.85) \end{pmatrix} = \begin{pmatrix} 1.85 \\ -3.25463 \end{pmatrix}$$

$$\mathbf{k}_{03} = \begin{pmatrix} 2+(-.16273) \\ .05-3(1.0925)-(.05)^2(1.83727) \end{pmatrix} = \begin{pmatrix} 1.83727 \\ -3.23209 \end{pmatrix}$$

$$\mathbf{k}_{04} = \begin{pmatrix} 2+(-.32321) \\ .1-3(1.18373)-(.1)^2(1.67679) \end{pmatrix} = \begin{pmatrix} 1.67679 \\ -3.46796 \end{pmatrix}$$

and thus

$x_1 = 1+(.1/6)[2 + 2(1.85)+2(1.83727)+(1.67679)]=1.18419$,
$y_1 = 2+(.1/6)[-3-2(3.25463)-2(3.23209)-3.46796]=1.67598$,

which are approximations to $x(.1)$ and $y(.1) = x'(.1)$.
In a similar fashion we find

$$\mathbf{k}_{11} = \begin{pmatrix} 1.67598 \\ -3.46933 \end{pmatrix} \qquad \mathbf{k}_{12} = \begin{pmatrix} 1.50251 \\ -3.68777 \end{pmatrix}$$

$$\mathbf{k}_{13} = \begin{pmatrix} 1.49159 \\ -3.66151 \end{pmatrix} \qquad \mathbf{k}_{14} = \begin{pmatrix} 1.30983 \\ -3.85244 \end{pmatrix}$$

and thus

$x_2 = x_1 + (.1/6)[1.67598 + 2(1.50251) + 2(1.49159) + 1.30983] = 1.33376$
$y_2 = y_1 - (.1/6)[3.46933 + 2(3.68777) + 2(3.66151) + 3.85244] = 1.30897.$

Three more steps must be taken in order to approximate
$x(.5)$ and $y(.5) = x'(.5)$. The intermediate steps yield
$x(.3) \approx 1.44489$, $y(.3) \approx .9093062$ and $x(.4) \approx 1.51499$,
$y(.4) \approx .4908795$.

CHAPTER 9

Section 9.1, Page 492

For Problems 1 through 16, once the eigenvalues have been
found, Table 9.1.1 will, for the most part, quickly yield the
type of critical point and the stability. In all cases it
can be easily verified that **A** is nonsingular.

1a. The eigenvalues are found from the equation det(**A**−r**I**)=0.
Substituting the values for **A** we have $\begin{vmatrix} 3-r & -2 \\ 2 & -2-r \end{vmatrix}$ =
$r^2 - r - 2 = 0$ and thus the eigenvalues are $r_1 = -1$ and
$r_2 = 2$. For $r_1 = -1$, we have $\begin{pmatrix} 4 & -2 \\ 2 & -1 \end{pmatrix}\begin{pmatrix} \xi_1 \\ \xi_2 \end{pmatrix} = \begin{pmatrix} 0 \\ 0 \end{pmatrix}$ and thus
$\xi^{(1)} = \begin{pmatrix} 1 \\ 2 \end{pmatrix}$ and for r_2 we have $\begin{pmatrix} 1 & -2 \\ 2 & -4 \end{pmatrix}\begin{pmatrix} \xi_1 \\ \xi_2 \end{pmatrix} = \begin{pmatrix} 0 \\ 0 \end{pmatrix}$ and thus
$\xi^{(2)} = \begin{pmatrix} 2 \\ 1 \end{pmatrix}$.

1b. Since the eigenvalues differ in sign, the critical point
is a saddle point and is unstable.

1d.

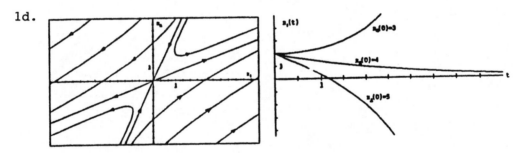

4a. Again the eigenvalues are given by $\begin{vmatrix} 1-r & -4 \\ 4 & -7-r \end{vmatrix}$ =
$r^2 + 6r + 9 = 0$ and thus $r_1 = r_2 = -3$. The eigenvectors
are solutions of $\begin{pmatrix} 4 & -4 \\ 4 & -4 \end{pmatrix}\begin{pmatrix} \xi_1 \\ \xi_2 \end{pmatrix} = \begin{pmatrix} 0 \\ 0 \end{pmatrix}$ and hence there is
just one eigenvector $\xi = \begin{pmatrix} 1 \\ 1 \end{pmatrix}$.

4b. Since the eigenvalues are negative, (0,0) is an improper
node which is asymptotically stable. If we had found
that there were two independent eigenvectors then (0,0)
would have been a proper node, as indicated in Case 3a.

4d.

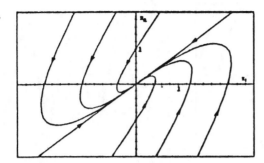

7a. In this case $\det(\mathbf{A} - r\mathbf{I}) = r^2 - 2r + 5$ and thus the eigenvalues are $r_{1,2} = 1 \pm 2i$. For $r_1 = 1 + 2i$ we have

$$\begin{pmatrix} 2-2i & -2 \\ 4 & -2-2i \end{pmatrix} \begin{pmatrix} \xi_1 \\ \xi_2 \end{pmatrix} = \begin{pmatrix} 2-2i & -2 \\ 8-8i & -8 \end{pmatrix} \begin{pmatrix} \xi_1 \\ \xi_2 \end{pmatrix} = \begin{pmatrix} 0 \\ 0 \end{pmatrix} \text{ and thus}$$

$\xi^{(1)} = \begin{pmatrix} 1 \\ 1-i \end{pmatrix}$ and hence $\xi^{(2)} = \begin{pmatrix} 1 \\ 1+i \end{pmatrix}$, which is the complex conjugate of $\xi^{(1)}$.

7b. Since the eigenvalues are complex with positive real part, we conclude that the critical point is a spiral point and is unstable.

7d.

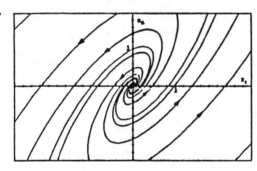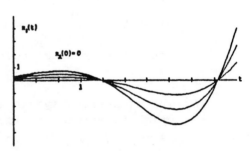

10a. Again, $\det(\mathbf{A}-r\mathbf{I}) = r^2 + 9$ and thus we have $r_{1,2} = \pm 3i$.

For $r_1 = 3i$ we have $\begin{pmatrix} 1-3i & 2 \\ -5 & -1-3i \end{pmatrix} \begin{pmatrix} \xi_1 \\ \xi_2 \end{pmatrix} = \begin{pmatrix} 0 \\ 0 \end{pmatrix}$ and thus

$\xi^{(1)} = \begin{pmatrix} 2 \\ -1+3i \end{pmatrix}$ and $\xi^{(2)} = \begin{pmatrix} 2 \\ -1-3i \end{pmatrix}$, which is the complex conjugate.

10b. Since the eigenvalues are pure imaginary the critical point is a center, which is stable.

10d.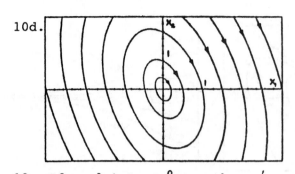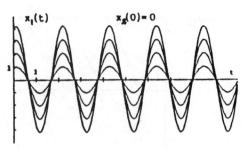

13. If we let $\mathbf{x} = \mathbf{x}^0 + \mathbf{u}$ then $\mathbf{x}' = \mathbf{u}'$ and thus the system

becomes $\mathbf{u}' = \begin{pmatrix} 1 & 1 \\ 1 & -1 \end{pmatrix} \mathbf{x}^0 + \begin{pmatrix} 1 & 1 \\ 1 & -1 \end{pmatrix} \mathbf{u} - \begin{pmatrix} 2 \\ 0 \end{pmatrix}$ which will be in

the form of Eq.(2) if $\begin{pmatrix} 1 & 1 \\ 1 & -1 \end{pmatrix} \mathbf{x}^0 = \begin{pmatrix} 2 \\ 0 \end{pmatrix}$. Using row

operations, this last set of equations is equivalent to
$\begin{pmatrix} 1 & 1 \\ 0 & -2 \end{pmatrix} \mathbf{x}^0 = \begin{pmatrix} 2 \\ -2 \end{pmatrix}$ and thus $x_1^0 = 1$ and $x_2^0 = 1$. Since

$\mathbf{u}' = \begin{pmatrix} 1 & 1 \\ 1 & -1 \end{pmatrix} \mathbf{u}$ has $(0,0)$ as the critical point, we

conclude that $(1,1)$ is the critical point of the original system. As in the earlier problems, the eigenvalues are

given by $\begin{vmatrix} 1-r & 1 \\ 1 & -1-r \end{vmatrix} = r^2 - 2 = 0$ and thus $r_{1,2} = \pm\sqrt{2}$.

Hence the critical point $(1,1)$ is an unstable saddle point.

17. The equivalent system is $dx/dt = y, dy/dt = -(k/m)x-(c/m)y$ which is written in the form of Eq.(2) as

$\dfrac{d}{dt}\begin{pmatrix} x \\ y \end{pmatrix} = \begin{pmatrix} 0 & 1 \\ -k/m & -c/m \end{pmatrix}\begin{pmatrix} x \\ y \end{pmatrix}$. The point $(0,0)$ is clearly a

critical point, and since $\mathbf{A}$ is nonsingular, it is the only one. The characteristic equation is $r^2+(c/m)r+k/m=0$ so $r_1,r_2 = [-c \pm (c^2 - 4km)^{1/2}]/2m$. In the underdamped case $c^2 - 4km < 0$, the characteristic roots are complex with negative real parts, since $c > 0$, and thus the critical point $(0,0)$ is an asymptotically stable spiral point. In the overdamped case $c^2 - 4km > 0$, the characteristic roots are real, unequal, and negative and hence the critical point $(0,0)$ is an asymptotically stable node. In the critically damped case $c^2 - 4km = 0$, the characteristic roots are equal and negative. As indicated in the solution to Prob.4, to determine whether this is an improper or proper node we must determine

whether there are one or two linearly independent
eigenvectors. The eigenvectors satisfy the equation
$\begin{pmatrix} c/2m & 1 \\ -k/m & -c/2m \end{pmatrix} \begin{pmatrix} \xi_1 \\ \xi_2 \end{pmatrix} = \begin{pmatrix} 0 \\ 0 \end{pmatrix}$, which has just one solution if
$c^2 - 4km = 0$. Thus the critical point $(0,0)$ is an
asymptotically stable improper node.

18a. If **A** has one zero eigenvalue then for $r = 0$ we have
     $\det(\mathbf{A}-r\mathbf{I}) = \det\mathbf{A} = 0$. Hence **A** is singular which means
     Ax = 0 has infinitely many solutions and consequently
     there are infinitely many critical points. Since **A** is
     2x2 the homogeneous equation **Ax = 0** will yield the
     solution $x_2 = cx_1$, which indicates that the critical
     points lie on a straight line through the origin.

18b. From Chapter 7, the solution is $\mathbf{x}(t) = c_1\xi^{(1)} + c_2\xi^{(2)}e^{r_2t}$,
     which can be written in scalar form as
     $x_1 = c_1\xi_1^{(1)} + c_2\xi_1^{(2)}e^{r_2t}$ and $x_2 = c_1\xi_2^{(1)} + c_2\xi_2^{(2)}e^{r_2t}$.
     Assuming $\xi_1^{(2)} \neq 0$, the first equation can be solved for
     $c_2e^{r_2t}$, which is then substituted into the second
     equation to yield $x_2 = c_1\xi_2^{(1)} + [\xi_2^{(2)}/\xi_1^{(2)}][x_1-c_1\xi_1^{(1)}]$.
     These are straight lines parallel to the vector $\xi^{(2)}$.
     Note that the family of lines is independent of c_2. If
     $\xi_1^{(2)} = 0$, then the lines are vertical. If $r_2 > 0$, the
     direction of motion will be in the same direction as
     indicated for $\xi^{(2)}$. If $r_2 < 0$, then it will be in the
     opposite direction.

19a. $\text{Det}(\mathbf{A}-r\mathbf{I}) = r^2 - (a_{11}+a_{22})r + a_{11}a_{22} - a_{21}a_{12} = 0$. If
     $a_{11} + a_{22} = 0$, then $r^2 = -(a_{11}a_{22} - a_{21}a_{12}) < 0$ if
     $a_{11}a_{22} - a_{21}a_{12} > 0$.

19b. Eq.(i) can be written in scalar form as
     $dx/dt = a_{11}x + a_{12}y$ and $dy/dt = a_{21}x + a_{22}y$, which then
     yields Eq.(iii). Ignoring the middle quotient in
     Eq.(iii), we can rewrite that equation as
     $(a_{21}x + a_{22}y)dx - (a_{11}x + a_{12}y)dy = 0$, which is exact
     since $a_{22} = -a_{11}$ from Eq.(ii)..

19c. Integrating $\phi_x = a_{21}x + a_{22}y$ we obtain
     $\phi = a_{21}x^2/2 + a_{22}xy + g(y)$ and thus
     $\phi_y = a_{22}x + g' = -a_{11}x - a_{12}y$ or $g' = -a_{12}y$ using Eq.(ii).

Hence $\phi(x,y) = a_{21}x^2/2 + a_{22}xy - a_{12}y^2/2 = k/2$ is the solution to Eq.(iii). The quadratic equation $Ax^2 + Bxy + Cy^2 = D$ is an ellipse provided $B^2 - 4AC < 0$. Hence for our problem if $a_{22}^2 + a_{21}a_{12} < 0$ then Eq.(iv) is an ellipse. From $a_{11} + a_{22} = 0$ we have $a_{22}^2 = -a_{11}a_{22}$ and hence the condition becomes $-a_{11}a_{22} + a_{21}a_{12} < 0$ or $a_{11}a_{22} - a_{21}a_{12} > 0$, which is true by Eqs.(ii). Thus Eq.(iv) is an ellipse under the conditions of Eqs.(ii).

20. The given system can be written as $\dfrac{d}{dt}\begin{pmatrix} x \\ y \end{pmatrix} = \begin{pmatrix} a_{11} & a_{12} \\ a_{21} & a_{22} \end{pmatrix}\begin{pmatrix} x \\ y \end{pmatrix}$.

Thus the eigenvalues are given by $r^2-(a_{11}+a_{22})r + a_{11}a_{22}-a_{12}a_{21} = 0$ and using the given definitions we rewrite this as $r^2 - pr + q = 0$ and thus $r_{1,2} = (p \pm \sqrt{p^2-4q})/2 = (p \pm \sqrt{\Delta})/2$. The results are now obtained using Table 9.1.1.

Section 9.2, Page 501

1. Solutions of the D.E. for x are y are $x = Ae^{-t}$ and $y = Be^{-2t}$ respectively. $x(0) = 4$ and $y(0) = 2$ yield $A = 4$ and $B = 2$, so $x = 4e^{-t}$ and $y = 2e^{-2t}$. Solving the first equation for e^{-t} and then substituting into the 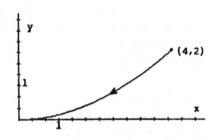 second yields $y = 2[x/4]^2 = x^2/8$, which is a parabola. From the original D.E., or from the parametric solutions, we find that $0 < x \le 4$ and $0 < y \le 2$ for $t \ge 0$ and thus only the portion of the parabola shown is the trajectory, with the direction of motion indicated.

3. Utilizing the approach indicated in Eq.(14), we have $dy/dx = -x/y$, which separates into $xdx + ydy = 0$. Integration then yields the circle $x^2 + y^2 = c^2$, where $c^2 = 16$ for both sets of I.C. The direction of motion can be found from the original D.E. and is counterclockwise for both I.C. To obtain the parametric equations, we write the system in the form

$$\frac{d}{dt}\begin{pmatrix} x \\ y \end{pmatrix} = \begin{pmatrix} 0 & -1 \\ 1 & 0 \end{pmatrix}\begin{pmatrix} x \\ y \end{pmatrix},$$ which has the characteristic

equation $\begin{vmatrix} -r & -1 \\ 1 & -r \end{vmatrix} = r^2 + 1 = 0$, or $r = \pm\,i$. Following

the procedures of Sect. 7.6, we find that one solution of

the above system is $\begin{pmatrix} 1 \\ -i \end{pmatrix}e^{it} = \begin{pmatrix} \cos t + i\sin t \\ \sin t - i\cos t \end{pmatrix}$ and thus two

real solutions are $\mathbf{u}(t) = \begin{pmatrix} \cos t \\ \sin t \end{pmatrix}$ and $\mathbf{v}(t) = \begin{pmatrix} \sin t \\ -\cos t \end{pmatrix}$. The

general solution of the system is then

$$\begin{pmatrix} x \\ y \end{pmatrix} = c_1\mathbf{u}(t) + c_2\mathbf{v}(t)$$ and hence the first I.C. yields

$c_1 = 4$, $c_2 = 0$, or $x = 4\cos t$, $y = 4\sin t$. The second I.C.
yields $c_1 = 0$, $c_2 = -4$, or $x = -4\sin t$, $y = 4\cos t$. Note
that both these parametric representations satsify the
form of the trajectories found in the first part of this
problem.

7a. The critical points are given by the solutions of
     $x(1-x-y) = 0$ and $y(1/2 - y/4 - 3x/4) = 0$. The solutions
     corresponding to either $x = 0$ or $y = 0$ are seen to be
     $x = 0$, $y = 0$; $x = 0$, $y = 2$; $x = 1$, $y = 0$. In addition,
     there is a solution corresponding to the intersection of
     the lines $1 - x - y = 0$ and $1/2 - y/4 - 3x/4 = 0$ which is
     the point $x = 1/2$, $y = 1/2$. Thus the critical points are
     $(0,0)$, $(0,2)$, $(1,0)$, and $(1/2,1/2)$.

7b.

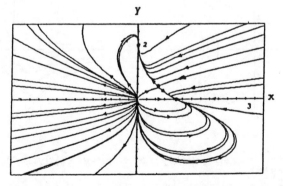

7c. For $(0,0)$, since all trajectories leave this point, this
     is an unstable node. For $(0,2)$ and $(1,0)$, since the
     trajectories tend to these points respectively, they are
     asymptotically stable nodes. For $(1/2,1/2)$, one
     trajectory tends to $(1/2,1/2)$ while all others tend to
     infinity, so this is an unstable saddle point.

12a. The critical points are given by y = 0 and
   x(1 - x^2/6 - y/5) = 0, so (0,0), ($\sqrt{6}$,0) and (-$\sqrt{6}$,0)
   are the only critical points.

12b.
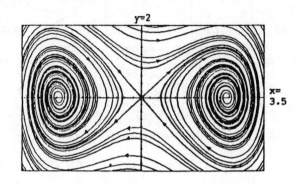

12c. Clearly ($\sqrt{6}$,0) and (-$\sqrt{6}$,0) are spiral points, and are
   asymptotically stable since the trajectories tend to each
   point, respectively. (0,0) is a saddle point, which is
   unstable, since the trajectories behave like the ones for
   (1/2,1/2) in Prob. 7.

15a. $\dfrac{dy}{dx} = \dfrac{dy/dt}{dx/dt} = \dfrac{8x}{2y}$, so 4xdx - ydy = 0 and thus 4x^2 - y^2 = c,
   which are hyperbolas for c ≠ 0 and straight lines y = ±2x for
   c = 0.

15b.
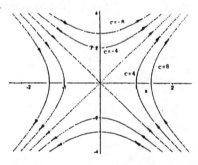

19b.
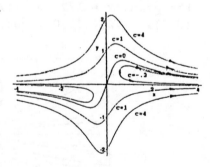

19a. $\dfrac{dy}{dx} = \dfrac{y-2xy}{-x+y+x^2}$, so (y-2xy)dx + (x-y-x^2)dy = 0, which is an
   exact D.E. Therefore ϕ(x,y) = xy - x^{2}y + g(y) and hence
   $\dfrac{\partial \phi}{\partial y}$ = x - x^2 + g$'$(y) = x - y - x^2, so g$'$(y) = -y and
   g(y) = -y^2/2. Thus 2x^{2}y - 2xy + y^2 = c (after multiplying by
   -2) is the desired solution.

21a. $\dfrac{dy}{dx} = \dfrac{-\sin x}{y}$, so $y\,dy + \sin x\,dx = 0$ and thus $y^2/2 - \cos x = c$.

21b.

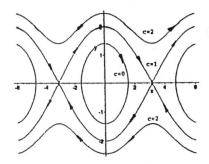

23. We know that $\phi'(t) = F[\phi(t),\psi(t)]$ and
    $\psi'(t) = G[\phi(t),\psi(t)]$ for $\alpha < t < \beta$. By direct
    substitution we have
    $\Phi'(t) = \phi'(t-s) = F[\phi(t-s),\psi(t-s)] = F[\Phi(t),\Psi(t)]$ and
    $\Psi'(t) = \psi'(t-s) = G[\phi(t-s),\psi(t-s)] = G[\Phi(t),\Psi(t)]$ for
    $\alpha < t - s < \beta$ or $\alpha + s < t < \beta + s$.

24. Suppose that $t_1 > t_0$. Let $s = t_1 - t_0$. Since the system
    is autonomous, the result of Prob. 23, with s replaced
    by $-s$ shows that $x = \phi_1(t+s)$ and $y = \psi_1(t+s)$ generates
    the same trajectory (C_1) as $x = \phi_1(t)$ and $y = \psi_1(t)$. But
    at $t = t_0$ we have $x = \phi_1(t_0+s) = \phi_1(t_1) = x_0$ and
    $y = \psi_1(t_0+s) = \psi_1(t_1) = y_0$. Thus the solution
    $x = \phi_1(t+s)$, $y = \psi_1(t+s)$ satisfies <u>exactly</u> the same
    initial conditions as the solution $x = \phi_0(t)$, $y = \psi_0(t)$
    which generates the trajectory C_0. Hence C_0 and C_1 are
    the same.

25. From the existence and uniqueness theorem we know that if
    the two solutions $x = \phi(t)$, $y = \psi(t)$ and $x = x_0$, $y = y_0$
    satisfy $\phi(a) = x_0$, $\psi(a) = y_0$ and $x = x_0$, $y = y_0$ at $t = a$,
    then these solutions are identical. Hence $\phi(t) = x_0$ and
    $\psi(t) = y_0$ for all t contradicting the fact that the
    trajectory generated by $[\phi(t),\psi(t)]$ started at a
    noncritical point.

26. By direct substitution
    $\Phi'(t) = \phi'(t+T) = F[\phi(t+T),\psi(t+T)] = F[\Phi(t),\Psi(t)]$ and
    $\Psi'(t) = \psi'(t+T) = G[\phi(t+T),\psi(t+T)], \; G[\Phi(t),\Psi(t)]$.
    Furthermore $\Phi(t_0) = x_0$ and $\Psi(t_0) = y_0$. Thus by the
    existence and uniqueness theorem $\Phi(t) = \phi(t)$ and
    $\Psi(t) = \psi(t)$ for all t.

Section 9.3, Page 511

In Problems 1 through 4, write the system in the form of
Eq.(4). Then if **g(0)** = **0** we may conclude that (0,0) is a
critical point. In addition, if **g** satisfies Eq.(5) or
Eq.(6), then the system is almost linear. In this case the
linear system, Eq.(1), will determine, in most cases, the
type and stability of the critical point (0,0) of the almost
linear system. These results are summarized in Table 9.3.1.

3. In this case the system can be written as

$$\frac{d}{dt}\begin{pmatrix} x \\ y \end{pmatrix} = \begin{pmatrix} 0 & 0 \\ -1 & 0 \end{pmatrix}\begin{pmatrix} x \\ y \end{pmatrix} + \begin{pmatrix} (1+x)\sin y \\ 1 - \cos y \end{pmatrix}.$$ However, the

coefficient matrix is singular and $g_1(x,y) = (1+x)\sin y$

does not satisfy Eq.(6). However, if we consider the
Taylor series for $\sin y$, we see that $(1+x)\sin y - y =$

$\sin y - y + x\sin y = -y^3/3! + y^5/5! + \cdots + x(y - y^3/3! + \cdots)$,
which does satisfy Eq.(6), using $x = r\cos\theta$, $y = r\sin\theta$.
Thus the first equation now becomes

$$\frac{dx}{dt} = y + [(1+x)\sin y - y] \text{ and hence}$$

$$\frac{d}{dt}\begin{pmatrix} x \\ y \end{pmatrix} = \begin{pmatrix} 0 & 1 \\ -1 & 0 \end{pmatrix}\begin{pmatrix} x \\ y \end{pmatrix} + \begin{pmatrix} (1+x)\sin y - y \\ 1 - \cos y \end{pmatrix}, \text{ where the}$$

coefficient matrix is now nonsingular and

$$\mathbf{g}(x,y) = \begin{pmatrix} (1+x)\sin y - y \\ 1 - \cos y \end{pmatrix} \text{ satisfies Eq.(6). The linear}$$

system, represented by the matrix $\mathbf{A} = \begin{pmatrix} 0 & 1 \\ -1 & 0 \end{pmatrix}$, has

eigenvalues $r_{1,2} = \pm i$, and thus the origin is a center
which is stable and the nonlinear system has either a
center or spiral point at the origin and the stablility is
indeterminent, from Table 9.3.1.

4. In this case the system can be written as

$$\frac{d}{dt}\begin{pmatrix} x \\ y \end{pmatrix} = \begin{pmatrix} 1 & 0 \\ 1 & 1 \end{pmatrix}\begin{pmatrix} x \\ y \end{pmatrix} + \begin{pmatrix} y^2 \\ 0 \end{pmatrix} \text{ and thus } \mathbf{A} = \begin{pmatrix} 1 & 0 \\ 1 & 1 \end{pmatrix} \text{ and}$$

$$\mathbf{g} = \begin{pmatrix} y^2 \\ 0 \end{pmatrix}. \text{ Since } \mathbf{g(0)} = \begin{pmatrix} 0 \\ 0 \end{pmatrix} \text{ we conclude that } (0,0) \text{ is a}$$

critical point. Following the procedure of Ex. 1, we let
$x = r\cos\theta$ and $y = r\sin\theta$ and thus

$g_1(x,y)/r = \dfrac{r^2 \sin^2 \theta}{r} \to 0$ as $r \to 0$ and thus the system is almost linear. Since $\det(\mathbf{A} - r\mathbf{I}) = (r-1)^2$, we find that the eigenvalues are $r_1 = r_2 = 1$. Since the roots are equal, we must determine whether there are one or two eigenvectors to classify the type of critical point. The eigenvectors are determined by $\begin{pmatrix} 0 & 0 \\ 1 & 0 \end{pmatrix} \begin{pmatrix} \xi_1 \\ \xi_2 \end{pmatrix} = \begin{pmatrix} 0 \\ 0 \end{pmatrix}$ and hence there is only one eigenvector $\xi = \begin{pmatrix} 0 \\ 1 \end{pmatrix}$. Thus the critical point for the linear system is an unstable improper node, from Sect. 9.1 From Table 9.3.1 we then conclude that the given system, which is almost linear, has a critical point near $(0,0)$ which is either a node or spiral point (depending on how the roots bifurcate) which is unstable.

6a. The critical points are the solutions of $x(1-x-y) = 0$ and $y(3-x-2y) = 0$. Solutions are $x = 0$, $y = 0$; $x = 0$, $3 - 2y = 0$ which gives $y = 3/2$; $y = 0$ and $1 - x = 0$ which give $x = 1$; and $1 - x - y = 0$, $3 - x - 2y = 0$ which give $x = -1$, $y = 2$. Thus the critical points are $(0,0)$, $(0,3/2)$, $(1,0)$ and $(-1,2)$.

6b, For the critical point $(0,0)$ the D.E. is already in the
6c. form of an almost linear system; and the corresponding linear system is $du/dt = u$, $dv/dt = 3v$ which has the eigenvalues $r_1 = 1$ and $r_2 = 3$. Thus the critical point $(0,0)$ is an unstable node. Each of the other three critical points is dealt with in the same manner; we consider only the critical point $(-1,2)$. In order to translate this critical point to the origin we set $x(t) = -1 + u(t)$, $y(t) = 2 + v(t)$ and substitute in the D.E. to obtain
$du/dt = -1 + u - (-1+u)^2 - (-1+u)(2+v) = u + v - u^2 - uv$
and
$dv/dt = 3(2+v) - (-1+u)(2+v) - 2(2+v)^2 = -2u - 4v - uv - 2v^2$.
   Writing this in the form of Eq.(4) we find that
$$\mathbf{A} = \begin{pmatrix} 1 & 1 \\ -2 & -4 \end{pmatrix} \text{ and } \mathbf{g} = -\begin{pmatrix} u^2 + uv \\ uv + 2v^2 \end{pmatrix} \text{ which is an almost}$$
linear system. The eigenvalues of the corresponding linear system are $r = (-3 \pm \sqrt{9 + 8})/2$ and hence the critical point $(-1,2)$, of the original system, is an unstable saddle point.

6d. The phase portrait near (-1,2) is shown.

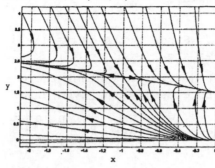

10a. The critical points are solutions of $x + x^2 + y^2 = 0$ and $y(1-x) = 0$, which yield $(0,0)$ and $(-1,0)$.

10b. For $(0,0)$ the D.E. is already in the form of an almost linear system and thus $du/dt = u$ and $dv/dt = v$ is the corresponding linear system. For $(-1,0)$ we let $u = x+1$, $v = y$ so that substituting $x = u-1$ and $y = v$ into the D.E. we obtain $\dfrac{du}{dt} = -u + u^2 + v^2$ and $\dfrac{dv}{dt} = 2v - uv$. Thus the corresponding linear system is $u' = -u$ and $v' = 2v$.

10c. For $(0,0)$ $\mathbf{A} = \begin{pmatrix} 1 & 0 \\ 0 & 1 \end{pmatrix}$ which has $r_1 = r_2 = 1$, so that $(0,0)$, for the nonlinear system, will be either a node or spiral point, depending on how the roots bifurcate. In any case, since r_1 and r_2 are positive, the system will be unstable. For $(-1,0)$ $\mathbf{A} = \begin{pmatrix} -1 & 0 \\ 0 & 2 \end{pmatrix}$ and thus $r_1 = -1$ and $r_2 = 2$, and hence the nonlinear system, from Table 9.3.1, has an unstable saddle point at $(-1,0)$.

18a. The system is $\dfrac{d}{dt} \begin{pmatrix} x \\ y \end{pmatrix} = \begin{pmatrix} 1 & 0 \\ 0 & -2 \end{pmatrix} \begin{pmatrix} x \\ y \end{pmatrix} + \begin{pmatrix} 0 \\ x^3 \end{pmatrix}$ and thus is almost linear using the procedures outlined in the earlier problems. The corresponding linear system has the eigenvalues $r_1 = 1$, $r_2 = -2$ and thus $(0,0)$ is an unstable saddle point for both the linear and almost linear systems.

18b. The trajectories of the linear system are the solutions of $dx/dt = x$ and $dy/dt = -2y$ and thus $x(t) = c_1 e^t$ and $y(t) = c_2 e^{-2t}$. To sketch these, solve the first equation for e^t and substitute into the second to obtain

$y = c_1^2 c_2/x^2$, $c_1 \neq 0$.
Several trajectories
are shown in the figure.
Since $x(t) = c_1 e^t$, we
must pick $c_1 = 0$ for
$x \to 0$ as $t \to \infty$. Thus
$x = 0$, $y = c_2 e^{-2t}$ (the
vertical axis) is the
only trajectory for
which $x \to 0$, $y \to 0$ as $t \to \infty$.

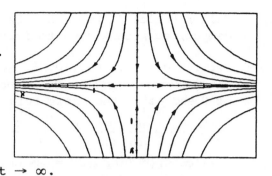

18c. For $x \neq 0$ we have $dy/dx = (dy/dt)/(dx/dt) = (-2y+x^3)/x$.
   This is a linear equation, and the general solution is
   $y = x^3/5 + k/x^2$, where k is an arbitrary constant.   In
   addition the system of equations has the solution $x = 0$,
   $y = Be^{-2t}$.   Any solution with its initial point on the
   y-axis ($x=0$) is given by the latter solution.   The
   trajectories corresponding to these solutions approach
   the origin as $t \to \infty$. The trajectory that passes

through the origin and
divides the family of
curves is given by $k = 0$,
namely $y = x^3/5$. This
trajectory corresponds to
the trajectory $y = 0$ for
the linear problem.
Several trajectories are
sketched in the figure.

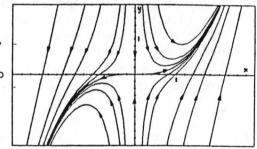

22a.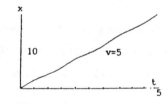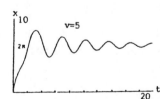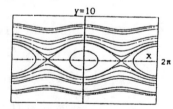

22b. From the graphs in part (a), we see that v_c is between $v = 2$
   and $v = 5$.   Using several values for v, we estimate $v_c \cong 4.00$.

23a.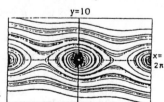

For $v = 2$, the motion is damped oscillatory about $x = 0$.
For $v = 5$, the pendulum swings all the way around once
and then is a damped oscillation about $x = 2\pi$ (after one

full rotation). For Prob. 22, this later case is not damped, so x continues to increase, as shown earlier.

27a. Setting c = 0 in Eq.(10) of Sect. 9.2 we obtain $mL^2 d^2\theta/dt^2 + mgL\sin\theta = 0$. Considering $d\theta/dt$ as a function of θ and using the chain rule we have

$$\frac{d}{dt}\left(\frac{d\theta}{dt}\right) = \frac{d}{d\theta}\left(\frac{d\theta}{dt}\right)\frac{d\theta}{dt} = \frac{1}{2}\frac{d}{d\theta}\left(\frac{d\theta}{dt}\right)^2 .$$ Thus

$(1/2)mL^2 d[(d\theta/dt)^2]/d\theta = -mgL\sin\theta$. Now integrate both sides from α to θ where $d\theta/dt = 0$ at $\theta = \alpha$: $(1/2)mL^2(d\theta/dt)^2 = mgL(\cos\theta - \cos\alpha)$. Thus $(d\theta/dt)^2 = (2g/L)(\cos\theta - \cos\alpha)$. Since we are releasing the pendulum with zero velocity from a positive angle α, the angle θ will initially be decreasing so $d\theta/dt < 0$. If we restrict attention to the range of θ from $\theta = \alpha$ to $\theta = 0$, we can assert $d\theta/dt = -\sqrt{2g/L}\,\sqrt{\cos\theta - \cos\alpha}$. Solving for dt gives $dt = -\sqrt{L/2g}\,d\theta/\sqrt{\cos\theta - \cos\alpha}$.

27b. Since there is no damping, the pendulum will swing from its initial angle α through 0 to $-\alpha$, then back through 0 again to the angle α in one period. It follows that $\theta(T/4) = 0$. Integrating the last equation and noting that as t goes from 0 to T/4, θ goes from α to 0 yields $T/4 = -\sqrt{L/2g}\int_\alpha^0 (1/\sqrt{\cos\theta - \cos\alpha})d\theta$.

28a. If $\dfrac{dx}{dt} = y$, then $\dfrac{d^2x}{dt^2} = \dfrac{dy}{dt} = -g(x) - c(x)y$.

28b. Under the given assumptions we have $g(x) = g(0) + g'(0)x + g''(\xi_1)x^2/2$ and $c(x) = c(0) + c'(\xi_2)x$, where $0 < \xi_1, \xi_2 < x$ and $g(0) = 0$. Hence

$\dfrac{dy}{dt} = -g'(0)x - c(0)y - [g''(\xi_1)x^2/2 - c'(\xi_2)xy]$ and thus the system can be written as

$$\frac{d}{dt}\begin{pmatrix} x \\ y \end{pmatrix} = \begin{pmatrix} 0 & 1 \\ -g'(0) & -c(0) \end{pmatrix}\begin{pmatrix} x \\ y \end{pmatrix} - \begin{pmatrix} 0 \\ -g''(\xi_1)x^2/2 - c'(\xi_2)xy \end{pmatrix},$$

from which the results follow.

Section 9.4, Page 525

3b. $x(1.5 - .5x - y) = 0$ and $y(2 - y - 1.125x) = 0$ yield $(0,0)$, $(0,2)$ and $(3,0)$ very easily. The fourth critical point is the intersection of .5x + y = 1.5 and 1.125x + y = 2, which is (.8,1.1).

3c. From Eq.(5) we get $\dfrac{d}{dt}\begin{pmatrix} u \\ v \end{pmatrix} = \begin{pmatrix} 1.5-x_0-y_0 & -x_0 \\ -1.125y_0 & 2-2y_0-1.125x_0 \end{pmatrix}\begin{pmatrix} u \\ v \end{pmatrix}$. For (0,0) we get u' = 1.5u and v' = 2v, so r = 3/2 and r = 2, and thus (0,0) is an unstable node. For (0,2) we have u' = -.5u and v' = -2.25u-2v, so r = -.5, -2 and thus (0,2) is an asymptotically stable node. For (3,0) we get u' = -1.5u-3v and v' = -1.375v, so r = -1.5, -1.375 and hence (3,0) is an symptotically stable node. For (.8,1.1) we have u' = -.4u -.8v and v' = -1.2375u - 1.1v which give r = -1.80475, .30475 and thus (.8,1.1) is an unstable saddle point.

3e.

3f. As in Ex.2, one species will die out, depending on the I.C. For an I.C. lying below the separatrix [not shown, but which is a curve starting at (0,0) and passing through (.8,1.1)], the species denoted by x will survive, while if the I.C. is above the separatrix the species denoted by y will survive.

5b. The critical points are found by setting dx/dt = 0 and dy/dt = 0 and thus we need to solve x(1 - x - y) = 0 and y(1.5 - y - x) = 0. The first yields x = 0 or y = 1 - x and the second yields y = 0 or y = 1.5 - x. Thus (0,0), (0,3/2) and (1,0) are the only critical points since the two straight lines do not intersect in the first quadrant (or anywhere in this case). This is an example of one of the cases shown in Figure 9.4.5 a or b.

5e.

5f. Only the species denoted by y will survive as t → ∞.

6b. The critical points are found by setting dx/dt = 0 and
     dy/dt = 0 and thus we need to solve x(1-x + y/2) = 0 and
     y(5/2 - 3y/2 + x/4) = 0. The first yields x = 0 or
     y = 2x - 2 and the second yields y = 0 or y = x/6 + 5/3.
     Thus we find the critical points (0,0), (1,0), (0,5/3)
     and (2,2). The last point is the intersection of the two
     straight lines, which will be used again in part d.

6c. For (0,0) the linearized system is $x' = x$ and $y' = 5y/2$,
     which has the eigenvalues $r_1 = 1$ and $r_2 = 5/2$. Thus the
     origin is an unstable node. For (2,2) we let
     x = u + 2 and y = v + 2 in the given system to find
     (since $x' = u'$ and $y' = v'$) that
     du/dt = (u+2)[1 - (u+2) + (v+2)/2] = (u+2)(-u+v/2) and
     dv/dt = (v+2)[5/2 - 3(v+2)/2 + (u+2)/4] = (v+2)(u/4 - 3v/2).
     Hence, keeping just the linear terms, the linearized

     equations are $\begin{pmatrix} u \\ v \end{pmatrix}' = \begin{pmatrix} -2 & 1 \\ 1/2 & -3 \end{pmatrix} \begin{pmatrix} u \\ v \end{pmatrix}$ which has the

     eigenvalues $r_{1,2} = (-5 \pm \sqrt{3})/2$. Since these are both
     negative we conclude that (2,2) is an asymptotically
     stable node. In a similar fashion for (1,0) we let x = u
     + 1 and y = v to obtain the linearized system
     $\begin{pmatrix} u \\ v \end{pmatrix}' = \begin{pmatrix} -1 & 1/2 \\ 0 & 11/4 \end{pmatrix} \begin{pmatrix} u \\ v \end{pmatrix}$. This has $r_1 = -1$ and $r_2 = 11/4$ as

     eigenvalues and thus (1,0) is an unstable saddle point.
     Likewise, for (0,5/3) we let x = u, y = v + 5/3 to find
     $\begin{pmatrix} u \\ v \end{pmatrix}' = \begin{pmatrix} 11/6 & 0 \\ 5/12 & -5/2 \end{pmatrix} \begin{pmatrix} u \\ v \end{pmatrix}$ as the corresponding linear system.

     Thus $r_1 = 11/6$ and $r_2 = -5/2$ and thus (0,5/3) is an
     unstable saddle point.

6d. To sketch the required trajectories, we must find the
     eigenvectors for each of the linearized systems and then
     analyze the behavior of the linear solution near the
     critical point. Using this approach we find that the

     solution near (0,0) has the form $\begin{pmatrix} x \\ y \end{pmatrix} = c_1 \begin{pmatrix} 1 \\ 0 \end{pmatrix} e^t +$

     $c_2 \begin{pmatrix} 0 \\ 1 \end{pmatrix} e^{5t/2}$ and thus the origin is approached only for

     large negative values of t. In this case e^t dominates
     $e^{5t/2}$ and hence in the neighborhood of the origin all
     trajectories are tangent to the x-axis except for one
     pair ($c_1 = 0$) that lies along the y-axis.

For (2,2) we find the eigenvector corresponding to
$r = (-5 + \sqrt{3})/2 = -1.63$ is given by $(1-\sqrt{3})\xi_1/2 + \xi_2 = 0$
and thus $\begin{pmatrix} 1 \\ (\sqrt{3}-1)/2 \end{pmatrix} = \begin{pmatrix} 1 \\ .37 \end{pmatrix}$ is one eigenvector. For
$r = (-5 - \sqrt{3})/2 = -3.37$ we have $(1 +\sqrt{3})\xi_1/2 + \xi_2 = 0$ and
thus $\begin{pmatrix} 1 \\ -(\sqrt{3}+1)/2 \end{pmatrix} = \begin{pmatrix} 1 \\ -1.37 \end{pmatrix}$ is the second eigenvector.
Hence the linearized solution is
$\begin{pmatrix} u \\ v \end{pmatrix} = c_1 \begin{pmatrix} 1 \\ .37 \end{pmatrix} e^{-1.63t} + c_2 \begin{pmatrix} 1 \\ -1.37 \end{pmatrix} e^{-3.37t}$. For large

positive values of t the first term is the dominant one
and thus we conclude that all trajectories but two
approach (2,2) tangent to the straight line with slope
.37. If $c_1 = 0$, we see that there are exactly two
($c_2 > 0$ and $c_2 < 0$) trajectories that lie on the straight
line with slope -1.37. In similar fashion, we find the
linearized solutions near (1,0) and (0,5/3) to be,
respectively,

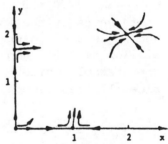

$\begin{pmatrix} u \\ v \end{pmatrix} = c_1 \begin{pmatrix} 1 \\ 0 \end{pmatrix} e^{-t} + c_2 \begin{pmatrix} 1 \\ 15/2 \end{pmatrix} e^{11t/4}$
and
$\begin{pmatrix} u \\ v \end{pmatrix} = c_1 \begin{pmatrix} 0 \\ 1 \end{pmatrix} e^{-5t/2} + c_2 \begin{pmatrix} 1 \\ 5/52 \end{pmatrix} e^{11t/6}$,

which, along with the above
analysis, yields the sketch shown.

6e. From the above sketch, it appears that $(x,y) \rightarrow (2,2)$ as
6f. $t \rightarrow \infty$ as long as (x,y) starts in the first quadrant.
To ascertain this, we need to prove that x and y cannot
become unbounded as $t \rightarrow \infty$. From the given system, we
can observe that, since $x > 0$ and $y > 0$, that dx/dt and
dy/dt have the same sign as the quantities $1 - x + y/2$
and $5/2 - 3y/2 + x/4$ respectively. If we set these
quantities equal to zero we get the straight lines
$y = 2x - 2$ and $y = x/6 + 5/3$, which divide the first
quadrant into the four sectors shown. The signs of
x' and y' are indicated,
from which it can be
concluded that x and y
must remain bounded [and
in fact approach (2,2)]
as $t \rightarrow \infty$. The discussion
leading up to Fig.9.4.4
is also useful here.

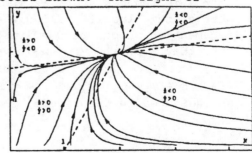

8a. Setting the right sides of the equations equal to zero
     gives the critical points $(0,0)$, $(0, \varepsilon_2/\sigma_2)$, $(\varepsilon_1/\sigma_1, 0)$
     [which are no fish, no bluegill or no redear respectively],
     and possibly
     $([\varepsilon_1\sigma_2 - \varepsilon_2\alpha_1]/[\sigma_1\sigma_2 - \alpha_1\alpha_2], [\varepsilon_2\sigma_1 - \varepsilon_1\alpha_2]/[\sigma_1\sigma_2 - \alpha_1\alpha_2])$.
     (The last point can be obtained from Eq.(36) also). The
     conditions $\varepsilon_2/\alpha_2 > \varepsilon_1/\sigma_1$ and $\varepsilon_2/\sigma_2 > \varepsilon_1/\alpha_1$ imply that
     $\varepsilon_2\sigma_1 - \varepsilon_1\alpha_2 > 0$ and $\varepsilon_1\sigma_2 - \varepsilon_2\alpha_1 < 0$. Thus either the x
     coordinate or the y coordinate of the last critical point
     is negative so both species cannot survive. The
     linearized system for $(0,0)$ is $x' = \varepsilon_1 x$ and $y' = \varepsilon_2 y$ and
     thus $(0,0)$ is an unstable equilibrium point. Similarly,
     it can be shown [by linearizing the given system or by
     using Eq.(35)] that $(0, \varepsilon_2/\sigma_2)$ is an asymptotically
     stable critical point and that $(\varepsilon_1\sigma_1, 0)$ is an unstable
     critical point. Thus the fish represented by y(redear)
     survive.

8b. The conditions $\varepsilon_1/\sigma_1 > \varepsilon_2/\alpha_2$ and $\varepsilon_1/\alpha_1 > \varepsilon_2/\sigma_2$ imply that
     $\varepsilon_2\sigma_1 - \varepsilon_1\alpha_2 < 0$ and $\varepsilon_1\sigma_2 - \varepsilon_2\alpha_1 > 0$ so again one of the
     coordinates of the fourth point in part (a) is negative
     and hence a mixed state is not possible. An analysis
     similar to that in part(a) shows that $(0,0)$ and $(0, \varepsilon_2/\sigma_2)$
     are unstable while $(\varepsilon_1/\sigma_1, 0)$ is stable. Hence the
     bluegill (represented by x) survive in this case.

9a. $x' = \varepsilon_1 x(1 - \dfrac{\sigma_1}{\varepsilon_1}x - \dfrac{\alpha_1}{\varepsilon_1}y) = \varepsilon_1 x(1 - \dfrac{1}{B}x - \dfrac{\gamma_1}{B}y)$

     $y' = \varepsilon_2 y(1 - \dfrac{\sigma_2}{\varepsilon_2}y - \dfrac{\alpha_2}{\varepsilon_2}x) = \varepsilon_2 y(1 - \dfrac{1}{R}y - \dfrac{\gamma_2}{R}x)$. The coexistence

     equilibrium point is given by $\dfrac{1}{B}x + \dfrac{\gamma_1}{B}y = 1$ and $\dfrac{\gamma_2}{R}x + \dfrac{1}{R}y = 1$.
     Solving these yields $X = (B - \gamma_1 R)/(1 - \gamma_1\gamma_2)$ and
     $Y = (R - \gamma_2 B)/(1 - \gamma_1\gamma_2)$.

9b. If B is reduced, it is clear from the answer to part(a)
     that X is reduced and Y is increased. To determine
     whether the bluegill will die out, we give an intuitive
     argument which can be confirmed by doing the analysis.
     Note that $B/\gamma_1 = \varepsilon_1/\alpha_1 > \varepsilon_2/\sigma_2 = R$ and
     $R/\gamma_2 = \varepsilon_2/\alpha_2 > \varepsilon_1/\sigma_1 = B$ so that the graph of the lines
     $1 - x/B - \gamma_1 y/B = 0$ and $1 - y/R - \gamma_2 x/R = 0$ must appear
     as indicated in the figure, where critical points are

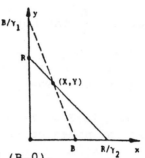

inidcated by heavy dots.
As B is decreased,
X decreases, Y increases
(as indicated above) and
the point of intersection
moves closer to (0,R). If
$B/\gamma_1 < R$ coexistence is not
possible, and the only
critical points are (0,0),(0,R)and (B,0).
It can be shown that (0,0) and (B,0) are unstable and
(0,R) is asymptotically stable. Hence we conlcude, when
coexistence is no longer possible, that $x \to 0$ and $y \to R$
and thus the bluegill population will die out.

13a. The nullclines are given by $x' = 0$ and $y' = 0$, or $y = 4x - x^2$
and $y = \dfrac{3\alpha}{2}$. Thus there are two critical points, where the

horizontal line $y = \dfrac{3\alpha}{2}$, for $\alpha < \dfrac{8}{3}$, intersects the parabola

$y = 4x - x^2$. As α increases the y value of the critical

points increases until $\alpha = \dfrac{8}{3}$.

13b. The critical points are determined by $4x - x^2 = \dfrac{3\alpha}{2}$ or

$x = 2 \pm \sqrt{4 - 3\alpha/2}$. Thus the critical points are
$(2 \pm \sqrt{4 - 3\alpha/2} , 3\alpha/2)$.

13c. For $\alpha = 2$ the critical points are (1,3) and (3,3). The
corresponding linear system has the coefficient matrix
$\begin{pmatrix} -4+2x_0 & 1 \\ 0 & -1 \end{pmatrix}$ and thus for (1,3) the eigenvalues are
$r_{1,2} = -1,-2$ which indicate (1,3) is an asymptotically stable
node. For (3,3), the eigenvalues are $r_{1,2} = -1,2$, so (3,3) is
an unstable saddle point.

13c.   13d.

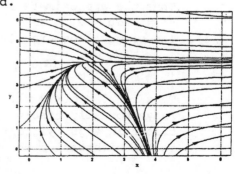

13d. For $\alpha = \dfrac{8}{3}$ the horizontal line $y = \dfrac{3\alpha}{2}$ is tangent to

$y = 4x - x^2$ and there is only one critical point $(2,4)$. Using the linear matrix of part (c) we find that $r_{1,2} = 0,-1$ are the eigenvalues. The phase portrait is shown.

13e. If $\alpha = 10/3$ we have $x' = -4x + y + x^2$ and $y' = 5 - y$, which has the phase portrait:

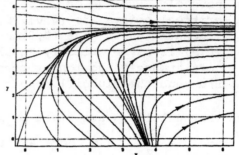

17a. Setting each equation equal to zero, we obtain $x = 0$ or $(4 - x - y) = 0$ and $y = 0$ or $(2 + 2\alpha - y - \alpha x) = 0$. Thus we have $(0,0)$, $(4,0)$, $(0,2 + 2\alpha)$, and the intersection of $x + y = 4$ and $\alpha x + y = 2 + 2\alpha$. If $\alpha \neq 1$, this yields $(2,2)$ as the fourth critical point. If $\alpha = 1$ there is a line of critical points, $x + y = 4$, since $x' = 0$ and $y' = 0$ for the entire line. This line, of course, includes $(2,2)$.

17b. For $\alpha = .75$ the linear system is $\begin{pmatrix} u \\ v \end{pmatrix}' = \begin{pmatrix} -2 & -2 \\ -1.5 & -2 \end{pmatrix}\begin{pmatrix} u \\ v \end{pmatrix}$, which

has the characteristic equation $r^2 + 4r + 1 = 0$ so that $r = -2 \pm \sqrt{3}$. Thus the critical point is an asymptotically

stable node. For $\alpha = 1.25$, we have $\begin{pmatrix} u \\ v \end{pmatrix}' = \begin{pmatrix} -2 & -2 \\ -2.5 & -2 \end{pmatrix}\begin{pmatrix} u \\ v \end{pmatrix}$, so

$r^2 + 4r -1 = 0$ and $r = -2 \pm \sqrt{5}$. Thus $(2,2)$ is an unstable saddle point.

17c. Letting $x = u+2$ and $y = v+2$ yields
$u' = (u+2)(4 - u-2 - v-2) = -2u - 2v - u^2 - uv$ and
$v' = (v+2)(2 + 2\alpha - v-2 - \alpha u - 2\alpha) = -2\alpha u - 2v - v^2 - \alpha uv$.
Thus the approximate linear system is $u' = -2u - 2v$ and
$v' = -2\alpha u - 2v$.

17d. The eigenvalues are given by
$\begin{vmatrix} -2-r & -2 \\ -2\alpha & -2-r \end{vmatrix} = r^2 + 4r + 4 - 4\alpha = 0$, or $r = -2 \pm 2\sqrt{\alpha}$.

Thus for $0 < \alpha < 1$ there are 2 negative real roots (asymptotially stable node) and for $\alpha > 1$ the real roots differ in sign, yielding an unstable saddle point. $\alpha = 1$ is the bifurcation point.

Section 9.5, Page 534

3b. We have x = 0 or (1 − .5x − .5y) = 0 and y = 0 or
(−.25 + .5x) = 0 and thus we have three critical points:
(0,0), (2,0) and (1/2,3/2).

3c. For (0,0) the linear system is dx/dt = x and
dy/dt = −.25y and hence $A = \begin{pmatrix} 1 & 0 \\ 0 & -1/4 \end{pmatrix}$ which has

eigenvalues $r_1 = 1$ and $r_2 = -1/4$ and corresponding

eigenvectors $\begin{pmatrix} 1 \\ 0 \end{pmatrix}$ and $\begin{pmatrix} 0 \\ 1 \end{pmatrix}$. Thus (0,0) is an unstable

saddle point.
For (2,0), we let x = 2 + u and y = v in the given

equations and obtain $\dfrac{du}{dt} = -(u+v) - \dfrac{1}{2}u(u+v)$ and

$\dfrac{dv}{dt} = \dfrac{3}{4}v + \dfrac{1}{2}uv$. The linear portion of this has matrix

$A = \begin{pmatrix} -1 & -1 \\ 0 & 3/4 \end{pmatrix}$, which has the eigenvalues $r_1 = -1$,

$r_2 = 3/4$ and corresponding eigenvectors $\begin{pmatrix} 1 \\ 0 \end{pmatrix}$ and $\begin{pmatrix} -4 \\ 7 \end{pmatrix}$.

Thus (2,0) is also an unstable saddle point.
For $\left(\dfrac{1}{2}, \dfrac{3}{2} \right)$ we let x = 1/2 + u and y = 3/2 + v in the

given equations, which yields $\dfrac{du}{dt} = -\dfrac{1}{4}u - \dfrac{1}{4}v$, $\dfrac{dv}{dt} = \dfrac{3}{4}u$

as the linear portion. Thus $A = \begin{pmatrix} -\dfrac{1}{4} & -\dfrac{1}{4} \\ \dfrac{3}{4} & 0 \end{pmatrix}$, which has

eigenvalues $r_{1,2} = (-1 \pm \sqrt{11}\,i)/8$. Thus $\left(\dfrac{1}{2}, \dfrac{3}{2} \right)$ is an

asymptotically stable spiral point since the eigenvalues
are complex with negative real part. Using
$r_1 = (-1 + \sqrt{11}\,i)/8$ we find that one eigenvector is
$\begin{pmatrix} -2 \\ 1 + \sqrt{11}\,i \end{pmatrix}$ and by Sect. 7.6 the second eigenvector is

the complex conjugate $\begin{pmatrix} -2 \\ 1 - \sqrt{11}\,i \end{pmatrix}$.

3e.

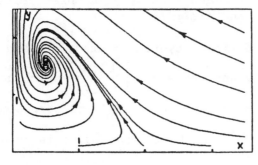

3f. For (x,y) above the line $x + y = 2$ we see that $x' < 0$ and thus x must remain bounded. For (x,y) to the right of $x = 1/2$, $y' > 0$ so it appears that y could grow large asymptotic to $x = $ constant. However, this implies a contradiction ($x = $ constant implies $x' = 0$, but as y gets larger, x' gets increasingly negative) and hence we conclude y must remain bounded and hence $(x,y) \to (1/2, 3/2)$ as $t \to \infty$, again assuming they start in the first quadrant.

7a. The amplitude ratio is $(cK/\gamma)/(\sqrt{ac}\,K/\alpha) = \alpha\sqrt{c}\,/\gamma\sqrt{a}$.

7b. From Eq.(2) $\alpha = .5$, $a = 1$, $\gamma = .25$ and $c = .75$, so the ratio is $.5\sqrt{.75}\,/.25\sqrt{1} = 2\sqrt{.75} = \sqrt{3} \cong 1.732$.

7c. A rough measurement of the amplitudes is $(6.2 - 1.2)/2 = 2.50$ and $(3.8 - .9)/2 = 1.45$ and thus the ratio is approximately 1.72. In this case the linear approximation is a good predictor.

11. The presence of a trapping company actually would require a modification of the equations, either by altering the coefficients or by including nonhomogeneous terms on the right sides of the D.E. The effects of indiscreminate trapping could decrease the populations of both rabbits and fox significantly or decrease the fox population which could possibly lead to a large increase in the rabbit population. Over the long run it makes sense for a trapping company to operate in such a way that a consistent supply of pelts is available and to disturb the predator-prey system as little as possible. Thus, the company should trap fox only when their population is increasing, trap rabbits only when their population is increasing, trap rabbits and fox only during the time

when both their populations are increasing, and trap neither during the time both their populations are decreasing. In this way the trapping company can have a moderating effect on the population fluctuations, keeping the trajectory close to the center.

13. The critical points of the system are the solutions of the algebraic equations $x(a - \sigma x - \alpha y) = 0$, and $y(-c + \gamma x) = 0$. the critical points are $x = 0$, $y = 0$; $x = a/\sigma$, $y = 0$; and $x = c/\gamma$, $y = a/\alpha - c\sigma/\alpha\gamma = \sigma A/\alpha$ where $A = a/\sigma - c/\gamma > 0$.

To study the critical point $(0,0)$ we discard the nonlinear terms in the system of D.E. to obtain the corresponding linear system $dx/dt = ax$, $dy/dt = -cy$. The characteristic equation is $r^2 - (a+c)r - ac = 0$ so $r_1 = a$, $r_2 = -c$. Thus the critical point $(0,0)$ is an unstable saddle point.

To study the critical point $(a/\sigma, 0)$ we let $x = (a/\sigma) + u$, $y = 0 + v$ and substitute in the D.E. to obtain the almost linear system $du/dt = -au - (a\alpha/\sigma)v - \sigma u^2 - \alpha uv$, $dv/dt = \gamma Av + \gamma uv$. The corresponding linear system is $du/dt = -au - (a\alpha/\sigma)v$, $dv/dt = \gamma Av$. The characteristic equation is $r^2 + (a - \gamma A)r - a\gamma A = 0$ so $r_1 = -a$, $r_2 = \gamma A$. Thus the critical point $(a/\sigma, 0)$ is an unstable saddle point.

To study the critical point $(c/\gamma, \sigma A/\alpha)$ we let $x = (c/\gamma) + u$, $y = (\sigma A/\alpha) + v$ and substitute in the D.E. to obtain the almost linear system

$$\frac{du}{dt} = -(c\sigma/\gamma)u - (ac/\gamma)v - \sigma u^2 - \alpha uv$$

$$\frac{dv}{dt} = (\sigma A\gamma/\alpha)u + \gamma uv$$

The corresponding linear system is $du/dt = -(c\sigma/\gamma)u - (\alpha c/\gamma)v$, $dv/dt = (\sigma A\gamma/\alpha)u$. The characteristic equation is $r^2 + (c\sigma/\gamma)r + c\sigma A = 0$, so $r_1, r_2 = [-(c\sigma/\gamma) \pm \sqrt{(c\sigma/\gamma)^2 - 4c\sigma A}]/2$. Thus, depending on the sign of the discriminant we have that $(c/\gamma, \sigma A/\alpha)$ is either an asymptotically stable spiral point or an asymptotically stable node. Thus for nonzero initial data $(x,y) \to (c/\gamma, \sigma A/\alpha)$ as $t \to \infty$.

Section 9.6, Page 544

1. Assuming that $V(x,y) = ax^2 + cy^2$ we find $V_x(x,y) = 2ax$, $V_y = 2cy$ and thus Eq.(7) yields

$$\dot{V}(x,y) = 2ax(-x^3 + xy^2) + 2cy(-2x^2y - y^3)$$
$$= -[2ax^4 + 2(2c-a)x^2y^2 + 2cy^4].$$ If we choose a

and c to be any positive real numbers with 2c > a, then $\dot{V}$
is a negative definite. By definition V is positive
definite, so by Theorem 9.6.1 the origin is an
asymptotically stable critical point.

3. Assuming the same form for V(x,y) as in Prob. 1, we have

    $$\dot{V}(x,y) = 2ax(-x^3 + 2y^3) + 2cy(-2xy^2) = -2ax^4 + 4(a-c)xy^3.$$

    If we choose a = c > 0, then $\dot{V}(x,y) = -2ax^4 \le 0$ in any

    neighborhood containing the origin and thus $\dot{V}$ is negative
    semidefinite and V is positive definite. Theorem 9.6.1
    then concludes that the origin is a stable critical
    point. Note that the origin may still be asymptotically
    stable, however, the V(x,y) used here is not sufficient
    to prove that.

6a. The correct system is dx/dt = y and dy/dt = -g(x). Since
    g(0) = 0, we conclude that (0,0) is a critical point.

6b. From the given conditions, the graph of g must be
    positive for 0 < x < k and negative for -k < x < 0. Thus
    if   0 < x < k   then $\int_0^x g(s)ds > 0,$

    if   -k < x < 0   then $\int_0^x g(s)ds = -\int_x^0 g(s)ds > 0.$

    Since V(0,0) = 0 it follows that $V(x,y) = y^2/2 + \int_0^x g(s)ds$

    is positive definite for -k < x < k, $-\infty < y < \infty$. Next,

    we have $\dot{V}(x,y) = V_x\dfrac{dx}{dt} + V_y\dfrac{dy}{dt} = g(x)y + y[-g(x)] = 0.$

    Since $\dot{V}(x,y)$ is never positive, we may conclude that it
    is negative semidefinite and hence by Theorem 9.6.1 (0,0)
    is at least a stable critical point.

7b. V is positive definite by Theorem 9.6.4. Since
    $V_x(x,y) = 2x$, $V_y(x,y) = 2y$, we obtain

    $$\dot{V}(x,y) = 2xy - 2y^2 - 2y\sin x = 2y[-y + (x - \sin x)].$$ If

    x < 0, then $\dot{V}(x,y) < 0$ for all y > 0. If x > 0, choose y

    so that $0 < y < x - \sin x$. Then $\dot{V}(x,y) > 0$. Hence V is
    not a Liapunov function.

7c. Since V(0,0) = 0, $1 - \cos x > 0$ for $0 < |x| < 2\pi$ and $y^2 > 0$
    for $y \ne 0$, it follows that V(x,y) is positive definite

in a neighborhood of the origin. Next $V_x(x,y) = \sin x$,

$V_y(x,y) = y$, so $\dot{V}(x,y) = (\sin x)(y) + y(-y - \sin x) = -y^2$.

Hence $\dot{V}$ is negative semidefinite and $(0,0)$ is a stable
critical point by Theorem 9.6.1.

7d. $V(x,y) = (x+y)^2/2 + x^2 + y^2/2 = 3x^2/2 + xy + y^2$ is
     positive definite by Theorem 9.6.4. Next
     $V_x(x,y) = 3x + y$, $V_y(x,y) = x + 2y$ so

$$\dot{V}(x,y) = (3x+y)y - (x+2y)(y+\sin x)$$
$$= 2xy - y^2 - (x+2y)\sin x$$
$$= 2xy - y^2 - (x+2y)(x - \alpha x^3/6) \quad \text{[from the hint]}$$
$$= -x^2 - y^2 + \alpha(x+2y)x^3/6$$
$$= -r^2 + \alpha r^4(\cos\theta + 2\sin\theta)(\cos^3\theta)/6$$
$$< -r^2 + r^4/2 = -r^2(1-r^2/2).$$ Thus $\dot{V}$ is negative

definite for $r < \sqrt{2}$. From Theorem 9.6.1 it follows
that the origin is an asymptotically stable critical point.

8.   Let $x = u$ and $y = du/dt$ to obtain the system $dx/dt = y$
     and $dy/dt = -c(x)y - g(x)$. Now consider
     $V(x,y) = y^2/2 + \int_0^x g(s)ds$, which yields

$$\dot{V} = g(x)y + y[-c(x)y - g(x)] = -y^2 c(x),$$ which is
negative semidefinite.

10b. Since $V_x(x,y) = 2Ax + By$, $V_y(x,y) = Bx + 2Cy$, we have

$$\dot{V}(x,y) = (2Ax + By)(a_{11}x + a_{12}y) + (Bx + 2Cy)(a_{21}x + a_{22}y)$$
$$= (2Aa_{11} + Ba_{21})x^2 + [2(Aa_{12} + Ca_{21}) + B(a_{11}+a_{22})]xy$$
$$+ (2Ca_{22} + Ba_{12})y^2.$$

We choose A, B, and C so that
$2Aa_{11} + Ba_{21} = -1$, $2(Aa_{12} + Ca_{21}) + B(a_{11}+a_{22}) = 0$, and
$2Ca_{22} + Ba_{12} = -1$. The first and third equations give us A
and C in terms of B, respectively. We substitute in the
second equation to find B and then calculate A and C. The
result is given in the text.

10c. Since $a_{11}a_{22} - a_{12}a_{21} > 0$ and $a_{11} + a_{22} < 0$, we see that
     $\Delta < 0$ and so $A > 0$. Using the expressions for A, B, and
     C found in part (b) we obtain
$$(4AC-B^2)\Delta^2 = [a_{21}^2+a_{22}^2 + (a_{11}a_{22}-a_{12}a_{21})][a_{11}^2+a_{12}^2 + (a_{11}a_{22}-a_{12}a_{21})]$$
$$- (a_{12}a_{22}+a_{11}a_{21})^2$$
$$= (a_{11}^2+a_{12}^2+a_{21}^2+a_{22}^2)(a_{11}a_{22}-a_{12}a_{21}) + (a_{11}^2+a_{12}^2)(a_{21}^2+a_{22}^2)$$
$$+ (a_{11}a_{22}-a_{12}a_{21})^2 - (a_{12}a_{22}+a_{11}a_{21})^2$$
$$= (a_{11}^2+a_{12}^2+a_{21}^2+a_{22}^2)(a_{11}a_{22}-a_{12}a_{21}) + 2(a_{11}a_{22}-a_{12}a_{21})^2.$$
Since $a_{11}a_{22} - a_{12}a_{21} > 0$ it follows that $4AC - B^2 > 0$.

11a. For $V(x,y) = Ax^2 + Bxy + Cy^2$ we have

$$\dot{V} = (2Ax + By)(a_{11}x+a_{12}y + F_1(x,y))+(Bx+2Cy)(a_{21}x+a_{22}y + G_1(x,y))$$
$$= (2Ax+By)(a_{11}x+a_{12}y)+(Bx+2Cy)(a_{21}x+a_{22}y)$$
$$+ (2Ax+By)F_1(x,y) + (Bx+2Cy)G_1(x,y)$$
$$= -x^2-y^2 + (2Ax+By)F_1(x,y) + (Bx+2Cy)G_1(x,y), \text{ if A,B and C are}$$

chosen as in Prob. 10.

11b. Substituting $x = r\cos\theta$, $y = r\sin\theta$ we find that

$$\dot{V}[x(r,\theta),y(r,\theta)] = -r^2+r(2A\cos\theta+B\sin\theta)F_1[x(r,\theta),y(r,\theta)]$$
$$+ r(B\cos\theta + 2C\sin\theta)G_1[x(r,\theta),y(r,\theta)].$$

Now we make use of the facts that: (1) there exists an M
such that $|2A| \le M$, $|B| \le M$, and $|2C| \le M$; and (2) given
any $\varepsilon > 0$ there exists a circle $r = R$ such that
$|F_1(x,y)| < \varepsilon r$ and $|G_1(x,y)| < \varepsilon r$ for $0 \le r < R$.   We have
$|2A\cos\theta + B\sin\theta| \le 2M$ and $|B\cos\theta + 2C\sin\theta| \le 2M$.   Hence

$$\dot{V}[x(r,\theta),y(r,\theta)] \le -r^2 + 2Mr(\varepsilon r)+2Mr(\varepsilon r) = -r^2(1 - 4M\varepsilon).$$

If we choose $\varepsilon = M/8$ we obtain $\dot{V}[x(r,\theta),y(r,\theta)] \le -r^2/2$

for $0 \le r < R$.   Hence $\dot{V}$ is negative definite in $0 \le r < R$
and from Prob.10c V is positive definite and thus V is a
Liapunov function for the almost linear system.

Section 9.7, Page 555

1.   Note that $r = 1$, $\theta = t + t_0$ satisfy the two equations for
all t and is thus a periodic solution. If $r < 1$, then
$dr/dt > 0$, and the direction of motion on a trajectory is
outward. If $r > 1$, then the direction of motion is
inward. It follows that the periodic solution $r = 1$,
$\theta = t + t_0$ is an asymptotically stable limit cycle.

2. $r = 1$, $\theta = -t + t_0$ is a periodic solution. If $r < 1$,
then $dr/dt > 0$, and the direction of motion on a
trajectory is outward. If $r > 1$, the $dr/dt > 0$, and the
direction of motion is still outward. It follows that
the solution $r = 1$, $\theta = -t + t_0$ is a semistable limit
cycle.

4. $r = 1$, $\theta = -t + t_0$ and $r = 2$, $\theta = -t + t_0$ are periodic
solutions. If $r < 1$, then $dr/dt < 0$, and the direction
of motion on a trajectory is inward. If $1 < r < 2$, then
$dr/dt > 0$, and the direction of motion is outward.
Similarly, if $r > 2$, the direction of motion is inward.
It follows that the periodic solution $r = 1$,
$\theta = -t + t_0$ is unstable and the periodic solution $r = 2$,

$\theta = -t + t_0$ is an asymptotically stable limit cycle.

7.   Differentiating x and y with respect to t we find that
$dx/dt = (dr/dt)\cos\theta - (r\sin\theta)d\theta/dt$ and
$dy/dt = (dr/dt)\sin\theta + (r\cos\theta)d\theta/dt$. Hence
$$ydx/dt - xdy/dt = (r\sin\theta\cos\theta)dr/dt - (r^2\sin^2\theta)d\theta/dt -$$
$$(r\cos\theta\sin\theta)dr/dt - (r^2\cos\theta)d\theta/dt$$
$$= -r^2 d\theta/dt.$$

8a. Multiplying the first equation by x and the second by y
and adding yields $xdx/dt + ydy/dt = (x^2+y^2)f(r)/r$, or
$rdr/dt = rf(r)$, as in the derivation of Eq.(8), and thus
$dr/dt = f(r)$. To obtain an equation for θ multiply the
first equation by y, the second by x and substract to
obtain $ydx/dt - xdy/dt = -x^2-y^2$, or $-r^2 d\theta/dt = -r^2$, using
the results of Prob. 7. Thus $d\theta/dt = 1$. It follows that
periodic solutions are given by $r = c$, $\theta = t + t_0$ where
$f(c) = 0$. Since $\theta = t + t_0$, the motion is
counterclockwise.

8b. First note that $f(r) = r(r-2)^2(r-3)(r-1)$. Thus $r = 1$,
$\theta = t + t_0$; $r = 2$, $\theta = t + t_0$; and $r = 3$, $\theta = t + t_0$ are
periodic solutions. If $r < 1$, then $dr/dt > 0$, and the
direction of motion on a trajectory is outward. If
$1 < r < 2$, then $dr/dt < 0$ and the direction of motion is
inward. Thus the periodic solution $r = 1$, $\theta = t + t_0$ is
an asymptotically stable limit cycle. If $2 < r < 3$, then
$dr/dt < 0$, and the direction of motion is inward. Thus
the periodic solution $r = 2$, $\theta = t + t_0$ is a semistable
limit cycle. If $r > 3$, then $dr/dt > 0$, and the direction
of motion is outward. Thus the periodic solution $r = 3$,
$\theta = t + t_0$ is unstable.

9.   Setting $x = r\cos\theta$, $y = r\sin\theta$ and using the techniques of
Prob. 8 the equations transform to $dr/dt = r^2 - 2$,
$d\theta/dt = -1$. This system has a periodic solution $r = \sqrt{2}$,
$\theta = -t + t_0$. If $r < \sqrt{2}$, then $dr/dt < 0$, and the
direction of motion along a trajectory is inward. If
$r > \sqrt{2}$, then $dr/dt > 0$, and the direction of motion is
outward. Thus the periodic solution $r = \sqrt{2}$, $\theta = -t + t_0$
is unstable.

11. If $F(x,y) = x+y+x^3-y^2$, $G(x,y) = -x+2y+x^2y+y^3/3$, then
$F_x(x,y) + G_y(x,y) = 1+3x^2+2+x^2+y^2 = 3+4x^2+y^2$. Since the
conditions of Theorem 9.7.2 are satisfied for all x and
y, and since $F_x + G_y > 0$ for all x and y, it follows that
the system has no periodic nonconstant solution.

13. Since $x = \phi(t)$, $y = \psi(t)$ is a solution of Eqs.(15), we
     have $d\phi/dt = F[\phi(t),\psi(t)]$, $d\psi/dt = G[\phi(t),\psi(t)]$. Hence
     on the curve C,
     $F(x,y)dy - G(x,y)dx = \phi'(t)\psi'(t)dt - \psi'(t)\phi'(t)dt = 0$. It
     follows that the line integral around C is zero.
     However, if $F_x + G_y$ has the same sign throughout D, then
     the double integral cannot be zero. This gives a
     contradiction. Thus either the solution of Eqs.(15) is
     not periodic or if it is, it cannot lie entirely in D.

18a. Setting $x' = 0$ and solving for y yields $y = x^3/3 - x + k$.
     Substituting this into $y'= 0$ then gives
     $W(x) = x +.8(x^3/3 - x + k) -.7 = .8x^3/3 + .2x + (.8k -.7) = 0$.
     Since $W'(x) = .8x^2 + .2$ is never zero, we conclude that W
     always has a positive slope and thus the cubic equation
     crosses the x axis only once for all values of k.

18b. Using an equation solver on the expression in part (a) we
     obtain $x = 1.1994$, $y = -.62426$ for $k = 0$ and $x = .80485$,
     $y = -.13106$ for $k = .5$. To determine the type of critical
     points these are, we use Eq.(13) of Section 9.3 to find the
     linear coefficient matrix to be $\mathbf{A} = \begin{pmatrix} 3(1-x_c^2) & 3 \\ -1/3 & -.8/3 \end{pmatrix}$, where x_c
     is the critical point. For $x_c = 1.1994$ we obtain complex
     conjugate eigenvalues with a negative real part, and
     therefore $k = 0$ yields an asymptotically stable spiral
     point. For $x_c = .80485$ the eigenvalues are also complex
     conjugates, but with positive real parts, so $k = .5$ yields
     an unstable spiral point.

18c. Letting $k = .1, .2, .3, .4$ in the cubic equation of part (a)
     and finding the corresponding eigenvalues from the matrix in
     part (b), we find that the real part of the eigenvalues
     changes sign between $k = .3$ and $k = .4$. Continuing to
     iterate in this fashion we find that for $k = .3464$ that the
     real part of the eigenvalue is $-.0002$ while for $k = .3465$
     the real part is $.00005$, which indicates $k_0 = .3465$ is the
     critical point for which the system changes from stable to
     unstable.

18e. Again, iterating as in part (c), we find for $k = 1.403$ that
     $x_c = -.9541$ and for $k = 1.404$ that $x_c = -.9549$. Substituting
     these values into the coefficient matrix of part (b) and
     finding the eigenvalues we find that the real part changes
     sign beween $k = 1.403$ and $k = 1.404$. Thus the critical point
     again becomes asymptotically stable.

Section 9.8, Page 565

1a. From Eq.(6), $\lambda_1 = -8/3$ is clearly one eigenvalue and the other two may be found from $\lambda^2 + 11\lambda - 10(r-1) = 0$ using the quadratic formula.

1b. For $\lambda = \lambda_1$ we have

$$\begin{pmatrix} -10+8/3 & 10 & 0 \\ r & -1+8/3 & 0 \\ 0 & 0 & 0 \end{pmatrix}\begin{pmatrix} \xi_1 \\ \xi_2 \\ \xi_3 \end{pmatrix} = \begin{pmatrix} 0 \\ 0 \\ 0 \end{pmatrix}, \text{ which requires } \xi_1 = \xi_2 = 0$$

and ξ_3 arbitrary and thus $\xi^{(1)} = (0,0,1)^T$.

For $\lambda = \lambda_3 = (-11 + \alpha)/2$, where $\alpha = \sqrt{81+40r}$, we have

$$\begin{pmatrix} -10+(11-\alpha)/2 & 10 & 0 \\ r & -1+(11-\alpha)/2 & 0 \\ 0 & 0 & -8/3+(11-\alpha)/2 \end{pmatrix}\begin{pmatrix} \xi_1 \\ \xi_2 \\ \xi_3 \end{pmatrix} = \begin{pmatrix} 0 \\ 0 \\ 0 \end{pmatrix}.$$

The last line implies $\xi_3 = 0$ and multiplying the first line by

$(-9+\alpha)/2$ we obtain $\begin{pmatrix} (81-\alpha^2)/4 & 10(-9+\alpha)/2 \\ r & (9-\alpha)/2 \end{pmatrix}\begin{pmatrix} \xi_1 \\ \xi_2 \end{pmatrix} = \begin{pmatrix} 0 \\ 0 \end{pmatrix}.$

Substituting $\alpha^2 = 81+40r$ we have

$$\begin{pmatrix} -10r & -10(9-\alpha)/2 \\ r & (9-\alpha)/2 \end{pmatrix}\begin{pmatrix} \xi_1 \\ \xi_2 \end{pmatrix} = \begin{pmatrix} 0 \\ 0 \end{pmatrix}. \text{ Thus } \xi^{(3)} = \begin{pmatrix} 9 - \sqrt{81+40r} \\ -2r \\ 0 \end{pmatrix},$$

which is proportional to the answer given in the text. Similar calculations give $\xi^{(2)}$.

1c. Simply substitute $r = 28$ into the answers in parts (a) and (b).

2a. The calculations are somewhat simplified if you let $x = \beta + u$, $y = \beta + v$, and $z = (r-1)+w$, where $\beta = \sqrt{8(r-1)/3}$. An alternate approach is to extend Eq.(13) of Section 9.3, which is:

$$\begin{pmatrix} u \\ v \\ w \end{pmatrix}' = \begin{pmatrix} F_x & F_y & F_z \\ G_x & G_y & G_z \\ H_x & H_y & H_z \end{pmatrix}_{(x_0,y_0,z_0)} \begin{pmatrix} u \\ v \\ w \end{pmatrix}.$$

In this example $F = -10x + 10y$, $G = rx - y - xz$ and $H = -8z/3 + xy$ and thus

$$\begin{pmatrix} u \\ v \\ w \end{pmatrix}' = \begin{pmatrix} -10 & 10 & 0 \\ r-z_0 & -1 & -x_0 \\ y_0 & x_0 & -8/3 \end{pmatrix} \begin{pmatrix} u \\ v \\ w \end{pmatrix}, \text{ which is Eq.(8) for } P_2 \text{ since}$$

$x_0 = y_0 = \sqrt{8(r-1)/3}$, $z_0 = r-1$.

2b. Eq.(9) is found by evaluating $\begin{vmatrix} -10-\lambda & 10 & 0 \\ 1 & -1-\lambda & -\beta \\ \beta & \beta & -8/3-\lambda \end{vmatrix} = 0.$

2c. If $r = 28$, then Eq.(9) is $3\lambda^3 + 41\lambda^2 + 304\lambda + 4320 = 0$, which has the roots -13.8546 and $.093956 \pm 10.1945i$.

3b. If r_1, r_2, r_3 are the three roots of a cubic polynomial, then the polynomial can be factored as $(x-r_1)(x-r_2)(x-r_3)$. Expanding this and equating to the given polynomial we have $A = -(r_1+r_2+r_3)$, $B = r_1r_2 + r_1r_3 + r_2r_3$ and $C = -r_1r_2r_3$. We are interested in the case when the real part of the complex conjugate roots changes sign. Thus let $r_2 = \alpha+i\beta$ and $r_3 = \alpha-i\beta$, which yields $A = -(r_1+2\alpha)$, $B = 2\alpha r_1 + \alpha^2 + \beta^2$ and $C = -r_1(\alpha^2+\beta^2)$. Hence, if $AB = C$, we have $-(r_1+2\alpha)(2\alpha r_1+\alpha^2+\beta^2) = -r_1(\alpha^2+\beta^2)$ or $-2\alpha[r_1^2 + 2\alpha r_1 + (\alpha^2+\beta^2)] = 0$ or $-2\alpha[(r_1+\alpha)^2+\beta^2] = 0$. Since the square bracket term is positive, we conclude that if $AB = C$, then $\alpha = 0$. That is, the conjugate complex roots are pure imaginary. Note that the converse is also true. That is, if the conjugate complex roots are pure imaginary then $AB = C$.

3c. Comparing Eq.(9) to that of part (b), we have $A = 41/3$, $B = 8(r+10)/3$ and $C = 160(r-1)/3$. Thus $AB = C$ yields $r = 470/19$.

4.   We have $\dot{V} = 2x[\sigma(-x+y)] + 2\sigma y[rx-y-xz] + 2\sigma z[-bz+xy]$
$= -2\sigma x^2 + 2\sigma xy + 2\sigma rxy - 2\sigma y^2 - 2\sigma bz^2$
$= 2\sigma\{-[x^2-(r+1)xy+y^2]-bz^2\}$. For $r < 1$, the
term in the square brackets remains positive for all

values of x and y, by Theorem 9.6.4, and thus $\dot{V}$ is
negative definite. Thus, by the extension of Theorem
9.6.1 to three equations, we conclude that the origin is
an asymptotically stable critical point.

5a. $V = rx^2 + \sigma y^2 + \sigma(z-2r)^2 = c > 0$ yields

$$\frac{dv}{dt} = 2rx[\sigma(-x+y)] + 2\sigma y(rx-y-xz) + 2\sigma(z-2r)(-bz+xy). \text{ Thus}$$

$$\dot{V} = -2\sigma[rx^2+y^2 + b(z^2 - 2rz)] = -2\sigma[rx^2 + y^2 + b(z-r)^2 - br^2].$$

5b. From the proof of Theorem 9.6.1, we find that we need to

show that $\dot{V}$, as found in part (a), is always negative as
it crosses $V(x,y,z) = c$. (Actually, we need to use the
extension of Theorem 9.6.1 to three equations, but the
proof is very similar using the vector calculus
approach.) From part (a) we see that

$\dot{V} < 0$ if $rx^2 + y^2 + b(z-r)^2 > br^2$, which holds if (x,y,z)

lies outside the ellipsoid $\dfrac{x^2}{br} + \dfrac{y^2}{br^2} + \dfrac{(z-r)^2}{r^2} = 1$, Eq.(i).

Thus we need to choose c such that $V = c$ lies outside
Eq.(i). Writing $V = c$ in the form of Eq.(i) we obtain the

ellipsoid $\dfrac{x^2}{c/r} + \dfrac{y^2}{c/\sigma} + \dfrac{(z-2r)^2}{c/\sigma} = 1$, Eq.(ii). Now let

$M = \max(\sqrt{br}, r\sqrt{b}, r)$, then the ellipsoid (i) is

contained inside the sphere S1: $\dfrac{x^2}{M^2} + \dfrac{y^2}{M^2} + \dfrac{(z-r)^2}{M^2} = 1.$

Let S2 be a sphere centered at $(0,0,2r)$ with radius

$M+r$: $\dfrac{x^2}{(M+r)^2} + \dfrac{y^2}{(M+r)^2} + \dfrac{(z-2r)}{(M+r^2)} = 1$, then S1 is contained

in S2. Thus, if we choose c, in Eq.(ii), such that

$\dfrac{c}{r} > (M+r)^2$ and $\dfrac{c}{\sigma} > (M+r)^2$, then $\dot{V} < 0$ as the trajectory

crosses $V(x,y,z) = c$. Note that this is a sufficient
condition and there may be many other "better" choices
using different techniques.

8b. Several cases are shown. Results may vary, particularly
for $r = 24$, due to the closeness of r to $r_3 \cong 24.06$.

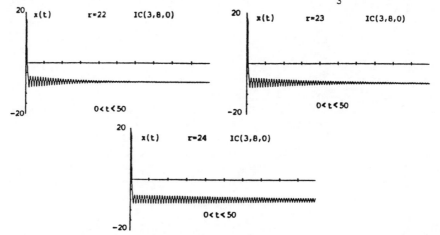

8b.

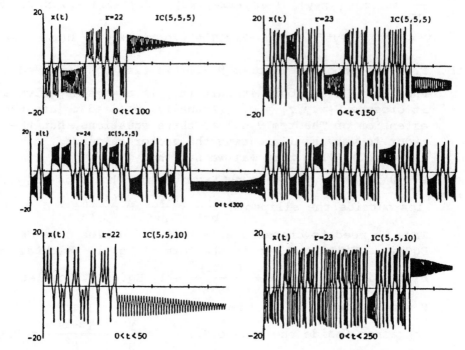

CHAPTER 10

2. $y(x) = c_1\cos\sqrt{2}\,x + c_2\sin\sqrt{2}\,x$ is the general solution of the
    D.E. Thus $y'(x) = -\sqrt{2}\,c_1\sin\sqrt{2}\,x + \sqrt{2}\,c_2\cos\sqrt{2}\,x$ and hence
    $y'(0) = \sqrt{2}\,c_2 = 1$, which gives $c_2 = 1/\sqrt{2}$. Now,
    $y'(\pi) = -\sqrt{2}\,c_1\sin\sqrt{2}\,\pi + \cos\sqrt{2}\,\pi = 0$ then yields
    $c_1 = \dfrac{\cos\sqrt{2}\,\pi}{\sqrt{2}\,\sin\sqrt{2}\,\pi} = \cot\sqrt{2}\,\pi/\sqrt{2}$. Thus the desired solution is
    $y = (\cot\sqrt{2}\,\pi\cos\sqrt{2}\,x + \sin\sqrt{2}\,x)/\sqrt{2}$.

3. We have $y(x) = c_1\cos x + c_2\sin x$ as the general solution and
    hence $y(0) = c_1 = 0$ and $y(L) = c_2\sin L = 0$. If $\sin L \neq 0$, then
    $c_2 = 0$ and $y(x) = 0$ is the only solution. If $\sin L = 0$, then
    $y(x) = c_2\sin x$ is a solution for arbitrary c_2.

7. $y(x) = c_1\cos 2x + c_2\sin 2x$ is the solution of the related
    homogeneous equation and $y_p(x) = \dfrac{1}{3}\cos x$ is a particular
    solution (using undetermined coefficients), yielding
    $y(x) = c_1\cos 2x + c_2\sin 2x + \dfrac{1}{3}\cos x$ as the general solution of
    the D.E. Thus $y(0) = c_1 + \dfrac{1}{3} = 0$ and $y(\pi) = c_1 - \dfrac{1}{3} = 0$ and
    hence there is no solution since there is no value of c_1 that
    will satisfy both boundary conditions.

12. This is an Euler Equation (Sect.5.5), so solutions of the
    homogeneous equation are of the form $y = x^\alpha$. Substituting this
    into the D.E. yields $(\alpha+1)^2 = 0$ and thus $y_c = (c_1 + c_2\ln x)x^{-1}$,
    from Theorem 5.5.1. For the particular solution we assume
    $y_p = Ax^2$, which upon substitution yields $A = 1/9$. Hence the
    general solution is $y(x) = (c_1 + c_2\ln x)x^{-1} + x^2/9$, so
    $y(1) = c_1 + 1/9 = 0$ and $y(e) = (c_1 + c_2)e^{-1} + e^2/9 = 0$. Solving
    for c_1 and c_2 yields the desired solution.

14. If $\lambda < 0$ set $\lambda = -\mu^2$, then the general solution of the D.E. is
    $y = c_1\sinh\mu x + c_2\cosh\mu x$. The two B.C. require that $c_2 = 0$ and
    $c_1 = 0$, so $\lambda < 0$ is not an eigenvalue. If $\lambda = 0$, the general
    solution of the D.E. is $y = c_1 + c_2 x$. The B.C. require that
    $c_1 = 0$, $c_2 = 0$, so again $\lambda = 0$ is not an eigenvalue.
    If $\lambda > 0$, the general solution of the D.E. is

$y = c_1\sin\sqrt{\lambda}\,x + c_2\cos\sqrt{\lambda}\,x$. The B.C. require that $c_2 = 0$ and $\sqrt{\lambda}\,c_1\cos\sqrt{\lambda}\,\pi = 0$. The second condition is satisfied for $\lambda \neq 0$ and $c_1 \neq 0$ if $\sqrt{\lambda}\,\pi = (2n-1)\pi/2$, $n = 1,2,\ldots$. Thus the eigenvalues are $\lambda_n = (2n-1)^2/4$, $n = 1,2,3\ldots$ with the corresponding eigenfunctions $y_n(x) = \sin[(2n-1)x/2]$, $n = 1,2,3\ldots$.

18. For $\lambda < 0$ there are no eigenvalues, using the same steps as in Prob. 14. For $\lambda = 0$ we have $y(x) = c_1 + c_2x$, so $y'(0) = c_2 = 0$ and $y'(L) = c_2 = 0$, and thus $\lambda = 0$ is an eigenvalue, with $y_0(x) = 1$ as the eigenfunction. For $\lambda > 0$ we again have $y(x) = c_1\sin\sqrt{\lambda}\,x + c_2\cos\sqrt{\lambda}\,x$, so $y'(0) = \sqrt{\lambda}\,c_1 = 0$ (so $c_1 = 0$) and $y'(L) = -c_2\sqrt{\lambda}\,\sin\sqrt{\lambda}\,L = 0$. We know $\lambda > 0$, in this case, so the eigenvalues are given by $\sin\sqrt{\lambda}\,L = 0$ or $\sqrt{\lambda}\,L = n\pi$. Thus $\lambda_n = (n\pi/L)^2$ and $y_n(x) = \cos(n\pi x/L)$ for $n = 1,2,3\ldots$.

21a. Multiplying by r^2 we obtain $r^2w'' + rw' = -\dfrac{Gr^2}{\mu}$, which is an Euler Eq. that has repeated roots given by $\alpha^2 = 0$. Thus, as in Prob. 12, we have $w(r) = (c_1 + c_2\ln r) - Gr^2/4\mu$. $w(r)$ bounded for $0 < r < R$ implies $c_2 = 0$ and $w(R) = 0$ implies $c_1 = GR^2/4\mu$, so $w(r) = G(R^2 - r^2)/4\mu$.

21b. Intergrating over a cross section, in polar coordinates, gives the integral $\int_0^{2\pi}\int_0^{R}w(r)\,r\,dr\,d\theta$, which when evaluated yields $Q = \pi R^4 G/8\mu$.

21c. Setting $R = .75$ we find $(.75)^4 = .3164$, which is the desired reduction since all other values are constant.

Section 10.2, Page 585

3. We look for values of T for which $\sinh 2(x+T) = \sinh 2x$ for all x. Expanding the left side of this equation gives $\sinh 2x\cosh 2T + \cosh 2x\sinh 2T = \sinh 2x$, which will be satisfied for all x if we can choose T so that $\cosh 2T = 1$ and $\sinh 2T = 0$. The only value of T satisfying these two constraints is $T = 0$. Since T is not positive we conclude that the function $\sinh 2x$ is not periodic.

5. We look for values of T for which $\tan\pi(x+T) = \tan\pi x$. Expanding the left side gives $\tan\pi(x+T) = (\tan\pi x + \tan\pi T)/(1-\tan\pi x\tan\pi T)$ which is equal

to tanπx only for tanπT = 0. The only positive solutions
of this last equation are T = 1,2,3... and hence tanπx is
periodic with fundamental period T = 1.

7.   To start, let n = 0, then $f(x) = \begin{cases} 0 & -1 \leq x < 0 \\ 1 & 0 \leq x < 1 \end{cases}$; for n = 1,

$f(x) = \begin{cases} 0 & 1 \leq x < 2 \\ 1 & 2 \leq x < 3 \end{cases}$; and for n = 2, $f(x) = \begin{cases} 0 & 3 \leq x < 4 \\ 1 & 4 \leq x < 5 \end{cases}$. By

continuing in this fashion, and drawing a graph, it can be
seen that f(x) is periodic with fundamental period T = 2.

10.   The graph of f(x) is:
We note that f(x) is
a straight line with
a slope of one in any
interval. Thus f(x) has the form x+b in any interval for
the correct value of b. Since f(x+2) = f(x), we may set
x = -1/2 to obtain f(3/2) = f(-1/2). Noting that 3/2 is
on the interval 1 < x < 2[f(3/2) = 3/2 + b] and that -1/2
is on the interval -1 < x < 0[f(-1/2) = -1/2 + 1], we
conclude that 3/2 + b = -1/2 + 1, or b = -1 for the
interval 1 < x < 2. For the interval 8 < x < 9 we have
f(x+8) = f(x+6) = ... = f(x) by successive applications
of the periodicity condition. Thus for x = 1/2 we have
f(17/2) = f(1/2) or 17/2 + b = 1/2 so b = -8 on the
interval 8 < x < 9.

In Problems 13 through 18 it is often necessary to use
integration by parts to evaluate the coefficients, although
all the details will not be shown here.

13a. The function represents
     a sawtooth wave. It is
     periodic with period 2L.

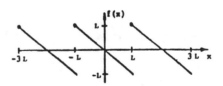

13b. The Fourier series is of the form

$f(x) = a_0/2 + \sum_{m=1}^{\infty} [a_m\cos(m\pi x/L) + b_m\sin(m\pi x/L)]$, where the

coefficients are computed from Eqs. (13) and (14).
Substituting for f(x) in these equations yields
$a_0 = (1/L)\int_{-L}^{L}(-x)dx = 0$ and $a_m = (1/L)\int_{-L}^{L}(-x)\cos(m\pi x/L)dx = 0$,
m = 1,2... (these can be shown by direct integration, or

using the fact that $\int_{-a}^{a} g(x)dx = 0$ when $g(x)$ is an odd function). Finally,

$$b_m = (1/L)\int_{-L}^{L}(-x)\sin(m\pi x/L)dx$$

$$= (x/m\pi)\cos(m\pi x/L)\Big|_{-L}^{L} - (1/m\pi)\int_{-L}^{L}\cos(m\pi x/L)dx$$

$$= (2L\cos m\pi)/m\pi - (L/m^2\pi^2)\sin(m\pi x/L)\Big|_{-L}^{L} = 2L(-1)^m/m\pi.$$

Substituting these terms in the above Fourier series for $f(x)$ yields the desired answer.

15a. See below.

15b. In this case $f(x)$ is periodic of period 2π and thus $L = \pi$ in Eqs. (9), (13,) and (14). The constant a_0 is found to be $a_0 = (1/\pi)\int_{-\pi}^{0} xdx = -\pi/2$ since $f(x)$ is zero on the interval $[0,\pi]$. Likewise $a_n = (1/\pi)\int_{-\pi}^{0} x\cos nxdx = [1 - (-1)^n]/n^2\pi$, using integration by parts and recalling that $\cos n\pi = (-1)^n$. Thus $a_n = 0$ for n even and $a_n = 2/n^2\pi$ for n odd, which may be written as $a_{2n-1} = 2/(2n-1)^2\pi$ since $2n-1$ is always an odd number. In a similar fashion $b_n = (1/\pi)\int_{-\pi}^{0} x\sin nxdx = (-1)^{n+1}/n$ and thus the desired solution is obtained. Notice that in this case both cosine and sine terms appear in the Fourier series for the given $f(x)$.

15a.   21a.

21b. $a_0 = \dfrac{1}{2}\displaystyle\int_{-2}^{2}\dfrac{x^2}{2}dx = \dfrac{1}{12}x^3\Big|_{-2}^{2} = \dfrac{4}{3}$, so $\dfrac{a_0}{2} = \dfrac{2}{3}$ and

$$a_n = \frac{1}{2}\int_{-2}^{2}\frac{x^2}{2}\cos\frac{n\pi x}{2}dx$$

$$= \frac{1}{4}[\frac{2x^2}{n\pi}\sin\frac{n\pi x}{2} + \frac{8x}{n^2\pi^2}\cos\frac{n\pi x}{2} - \frac{16}{n^3\pi^3}\sin\frac{n\pi x}{2}]\Big|_{-2}^{2}$$

$$= (8/n^2\pi^2)\cos(n\pi) = (-1)^n 8/n^2\pi^2$$

where the second line for a_n is found by integration by parts or a computer algebra system. Similarly,

$$b_n = \frac{1}{2}\int_{-2}^{2} \frac{x^2}{2}\sin\frac{n\pi x}{2}dx = 0, \text{ since } x^2\sin\frac{n\pi x}{2} \text{ is an odd}$$

function. Thus $f(x) = \frac{2}{3} + \frac{8}{\pi^2}\sum_{n=1}^{\infty} \frac{(-1)^n}{n^2}\cos\frac{n\pi x}{2}.$

21c. As in Eq. (27), we have $s_m(x) = \frac{2}{3} + \frac{8}{\pi^2}\sum_{n=1}^{m} \frac{(-1)^n}{n^2}\cos\frac{n\pi x}{2}$

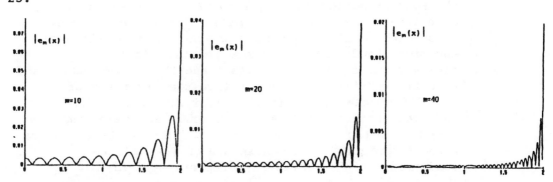

21d. Observing the graphs we see that the Fourier series
converges quite rapidly, except, at $x = -2$ and $x = 2$,
where there is a sharp "point" in the periodic function.

25.

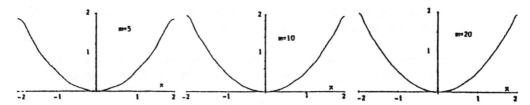

Clearly the maximum error occurs at $x = 2$, so use a
computer algebra system to evaluate $e_m(2)$ to show that at
least 81 terms are needed for $|e_m(2)| \le .01$.

27a. First we have $\int_{T}^{a+T}g(x)dx = \int_{0}^{a}g(s)ds$ by letting $x = s + T$
in the left integral. Now, if $0 \le a \le T$, then from
elementary calculus we know that
$\int_{a}^{a+T}g(x)dx = \int_{a}^{T}g(x)dx + \int_{T}^{a+T}g(x)dx = \int_{a}^{T}g(x)dx + \int_{0}^{a}g(x)dx$
using the equality derived above. This last sum is
$\int_{0}^{T}g(x)dx$ and thus we have the desired result.

Section 10.3, Page 592

2a. Substituting for f(x) in Eqs.(2) and (3) with L = π
     yields $a_0 = (1/\pi)\int_0^\pi x\,dx = \pi/2$;

     $a_m = (1/\pi)\int_0^\pi x\cos mx\,dx = (\cos m\pi - 1)/\pi m^2 = 0$ for m even and

        $= -2/\pi m^2$ for m odd; and

     $b_m = (1/\pi)\int_0^\pi x\sin mx\,dx = -(\pi\cos m\pi)/m\pi = (-1)^{m+1}/m$,

     m = 1,2... . Substituting these values into Eq.(1) with
     L = π yields the desired solution.

2b. The function to which the series converges is indicated
     in the figure and is periodic with period 2π. Note that

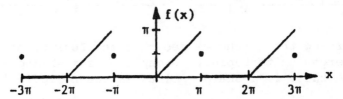

     the Fourier series converges to π/2 at x = -π, π, etc.,
     even though the function is defined to be zero there.
     This value (π/2) represents the mean value of the left
     and right hand limits at those points. In (-π, 0),
     f(x) = 0 and f'(x) = 0 so both f and f' are continuous and
     have finite limits as x → -π from the right and as
     x → 0 from the left. In (0, π), f(x) = x and f'(x) = 1
     and again both f and f' are continuous and have limits as
     x → 0 from the right and as x → π from the left. Since
     f and f' are piecewise continuous on [-π, π] the
     conditions of the Fourier theorem are satisfied.

4a. Substituting for f(x) in Eqs.(2) and (3), with L = 1

     yields $a_0 = \int_{-1}^1 (1-x^2)\,dx = 4/3$;

     $a_n = \int_{-1}^1 (1-x^2)\cos n\pi x\,dx = (2/n\pi)\int_{-1}^1 x\sin n\pi x\,dx$

        $= (-2/n^2\pi^2)[x\cos n\pi x\Big|_{-1}^1 - \int_{-1}^1 \cos n\pi x\,dx]$

        $= 4(-1)^{n+1}/n^2\pi^2$; and

     $b_n = \int_{-1}^1 (1-x^2)\sin n\pi x\,dx = 0$. Substituting these values
     into Eq.(1) gives the desired series.

4b. The function to which the series converges is shown in
     the figure and is periodic of fundamental period 2. In
     [-1,1] f(x) and f'(x) = -2x are both continuous and have

finite limits as the endpoints of the interval are approached from within the interval.

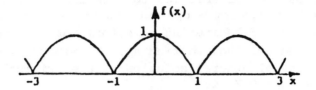

7a. As in Prob. 15, Sect. 10.2, we have

$$f(x) = -\frac{\pi}{4} + \sum_{n=1}^{\infty}[\frac{2\cos(2n-1)x}{\pi(2n-1)^2} + \frac{(-1)^{n+1}\sin nx}{n}].$$

7b. $e_n(x) = f(x) + \frac{\pi}{4} - \sum_{k=1}^{n}[\frac{2\cos(2k-1)x}{\pi(2k-1)^2} + \frac{(-1)^{k+1}\sin kx}{k}].$

Using a computer algebra system, we find that for $n = 5$, 10 and 20 the maximum error occurs at $x = -\pi$ in each case and is 1.6025, 1.5867 and 1.5787 respectively. Note that the author's n values are 10, 20 and 40, since he has included the zero cosine coefficient terms and the sine terms are all zero at $x = -\pi$.

7c. It's not possible in this case, due to Gibb's phenomenon, to satisfy $|e_n(x)| \leq 0.01$ for all x.

12a. $a_0 = \int_{-1}^{1}(x-x^3)dx = 0$ and $a_n = \int_{-1}^{1}(x-x^3)\cos n\pi x dx = 0$ since $(x-x^3)$ and $(x-x^3)\cos n\pi x$ are odd functions.

$b_n = \int_{-1}^{1}(x-x^3)\sin n\pi x dx$

$= [\frac{x^3}{n\pi}\cos n\pi x - \frac{3x^2}{n^2\pi^2}\sin n\pi x - \frac{(n^2\pi^2+6)}{n^3\pi^3}x\cos n\pi x + \frac{(n^2\pi^2+6)}{n^4\pi^4}\sin n\pi x]_{-1}^{1}$

$= \frac{-12}{n^3\pi^3}\cos n\pi$, so $f(x) = -\frac{12}{\pi^3}\sum_{n=1}^{\infty}\frac{(-1)^n}{n^3}\sin n\pi x.$

12b. $e_n(x) = f(x) + \frac{12}{\pi^3}\sum_{k=1}^{n}\frac{(-1)^k}{k^3}\sin k\pi x.$ These errors will be

much smaller than in the earlier problems due to the n^3 factor in the denominator. Convergence is much more rapid in this case.

14. The solution to the corresponding homogeneous equation is found by methods presented in Sect. 3.4 and is $y(t) = c_1\cos\omega t + c_2\sin\omega t$. For the nonhomogeneous terms we use the method of superposition and consider the

sequence of equations $y_n'' + \omega^2 y_n = b_n \sin nt$ for
$n = 1, 2, 3...$. If $\omega > 0$ is not equal to an integer,
then the particular solution to this last equation has
the form $Y_n = a_n \cos nt + d_n \sin nt$, as previously discussed
in Sect. 3.6. Substituting this form for Y_n into the
equation and solving, we find $a_n = 0$ and $d_n = b_n/(\omega^2 - n^2)$.
Thus the formal general solution of the original
nonhomogeneous D.E. is

$$y(t) = c_1 \cos \omega t + c_2 \sin \omega t + \sum_{n=1}^{\infty} b_n (\sin nt)/(\omega^2 - n^2), \text{ where we}$$

have superimposed all the Y_n terms to obtain the infinite
sum. To evalute c_1 and c_2 we set $t = 0$ to obtain

$$y(0) = c_1 = 0 \text{ and } y'(0) = \omega c_2 + \sum_{n=1}^{\infty} nb_n/(\omega^2 - n^2) = 0 \text{ where}$$

we have formally differentiated the infinite series term
by term and evaluated it at $t = 0$. (Differentiation of a
Fourier Series has not been justified yet and thus we can
only consider this method a formal solution at this

point). Thus $c_2 = -(1/\omega) \sum_{n=1}^{\infty} nb_n/(\omega^2 - n^2)$, which when

substituted into the above series yields the desired
solution.

If $\omega = m$, a positive integer, then the particular
solution of $y_m'' + m^2 y_m = b_m \sin mt$ has the form
$Y_m = t(a_m \cos mt + d_m \sin mt)$ since $\sin mt$ is a solution of
the related homogeneous D.E.[all other particular
solutions are the same as previously, with $\omega = m$].
Substituting Y_m into the D.E. yields $a_m = -b_m/2m$ and
$d_m = 0$ and thus the general solution of the D.E.
(when $\omega = m$) is now

$$y(t) = c_1 \cos mt + c_2 \sin mt - b_m t (\cos mt)/2m + \sum_{n=1, n \neq m}^{\infty} b_n (\sin nt)/(m^2 - n^2).$$

To evaluate c_1 and c_2 we set $y(0) = 0 = c_1$ and

$$y'(0) = c_2 m - b_m/2m + \sum_{n=1, n \neq m}^{\infty} b_n n/(m^2 - n^2) = 0. \text{ Thus}$$

$$c_2 = b_m/2m^2 - \sum_{n=1, n \neq m}^{\infty} b_n n/m(m^2 - n^2), \text{ which when substituted}$$

into the equation for $y(t)$ yields the desired solution.

15. In order to use the results of Prob. 14, we must first

find the Fourier series for the given f(t). From Eqs.(2) and (3) with L = π, we find that [Using the result of Prob. 27, Sect 10.2] $a_0 = (1/\pi)\int_0^\pi dx - (1/\pi)\int_\pi^{2\pi} dx = 0$;

$$a_n = (1/\pi)\int_0^\pi \cos nx\, dx - (1/\pi)\int_\pi^{2\pi} \cos nx\, dx = 0; \text{ and}$$

$b_n = (1/\pi)\int_0^\pi \sin nx\, dx - (1/\pi)\int_\pi^{2\pi}\sin nx\, dx = 0$ for n even and $= 4/n\pi$ for n odd.   Thus

$f(t) = (4/\pi)\sum_{n=1}^\infty \sin(2n-1)t/(2n-1)$.   Comparing this to the forcing function of Prob. 14 we see that b_n of Prob. 14 has the specific values $b_{2n} = 0$ and $b_{2n-1} = (4/\pi)/(2n-1)$ in this example.   Substituting these into the answer to Prob. 14 yields the desired solution. Note that we have asumed ω is not a positive integer.   Note also, that if the solution to Prob. 14 is not available, the procedure for solving this problem would be exactly the same as shown in Prob. 14.

16. From Prob. 8, the Fourier series for f(t) is given by

$$f(t) = 1/2 + (4/\pi^2)\sum_{n=1}^\infty \cos(2n-1)\pi t/(2n-1)^2 \text{ and thus we may}$$

not use the form of the answer in Prob. 14. The procedure outlined there, however, is applicable and will yield the desired solution.

18a. We will assume f(x) is continuous for this part. For the case where f(x) has jump discontinuities, a more detailed proof can be developed, as shown in part (b). From Eq.(3) we have $b_n = \dfrac{1}{L}\int_{-L}^L f(x)\sin\dfrac{n\pi x}{L}dx$. If we let u = f(x) and

$dv = \sin\dfrac{n\pi x}{L}dx$, then $du = f'(x)dx$ and $v = \dfrac{-L}{n\pi}\cos\dfrac{n\pi x}{L}$.
Thus

$$b_n = \frac{1}{L}[\frac{-L}{n\pi}f(x)\cos\frac{n\pi x}{L}\Big|_{-L}^L + \frac{L}{n\pi}\int_{-L}^L f'(x)\cos\frac{n\pi x}{L}dx]$$

$$= -\frac{1}{n\pi}[f(L)\cos n\pi - f(-L)\cos(-n\pi)] + \frac{1}{n\pi}\int_{-L}^L f'(x)\cos\frac{n\pi x}{L}dx$$

$$= \frac{1}{n\pi}\int_{-L}^L f'(x)\cos\frac{n\pi x}{L}dx, \text{ since } f(L) = f(-L) \text{ and}$$

$\cos(-n\pi) = \cos n\pi$.   Hence $nb_n = \dfrac{1}{\pi}\int_{-L}^L f'(x)\cos\dfrac{n\pi x}{L}dx$,   which exists for all n since $f'(x)$ is piecewise continuous. Thus nb_n is bounded as $n \to \infty$. Likewise, for a_n, we obtain $na_n = -\frac{1}{\pi}\int_{-L}^{L}f'(x)\sin\frac{n\pi x}{L}dx$ and hence na_n is also bounded as $n \to \infty$.

18b. Note that f and f' are continuous at all points where f″ is continuous. Let $x_1, \ldots, x_m$ be the points in $(-L,L)$ where f″ is not continuous. By splitting up the interval of integration at these points, and integrating Eq.(3) by parts twice, we obtain

$$n^2 b_n = \frac{n}{\pi}\sum_{i=1}^{m}[f(x_i+)-f(x_i-)]\cos\frac{n\pi x_i}{L} - \frac{n}{\pi}[f(L-)-f(-L+)]\cos n\pi$$

$$-\frac{L}{\pi^2}\sum_{i=1}^{m}[f'(x_i+)-f'(x_1-)]\sin\frac{n\pi x_i}{L} - \frac{L}{\pi^2}\int_{-L}^{L}f''(x)\sin\frac{n\pi x}{L}dx, \text{ where}$$

we have used the fact that cosine is continuous. We want the first two terms on the right side to be zero, for otherwise they grow in magnitude with n. Hence f must be continuous throughout the closed interval [-L,L]. The last two terms are bounded, by the hypotheses on f' and f″. Hence $n^2 b_n$ is bounded as $n \to \infty$; similarly $n^2 a_n$ is bounded as $n \to \infty$.

If $f(x)$ is continuous on $[-L,L]$ and $f(-L) = f(L)$ [since f is periodic of period 2L], then the periodic extension of $f(x)$ is continuous for all x. Thus there will be no jump discontinuities, as might be the case in part (a).

18c. If $n^2 a_n$ is bounded as as $n \to \infty$ then $|a_n| \le \frac{M}{n^2}$ as $n \to \infty$ and hence $\sum_{n=1}^{\infty}|a_n|$ converges by the comparison test using $\sum_{n=1}^{\infty}\frac{1}{n^2}$.

18d. Since $|a_n\cos(n\pi x/L) + b_n\sin(n\pi x/L)| \le |a_n| + |b_n|$ and since $\sum_{n=1}^{\infty}|a_n|$ and $\sum_{n=1}^{\infty}|b_n|$ converge from part (c), we conclude that the Fourier Series (4) converges absolutely.

Section 10.4, Page 600

3. Let $f(x) = \tan 2x$, then

$$f(-x) = \tan(-2x) = \frac{\sin(-2x)}{\cos(-2x)} = \frac{-\sin(2x)}{\cos(2x)} = -\tan 2x = -f(x)$$

and thus tan2x is an odd function.

6. Let $f(x) = e^{-x}$, then $f(-x) = e^x$ so that $f(-x) \neq f(x)$ and $f(-x) \neq -f(x)$ and thus e^{-x} is neither even nor odd.

7.

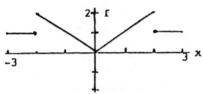
Even Extension

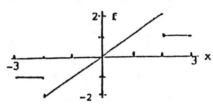
Odd Extension

10.

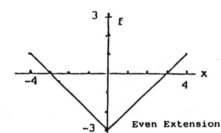
Even Extension   Odd Extension

13. By the hint $f(-x) = g(-x) + h(-x) = g(x) - h(x)$, since g is an even function and f is an odd function. Thus $f(x) + f(-x) = 2g(x)$ and hence $g(x) = [f(x) + f(-x)]/2$ defines $g(x)$. Likewise $f(x)-f(-x) = g(x)-g(-x) + h(x)-h(-x) = 2h(x)$ and thus $h(x) = [f(x) - f(-x)]/2$

All functions and their derivatives in Problems 14 through 30 are piecewise continuous on the given intervals and their extensions. Thus the Fourier Theorem applies in all cases.

14. For the cosine series we use the even extension of the function given in Eq.(13) and hence
$$f(x) = \begin{cases} 0 & -2 \leq x < -1 \\ 1+x & -1 \leq x < 0 \end{cases} \quad \text{on the interval } -2 \leq x < 0.$$
However, we don't really need this, as the coefficients in this case are given by Eqs.(7), which just use the original values for $f(x)$ on $0 < x \leq 2$. Applying Eqs.(7) we have $L = 2$ and thus
$$a_0 = (2/2)\int_0^1 (1-x)dx + (2/2)\int_1^2 0\,dx = 1/2. \quad \text{Similarly,}$$
$$a_n = (2/2)\int_0^1 (1-x)\cos(n\pi x/2)dx = 4[1-\cos(n\pi/2)]/n^2\pi^2 \text{ and}$$
$$b_n = 0. \quad \text{Substituting these values in the Fourier series}$$

yields the desired results.

For the sine series, we use Eqs.(8) with $L = 2$. Thus $a_n = 0$ and

$b_n = (2/2)\int_0^1 (1-x)\sin(n\pi x/2)dx = 4[n\pi/2 - \sin(n\pi/2)]/n^2\pi^2.$

15. The graph of the function to which the series converges is shown in the figure. Using Eqs.(7) with $L = 2$ we have $a_0 = \int_0^1 dx = 1$ and $a_n = \int_0^1 \cos(n\pi x/2)dx = 2\sin(n\pi/2)/n\pi.$ Thus $a_n = 0$ for n even, $a_n = 2/n\pi$ for $n = 1,5,9,\ldots$ and $a_n = -2/n\pi$ for $n = 3,7,11,\ldots$. Hence we may write $a_{2n} = 0$ and $a_{2n-1} = 2(-1)^{n+1}/(2n-1)\pi$, which when substituted into the series gives the desired answer. This is the same series as in Prob. 5 of Sect. 10.3.

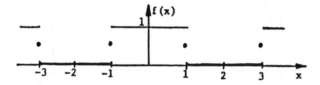

18. The graph of the function to which the series converges is indicated in the figure.

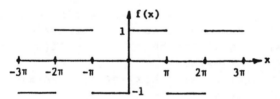

Since we want a sine series, we use Eqs.(8) to find, with $L = \pi$, that $b_n = (2/\pi)\int_0^\pi \sin nx\, dx = 2[1-(-1)^n]/n\pi$ and thus $b_n = 0$ for n even and $b_n = 4/n\pi$ for n odd.

20. The graph of the function to which the series converges is shown in the figure.

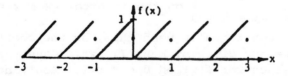

We note that $f(x)$ is specified over its entire fundamental period ($T = 1$) and hence we cannot extend f to make it either an odd or an even function. Using Eqs.(2) and (3) from Sect. 10.3 we have ($L = 1/2$)

$a_0 = 2\int_0^1 x\,dx = 1$, $a_n = 2\int_0^1 x\cos(2n\pi x)dx = 0$ and

$b_n = 2\int_0^1 x\sin(2n\pi x)dx = -1/n\pi$. [Note: We have used the results of Prob. 27 of Sect. 10.2 in writing these integrals. That is, if f(x) is periodic of period T, then every integral of f over an interval of length T has the same value. Thus we integrate from 0 to 1 here, rather than -1/2 to 1/2.] Substituting the above values into Eq.(1) of Sect. 10.3 yields the desired solution. It can also be observed from the above graph that g(x) = f(x) - 1/2 is an odd function. If Eqs.(8) are used with g(x), then it is found that

$$g(x) = (-1/\pi)\sum_{n=1}^{\infty}\sin(2n\pi x)/n \text{ and thus we obtain the same}$$

series for f(x) as found above.

25a. $b_n = \dfrac{2}{2}\int_0^2 (2-x^2)\sin\dfrac{n\pi x}{2}dx$

$= [\dfrac{-2}{n\pi}(2-x^2)\cos\dfrac{n\pi x}{2} - \dfrac{8x}{n^2\pi^2}\sin\dfrac{n\pi x}{2} - \dfrac{16}{n^3\pi^3}\cos\dfrac{n\pi x}{2}]\Big|_0^2$

$= \dfrac{4}{n\pi}(1+\cos n\pi) + \dfrac{16}{n^3\pi^3}(1-\cos n\pi)$ and thus

$$f(x) = \sum_{n=1}^{\infty}(\dfrac{4n^2\pi^2(1+\cos n\pi) + 16(1-\cos n\pi)}{n^3\pi^3})\sin\dfrac{n\pi x}{2}$$

25b.

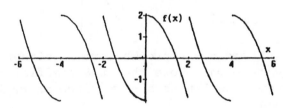

25c.

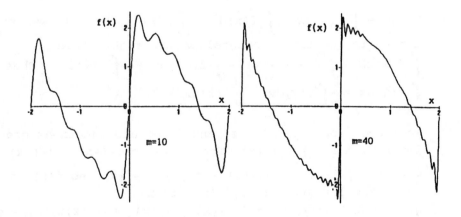

28b.For the cosine series(even extension) we have

$$a_0 = \frac{2}{2}\int_0^1 x\,dx = \frac{1}{2}$$

$$a_n = \frac{2}{2}\int_0^1 x\cos\frac{n\pi x}{2}\,dx = [\frac{2x}{n\pi}\sin\frac{n\pi x}{2} + \frac{4}{n^2\pi^2}\cos\frac{n\pi x}{2}]\Big|_0^1$$

$$= \frac{2}{n\pi}\sin\frac{n\pi}{2} + \frac{4}{n^2\pi^2}\cos\frac{n\pi}{2} - \frac{4}{n^2\pi^2}, \text{ so}$$

$$g(x) = \frac{1}{4} + \sum_{n=1}^{\infty}\frac{4\cos(n\pi/2) + 2n\pi\sin(n\pi/2) - 4}{n^2\pi^2}\cos\frac{n\pi x}{2}.$$

For the sine series (odd extension) we have

$$b_n = \frac{2}{2}\int_0^1 x\sin\frac{n\pi x}{2}\,dx = [\frac{-2x}{n\pi}\cos\frac{n\pi x}{2} + \frac{4}{n^2\pi^2}\sin\frac{n\pi x}{2}]\Big|_0^1$$

$$= \frac{-2}{n\pi}\cos\frac{n\pi}{2} + \frac{4}{n^2\pi^2}\sin\frac{n\pi}{2}, \text{ so}$$

$$h(x) = \sum_{n=1}^{\infty}\frac{4\sin(n\pi/2) - 2n\pi\cos(n\pi/2)}{n^2\pi^2}\sin\frac{n\pi x}{2}.$$

28c.

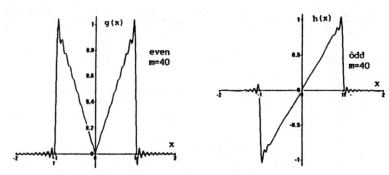

28d. The maximum error does not approach zero in either case,
due to Gibb's phenomenon. Note that the coefficients in
both series behave like 1/n as n → ∞ since there is an
n in the numerator.

31. We have $\int_{-L}^{L}f(x)\,dx = \int_{-L}^{0}f(x)\,dx + \int_0^L f(x)\,dx$. Now, if we let
x = -y in the first integral on the right, then
$\int_{-L}^{0}f(x)\,dx = \int_L^0 f(-y)(-dy) = \int_0^L f(-y)\,dy = -\int_0^L f(y)\,dy$. Thus
$\int_{-L}^{L}f(x)\,dx = -\int_0^L f(y)\,dy + \int_0^L f(x)\,dx = 0.$

32. To prove property 2 let f_1 and f_2 be odd functions and
let $f(x) = f_1(x) \pm f_2(x)$. Then $f(-x) = f_1(-x) \pm f_2(-x) =$
$-f_1(x) \pm [-f_2(x)] = -f_1(x) \mp f_2(x) = -f(x)$, so $f(x)$ is
odd. Now let $g(x) = f_1(x)f_2(x)$, then
$g(-x) = f_1(-x)f_2(-x) = [-f_1(x)][-f_2(x)] = f_1(x)f_2(x) = g(x)$

and thus g(x) is even. Finally, let h(x) = $f_1(x)/f_2(x)$
and hence h(-x) = $f_1(-x)/f_2(-x) = [-f_1(x)]/[-f_2(x)]$ =
$f_1(x)/f_2(x)$ = h(x), which says h(x) is also even.
Property 3 is proven in a similar manner.

34. Since $F(x) = \int_0^x f(t)dt$ we have

$F(-x) = \int_0^{-x} f(t)dt = -\int_0^x f(-s)ds$ by letting t = -s. If f
is an even function, f(-s) = f(s), we then have
$F(-x) = -\int_0^x f(s)ds = -F(x)$ from the original definition of
F. Thus F(x) is an odd function. The argument is
similar if f is odd.

35. Set x = L/2 in Eq.(6) of Sect. 10.3. Since we know f is
continuous at L/2, we may conclude, by the Fourier
theorem, that the series will converge to
f(L/2) = L at this point. Thus we have

$L = L/2 + (2L/\pi) \sum_{n=1}^{\infty} (-1)^{n+1}/(2n-1)$, since

$\sin[(2n-1)\pi/2] = (-1)^{n+1}$. Dividing by L and simplifying

yields $\dfrac{\pi}{4} = \sum_{n=1}^{\infty} \dfrac{(-1)^{n+1}}{(2n-1)} = \sum_{n=0}^{\infty} \dfrac{(-1)^n}{2n+1}$, by changing the

summation index.

37a. Multiplying both sides of the equation by f(x) and
integrating from 0 to L gives

$$\int_0^L [f(x)]^2 dx = \int_0^L \left[f(x) \sum_{n=1}^{\infty} b_n \sin(n\pi x/L) \right] dx$$

$$= \sum_{n=1}^{\infty} b_n \int_0^L f(x)\sin(n\pi x/L)dx = (L/2) \sum_{n=1}^{\infty} b_n^2, \text{ by Eq.(8).}$$

This result is identical to that of Prob. 17 of Sect. 10.3
if we set $a_n = 0$, n = 0,1,2,... , since
$\dfrac{1}{L}\int_{-L}^{L} [f(x)]^2 dx = \dfrac{2}{L}\int_0^L [f(x)]^2 dx$. In a similar manner, it can

be shown that $(2/L)\int_0^L [f(x)]^2 dx = a_0^2/2 + \sum_{n=1}^{\infty} a_n^2$.

37b. From Eq.(9) we have $b_n = 2L(-1)^{n+1}/n\pi$, so

$$\sum_{n=1}^{\infty} b_n^2 = \sum_{n=1}^{\infty} 4L^2/n^2\pi^2 = 4L^2/\pi^2 \sum_{n=1}^{\infty} (1/n^2). \text{ Also } f(x) = x, \text{ so}$$

$(2/L)\int_0^L [f(x)]^2 dx = (2/L)\int_0^L x^2 dx = 2L^2/3$. From part (a)

then, $2L^2/3 = 4L^2/\pi^2 \sum_{n=1}^{\infty} (1/n^2)$, or $\pi^2/6 = \sum_{n=1}^{\infty} (1/n^2)$.

38. We assume that the extensions of f and f' are piecewise
    continuous on $[-2L, 2L]$. Since f is an odd periodic
    function of fundamental period 4L it follows from
    properties 2 and 3 that $f(x)\cos(n\pi x/2L)$ is odd and
    $f(x)\sin(n\pi x/2L)$ is even. Thus the Fourier coefficients
    of f are given by Eqs.(8) with L replaced by 2L; that is
    $a_n = 0$, $n = 0,1,2,\ldots$ and
    $b_n = (2/2L)\int_0^{2L} f(x)\sin(n\pi x/2L) dx$, $n = 1,2,\ldots$. The

    Fourier sine series for f is $f(x) = \sum_{n=1}^{\infty} b_n \sin(n\pi x/2L)$.

39. From Prob. 38 we have $b_n = (1/L)\int_0^{2L} f(x)\sin(n\pi x/2L) dx$

    $= (1/L)\int_0^L f(x)\sin(n\pi x/2L) dx + (1/L)\int_L^{2L} f(2L-x)\sin(n\pi x/2L) dx$

    $= (1/L)\int_0^L f(x)\sin(n\pi x/2L) dx - (1/L)\int_L^0 f(s)\sin[n\pi(2L-s)/2L] ds$

    $= (1/L)\int_0^L f(x)\sin(n\pi x/2L) dx - (1/L)\int_L^0 f(s)\cos(n\pi)\sin(n\pi s/2L) ds$

    and thus $b_n = 0$ for n even and
    $b_n = (2/L)\int_0^L f(x)\sin(n\pi x/2L) dx$ for n odd. The Fourier
    series for f is given in Prob. 38, where the b_n are given
    above.

Section 10.5, Page 610

3. We seek solutions of the form $u(x,t) = X(x)T(t)$.
    Substituting into the P.D.E. yields $X''T + X'T' + XT' =$
    $X''T + (X' + X)T' = 0$. Formally dividing by the quantity
    $(X' + X)T$ gives the equation $X''/(X' + X) = -T'/T$ in which
    the variables are separated. Using the same concepts as
    those leading to Eq.(18) we conclude that in order for
    this equation to be valid on the domain of u it is
    necessary that both sides be equal to the same constant
    λ. Hence $X''/(X' + X) = -T'/T = \lambda$ or equivalently,
    $X'' - \lambda(X' + X) = 0$ and $T' + \lambda T = 0$.

5. We seek solutions of the form $u(x,y) = X(x)Y(y)$.
    Substituting into the P.D.E. yields $X''Y + (x+y)XY'' = 0$.
    Formally dividing by XY yields $X''/X + (x + y)Y''/Y = 0$.
    From this equation we see that the presence of the

independent variable x multiplying the term u_{yy} in the
original equation leads to the term xY''/Y when we attempt
to separate the variables. It follows that the argument
for a separation constant does not carry through and we
cannot replace the P.D.E. by two O.D.E.

8. Following the procedures of Eqs.(5) through (8), we set
    $u(x,y) = X(x)T(t)$ in the P.D.E. to obtain $X''T = 4XT'$, or
    $X''X = 4T'/T$, which must be a constant. As stated in the text
    this separation constant must be $-\lambda^2$ (we choose $-\lambda^2$ so that
    when a square root is used later, the symbols are simpler)
    and thus $X'' + \lambda^2X = 0$ and $T' + (\lambda^2/4)T = 0$. Now $u(0,t) =$
    $X(0)T(t) = 0$, for all $t > 0$, yields $X(0) = 0$, as discussed
    after Eq.(11) and similarly $u(2,t) = X(2)T(t) = 0$, for all
    $t > 0$, implies $X(2) = 0$. The D.E. for X has the solution
    $X(x) = C_1\cos\lambda x + C_2\sin\lambda x$ and $X(0) = 0$ yields $C_1 = 0$. Setting
    $x = 2$ in the remaining form of X yields $X(2) = C_2\sin2\lambda = 0$,
    which has the solutions $2\lambda = n\pi$ or $\lambda = n\pi/2$, $n = 1,2,\ldots$.
    Note that we exclude $n = 0$ since then $\lambda = 0$ would yield
    $X(x) = 0$, which is unacceptable. Hence
    $X(x) = \sin(n\pi x/2)$, $n = 1,2,\ldots$. Finally, the solution of the
    D.E. for T yields $T(t) = \exp(-\lambda^2t/4) = \exp(-n^2\pi^2t/16)$. Thus we
    have found $u_n(x,t) = \exp(-n^2\pi^2t/16)\sin(n\pi x/2)$, which is
    Eq.(17) for this problem. Setting $t = 0$ in this last
    expression indicates that $u_n(x,0)$ has, for $n = 1,2$ and 4, the
    same form as the terms in $u(x,0)$, the initial condition.
    Using the principle of superposition we know that
    $u(x,t) = c_1u_1(x,t) + c_2u_2(x,t) + c_4u_4(x,t)$ satisfies the
    P.D.E. and the B.C. and hence we let $t = 0$ to obtain
    $u(x,0) = c_1u_1(x,0) + c_2u_2(x,0) + c_4u_4(x,0)$
    $= c_1\sin n\pi x/2 + c_2\sin n\pi x + c_4\sin2\pi x$. If we choose
    $c_1 = 2$, $c_2 = -1$ and $c_4 = 4$ then $u(x,0)$ here will match
    the given initial condition, and hence substituting these
    values in $u(x,t)$ above then gives the desired solution.

10. Since the B.C. for this heat conduction problem are
    $u(0,t) = u(40,t) = 0$, $t > 0$, the solution $u(x,t)$ is given
    by Eq.(19) with $\alpha^2 = 1$ cm^2/sec, $L = 40$ cm, and the
    coefficients c_n determined by Eq.(21) with the I.C.

$$u(x,0) = f(x) = \begin{cases} x & 0 \le x < 20 \\ 40 - x & 20 \le x \le 40 \end{cases}. \text{ Thus}$$

$$c_n = \frac{1}{20}\left[\int_0^{20} x\sin\frac{n\pi x}{40}dx + \int_{20}^{40}(40-x)\sin\frac{n\pi x}{40}dx\right] = \frac{160}{n^2\pi^2}\sin\frac{n\pi}{2}.$$

It follows that $u(x,t) = \dfrac{160}{\pi^2}\displaystyle\sum_{n=1}^{\infty}\dfrac{\sin(n\pi/2)}{n^2}e^{-n^2\pi^2t/1600}\sin\dfrac{n\pi x}{40}.$

15a.

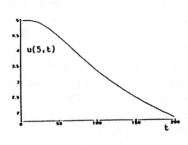

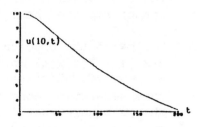

15b.

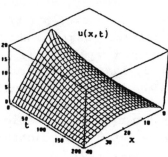

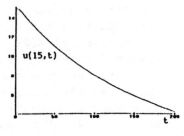

15c.

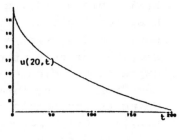

15d. As in Ex. 1, the maximum temperature will be at the
midpoint, x = 20. We use just the first term (n=1) of the
series found in Prob. 10, since the others will be
negligible for this temperature, since t is large. Thus

$$u(20,t) = 1 = \frac{160}{\pi^2}\sin(\pi/2)e^{-\pi^2 t/1600}\sin(20\pi/40).$$ Solving

for t, we obtain $e^{-\pi^2 t/1600} = \pi^2/160$, or

$$t = \frac{1600}{\pi^2}\ln\frac{160}{\pi^2} = 451.60 \text{ sec.}$$ If t is not so large in

other problems, then two or more terms must be used,
which would require an equation solver or numerical
solution to find t.

18. From the problem statement we conclude that the I.C. is
u(x,0) = 100°C and the B.C. are u(0,t) = u(20,t) = 0,

t > 0. The solution $u(x,t)$ is given by Eq.(19) with
L = 20 cm, and the coefficients c_n determined by Eq.(21)
with $f(x) = 100$. Thus
$c_n = (1/10)\int_0^{20}(100)\sin(n\pi x/20)dx = -200[(-1)^n-1]/n\pi$ and hence
$c_{2n} = 0$ and $c_{2n-1} = 400/(2n-1)\pi$. Substituting these values
into Eq.(19) yields
$$u(x,t) = \frac{400}{\pi}\sum_{n=1}^{\infty}\frac{e^{-(2n-1)^2\pi^2\alpha^2 t/400}}{2n-1}\sin\frac{(2n-1)\pi x}{20}.$$

18b. For aluminum, we have $\alpha^2 = .86$ cm^2/sec (Table 10.5.1) and
thus the first two terms give
$$u(10,30) = \frac{400}{\pi}\left\{e^{-\pi^2(.86)30/400} - \frac{1}{3}e^{-9\pi^2(.86)30/400}\right\}$$
$$= 67.228^\circ C.$$
If an additional term is used, the temperature is
increased by $\frac{80}{\pi}e^{-25\pi^2(.86)30/400} = 3\times10^{-6}$ degrees C.

19b. Using only one term in the series for $u(x,t)$ of Prob. 18,
we must solve the equation $(400/\pi)\exp[-\pi^2(.86)t/400] = 5$
for t. Taking the logarithm of both sides and solving
for t yields $t \cong 400\ln(80/\pi)/\pi^2(.86) = 152.56$ sec.

20. Applying the chain rule to partial differentiation of u
with respect to x we see that $u_x = u_\xi\xi_x = u_\xi(1/L)$ and
$u_{xx} = u_{\xi\xi}(1/L)^2$. Substituting $u_{\xi\xi}/L^2$ for u_{xx} in the heat
equation gives $\alpha^2 u_{\xi\xi}/L^2 = u_t$ or $u_{\xi\xi} = (L^2/\alpha^2)u_t$. In a
similar manner, $u_t = u_\tau\tau_t = u_\tau(\alpha^2/L^2)$ and hence $\frac{L^2}{\alpha^2}u_t = u_\tau$
and thus $u_{\xi\xi} = u_\tau$.

22. Substituting $u(x,y,t) = X(x)Y(y)T(t)$ in the P.D.E. yields
$\alpha^2(X''YT + XY''T) = XYT'$, which is equivalent to
$\frac{X''}{X} + \frac{Y''}{Y} = \frac{T'}{\alpha^2 T}$. By keeping the independent variables x
and y fixed and varying t we see that $T'/\alpha^2 T$ must equal
some constant σ_1 since the left side of the equation is
fixed. Hence, $X''/X + Y''/Y = T'/\alpha^2 T = \sigma_1$, or
$X''/X = \sigma_1 - Y''/Y$ and $T' - \sigma_1\alpha^2 T = 0$. By keeping x fixed
and varying y in the equation involving X and Y we see
that $\sigma_1 - Y''/Y$ must equal some constant σ_2 since the left side
of the equation is fixed. Hence,
$X''/X = \sigma_1 - Y''/Y = \sigma_2$ so $X'' - \sigma_2 X = 0$ and $Y'' - (\sigma_1 - \sigma_2)Y = 0$.

For $T' - \sigma_1\alpha^2 T = 0$ to have solutions that remain bounded as $t \to \infty$ we must have $\sigma_1 < 0$. Thus, setting $\sigma_1 = -\lambda^2$, we have $T' + \alpha^2\lambda^2 T = 0$. For $X'' - \sigma_2 X = 0$ and homogeneous B.C., we conclude, as in Sect. 10.1, that $\sigma_2 < 0$ and, if we let $\sigma_2 = -\mu^2$, then $X'' + \mu^2 X = 0$. With these choices for σ_1 and σ_2 we then have $Y'' + (\lambda^2 - \mu^2)Y = 0$.

Section 10.6, Page 620

3. The steady-state temperature distribution $v(x)$ must satisfy Eq.(9) and also satsify the B.C. $v_x(0) = 0$, $v(L) = 0$. The general solution of $v'' = 0$ is $v(x) = Ax + B$. The B.C. $v_x(0) = 0$ implies $A = 0$ and then $v(L) = 0$ implies $B = 0$, so the steady state solution is $v(x) = 0$.

7. Again, $v(x)$ must satisfy $v'' = 0$, $v'(0) - v(0) = 0$ and $v(L) = T$. The general solution of $v'' = 0$ is $v(x) = ax + b$, so $v(0) = b$, $v'(0) = a$ and $v(L) = T$. Thus $a - b = 0$ and $aL + b = T$, which give $a = b = T/(L+1)$. Hence $v(x) = T(x+1)/(L+1)$.

9a. Since the B.C. are not homogeneous, we must first find the steady state solution. Using Eqs.(9) and (10) we have $v'' = 0$ with $v(0) = 0$ and $v(20) = 60$, which has the solution $v(x) = 3x$. Thus the transient solution $w(x,t)$ satisfies the equations $\alpha^2 w_{xx} = w_t$, $w(0,t) = 0$, $w(20,t) = 0$ and $w(x,0) = 25 - 3x$, which are obtained from Eqs.(13) to (15). The solution of this problem is given by Eq.(4) with the c_n given by Eq.(6):

$$c_n = \frac{1}{10}\int_0^{20}(25-3x)\sin\frac{n\pi x}{20}dx = (70\cos n\pi + 50)/n\pi, \text{ and thus}$$

$$u(x,t) = 3x + \sum_{n=1}^{\infty}\frac{70\cos n\pi + 50}{n\pi}e^{-0.86n^2\pi^2 t/400}\sin\frac{n\pi x}{20} \text{ since}$$

$\alpha^2 = .86$ for aluminum.

9b.   9c.

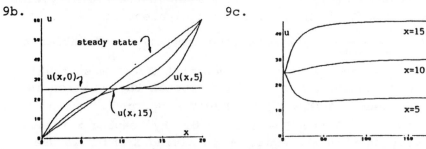

9d. Using just the first term of the sum, we have

$$u(5,t) = 15 - \frac{20}{\pi}e^{-0.86\pi^2 t/400}\sin\frac{\pi}{4} = 15 - .15. \quad \text{Thus}$$

$\frac{20}{\pi}e^{-0.86\pi^2 t/400}\sin\frac{\pi}{4} = .15$, which yields t = 160.30 sec.

To obtain the answer in the text, the first two terms of the sum must be used, which requires an equation solver to solve for t. Note that this reduces t by only .01 seconds.

12a. Since the B.C. are $u_x(0,t) = u_x(L,t) = 0$, t > 0, the solution u(x,t) is given by Eq.(35) with the coefficients c_n determined by Eq.(37). Substituting the I.C. u(x,0) = f(x) = sin(πx/L) into Eq.(37) yields

$c_0 = (2/L)\int_0^L \sin(\pi x/L)dx = 4/\pi,$

$c_1 = (2/L)\int_0^L \sin(\pi x/L)\cos(\pi x/L)dx = (1/\pi)(\sin(\pi x/L)^2\big|_0^L = 0,$

$c_n = (2/L)\int_0^L \sin(\pi x/L)\cos(n\pi x/L)dx$

$= (1/L)\int_0^L \{\sin[(n+1)\pi x/L] - \sin[(n-1)\pi x/L]\}dx$

$= (1/\pi)\{[1 - \cos(n+1)\pi]/(n+1) - [1 - \cos(n-1)\pi]/(n-1)\},$

for n ≠ 1. Thus $c_n = \begin{cases} 0 & \text{n odd} \\ -4/(n^2-1)\pi & \text{n even} \end{cases}$. Hence

$u(x,t) = 2/\pi - (4/\pi)\sum_{n=1}^{\infty}\exp[-4n^2\pi^2\alpha^2 t/L^2]\cos(2n\pi x/L)/(4n^2-1)$

where we are now summing over even terms by setting n = 2n.

12b. As t → ∞ we see that all terms in the series decay to zero except the constant term, 2/π. Hence
$\lim_{t\to\infty} u(x,t) = 2/\pi.$

12c.

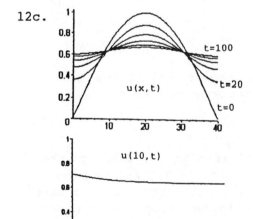

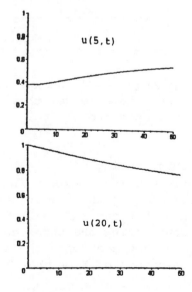

12d. The original heat in the rod is redistributed to give the final temperature distribution, since no heat is lost.

14a. Since the ends are insulated, the solution to this problem is given by Eq.(35) and Eq.(37), with $\alpha^2 = 1$ and $L = 30$. Thus

$$u(x,t) = \frac{c_0}{2} + \sum_{n=1}^{\infty} c_n \exp(-n^2\pi^2 t/900)\cos(n\pi x/30), \text{ where}$$

$$c_0 = \frac{2}{30}\int_0^{30} f(x)dx = \frac{1}{15}\int_5^{10} 25dx = \frac{25}{3} \text{ and}$$

$$c_n = \frac{2}{30}\int_0^{30} f(x)\cos\frac{n\pi x}{30}dx = \frac{1}{15}\int_5^{10} 25\cos\frac{n\pi x}{30}dx = \frac{50}{n\pi}[\sin\frac{n\pi}{3} - \sin\frac{n\pi}{6}].$$

14b.

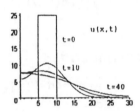
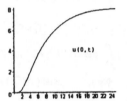
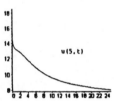
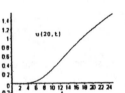
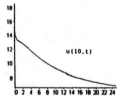

14c.

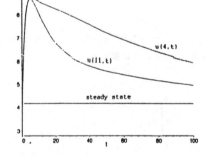

Although $x = 4$ and $x = 11$ are symmetrical to the initial temperature pulse, they are not symmetrical to the insulated end points.

15a. Substituting $u(x,t) = X(x)T(t)$ into Eq.(1) leads to the two O.D.E. $X'' - \sigma X = 0$ and $T' - \alpha^2\sigma T = 0$. An argument similar to the one in the text implies that we must have $X(0) = 0$ and $X'(L) = 0$. Also, by assuming σ is real and considering the three cases $\sigma < 0$, $\sigma = 0$, and $\sigma > 0$ we can show that only the case $\sigma < 0$ leads to nontrivial solutions of $X'' - \sigma X = 0$ with $X(0) = 0$ and $X'(L) = 0$.

Setting $\sigma = -\lambda^2$, we obtain $X(x) = k_1\sin\lambda x + k_2\cos\lambda x$.
Now, $X(0) = 0 \Rightarrow k_2 = 0$ and thus $X(x) = k_1\sin\lambda x$.
Differentiating and setting $x = L$ yields $\lambda k_1\cos\lambda L = 0$.
Since $\lambda = 0$ and $k_1 = 0$ lead to $u(x,t) \equiv 0$, we must choose
λ so that $\cos\lambda L = 0$, or $\lambda = (2n-1)\pi/2L$, $n = 1,2,3,\ldots$ and
thus $X_n = \sin[(2n-1)\pi x/2L]$. The values for λ imply that
$\sigma = -(2n-1)^2\pi^2/4L^2$ so the solutions $T(t)$ of $T' - \alpha^2\sigma T = 0$
are proportional to $\exp[-(2n-1)^2\pi^2\alpha^2 t/4L^2]$. Combining
the above results leads to the desired set of fundamental
solutions.

15b. In order to satisfy the I.C. $u(x,0) = f(x)$ we assume that
$u(x,t)$ has the form

$$u(x,t) = \sum_{n=1}^{\infty} c_n\exp[-(2n-1)^2\pi^2\alpha^2 t/4L^2]\sin[(2n-1)\pi x/2L].$$ The

coefficients c_n are determined by the requirement that

$$u(x,0) = \sum_{n=1}^{\infty} c_n\sin[(2n-1)\pi x/2L] = f(x).$$ Referring to

Prob. 39 of Sect. 10.4 reveals that such a representation
for $f(x)$ is possible if we choose the coefficients
$c_n = (2/L)\int_0^L f(x)\sin[(2n-1)\pi x/2L]dx$.

19. We must solve $v_1''(x) = 0$, $0 \le x \le a$ and $v_2''(x) = 0$,
$a \le x \le L$ subject to the B.C. $v_1(0) = 0$, $v_2(L) = T$ and
the continuity conditions at $x = a$. For the temperature
to be continuous at $x = a$ we must have $v_1(a) = v_2(a)$ and
for the rate of heat flow to be continuous we must have
$-\kappa_1 A_1 v_1'(a) = -\kappa_2 A_2 v_2'(a)$, from Eq.(2) of Appendix A. The
general solutions to the two O.D.E. are $v_1(x) = C_1 x + D_1$
and $v_2(x) = C_2 x + D_2$. By applying the boundary and
continuity conditions we may solve for C_1, D_1, and C_2 and
D_2 to obtain the desired solution.

21. For the steady state solution we set $u(x,t) = v(x)$ which
gives $u_{xx} = v''$, $u_t = 0$, $u(0,t) = v(0)$, and $u(L,t) = v(L)$.
Substituting these into the original problem gives
$0 = \alpha^2 v'' + s(x)$, $v(0) = T_1$ and $v(L) = T_2$. For the
transient solution we set $u(x,t) = v(x) + w(x,t)$ which
gives $u_{xx} = v'' + w_{xx}$, $u_t = w_t$, $u(0,t) = T_1 + w(0,t) = T_1$,
$u(L,t) = T_2 + w(L,t) = T_2$, and $u(x,0) = v(x) + w(x,0) = f(x)$.
Using these expressions for u in the original equations
yields $w_t = \alpha^2(v'' + w_{xx}) + s(x) = \alpha^2 w_{xx}$, $w(0,t) = 0$,
$w(L,t) = 0$ and $w(x,0) = f(x) - v(x)$.

22a. From Prob. 21 $v'' = -k$ and thus $v(x) = c_1 + c_2 x - kx^2/2$.
     $V(0) = T_1 \Rightarrow c_1 = T_1$ and $v(L) = T_2 \Rightarrow c_2 = (T_2 - T_1)/L + kL/2$
     and hence $v(x) = T_1 + (T_2 - T_1)x/L + kLx/2 - kx^2/2$.

22b. From Prob. 21 $w(x,t)$ satisfies $w_t = w_{xx}$, $w(0,t) = 0$, $w(20,t) = 0$
     and $w(x,0) = -v(x) = (x^2 - 20x)/10$, using the givn values

     for T_1, T_2, L and k. Thus $w(x,t) = \sum_{n=1}^{\infty} c_n e^{-n^2\pi^2 t/400} \sin\left(\frac{n\pi x}{20}\right)$,

     where $c_n = \frac{2}{20}\int_0^{20} \frac{x^2-20x}{10} \sin\left(\frac{n\pi x}{20}\right)dx$

     $= \frac{1}{100}\int_0^{20}(x^2-20x)\sin\left(\frac{n\pi x}{20}\right)dx = \begin{cases} 0 & n \text{ even} \\ \dfrac{-320}{n^3\pi^3} & n \text{ odd} \end{cases}$.

     Thus $u(x,t) = v(x) + w(x,t)$

     $= \frac{20x - x^2}{10} - \frac{320}{n^3\pi^3} \sum_{n=1}^{\infty} \frac{e^{-(2n-1)^2\pi^2 t/400}}{(2n-1)^3} \sin\frac{(2n-1)\pi x}{20}$.

     Graphs of $u(x,t)$ are shown for $t = 10,40,100,150$. In all
     cases six terms of the series were used.

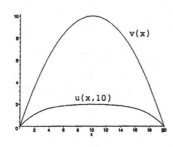
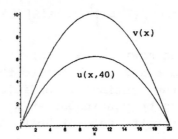
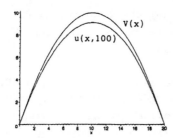
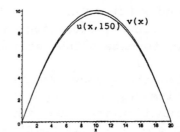

Section 10.7, Page 632

1a. Since the initial velocity is zero, the solution is given
    by Eq.(20) with the coefficients c_n given by Eq.(22).
    Substituting $f(x)$ into Eq.(22) yields

$$c_n = \frac{2}{L} \left[\int_0^{L/2} \frac{2x}{L} \sin\frac{n\pi x}{L} dx + \int_{L/2}^{L} \frac{2(L-x)}{L} \sin\frac{n\pi x}{L} dx \right]$$

$$= \frac{8}{n^2\pi^2} \sin\frac{n\pi}{2}. \quad \text{Thus Eq. (20) becomes}$$

$$u(x,t) = \frac{8}{\pi^2} \sum_{n=1}^{\infty} \frac{1}{n^2} \sin\frac{n\pi}{2} \sin\frac{n\pi x}{L} \cos\frac{n\pi at}{L}.$$

1b.

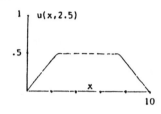
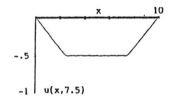
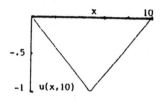

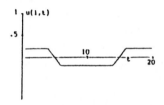
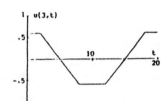
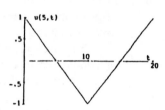

1c.

1e. The graphs in part (b) can best be understood using
    Eq.(28) (or the results of Probs 13 and 14). The
    original triangular shape is composed of two similar
    triangles of 1/2 the height, one of which moves to the
    right, h(x-at), and the other to the left, h(x+at).
    Recalling that h(x) extends f(x) periodically for all x
    and that h(x+at) is a wave moving left and h(x-at) is a
    wave moving right then gives the results shown. The
    graphs in part (c) can then be visualized from those in
    part (b).

6a. The motion is governed by Eqs.(1), (3) and (31), and thus
    the solution is given by Eq.(34) where the k_n are given
    by Eq.(36):

$$k_n = \frac{2}{n\pi a} \left[\int_0^{L/4} \frac{4x}{L} \sin\frac{n\pi x}{L} dx + \int_{L/4}^{3L/4} \sin\frac{n\pi x}{L} dx + \int_{3L}^{L} \frac{4(L-x)}{L} \sin\frac{n\pi x}{L} dx \right]$$

$$= \frac{8L}{n^3\pi^3 a} \left(\sin\frac{n\pi}{4} + \sin\frac{3n\pi}{4} \right). \quad \text{Substituting this in Eq.(34)}$$

$$\text{yields } u(x,t) = \frac{8L}{a\pi^3} \sum_{n=1}^{\infty} \frac{\sin\frac{n\pi}{4} + \sin\frac{3n\pi}{4}}{n^3} \sin\frac{n\pi x}{L} \sin\frac{n\pi at}{L}.$$

6b.

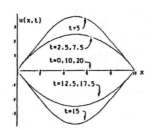

6c.

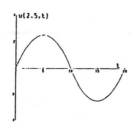

6e. Note that again this is a vibrating system. Of particular importance, however, is that the solution does not have any 'corners', as does the initial velocity. This is due to the integration of the initial velocity, Eq.(36), which improves the convergence of the series and consequently gives a smoothing effect.

9. Assuming that $u(x,t) = X(x)T(t)$ and substituting for u in Eq.(1) leads to the pair of O.D.E. $X'' + \sigma X = 0$, $T'' + a^2\sigma T = 0$. Applying the B.C. $u(0,t) = 0$ and $u_x(L,t) = 0$ to $u(x,t)$ we see that we must have $X(0) = 0$ and $X'(L) = 0$. By considering the three cases $\sigma < 0$, $\sigma = 0$, and $\sigma > 0$ it can be shown that nontrivial solutions of the problem $X'' + \sigma X = 0$, $X(0) = 0$, $X'(L) = 0$ are possible if and only if $\sigma = (2n-1)^2\pi^2/4L^2$, $n = 1,2,\ldots$ and the corresponding solutions for $X(x)$ are proportional to $\sin[(2n-1)\pi x/2L]$. Using these values for σ we find that $T(t)$ is a linear combination of $\sin[(2n-1)\pi at/2L]$ and $\cos[(2n-1)\pi at/2L]$. Now, the I.C. $u_t(x,0)$ implies that $T'(0) = 0$ and thus functions of the form
$u_n(x,t) = \sin[(2n-1)\pi x/2L]\cos[(2n-1)\pi at/2L]$, $n = 1,2,\ldots$
satsify the P.D.E. (1), the B.C. $u(0,t) = 0$, $u_x(L,t) = 0$, and the I.C. $u_t(x,0) = 0$. We now seek a superposition of these fundamental solutions u_n that also satisfies the I.C. $u(x,0) = f(x)$. Thus we assume that

$$u(x,t) = \sum_{n=1}^{\infty} c_n \sin[(2n-1)\pi x/2L]\cos[(2n-1)\pi at/2L].$$ The

I.C. now implies that we must have

$$f(x) = \sum_{n=1}^{\infty} c_n \sin[(2n-1)\pi x/2L].$$ From Prob. 39 of Sect. 10.4

we see that $f(x)$ can be represented by such a series and that $c_n = (2/L)\int_0^L f(x)\sin[(2n-1)\pi x/2L]dx$, $n = 1, 2, \ldots$. Substituting these values into the above series for $u(x,t)$ yields the desired solution.

10a. From Prob. 9 we have

$$c_n = \frac{2}{L}\int_{(L-2)/2}^{(L+2)/2} \sin\frac{(2n-1)\pi x}{2L}dx$$

$$= \frac{-4}{(2n-1)\pi}\left[\cos\left(\frac{(2n-1)\pi(L+2)}{4L}\right) - \cos\left(\frac{(2n-1)\pi(L-2)}{4L}\right)\right]$$

$$= \frac{8}{(2n-1)\pi}\sin\frac{(2n-1)\pi}{4}\sin\frac{(2n-1)\pi}{2L} \text{ using the}$$

trigonometric relations for $\cos(A \pm B)$. Substituting this value of c_n into $u(x,t)$ in Prob. 9 yields the desired solution.

10b.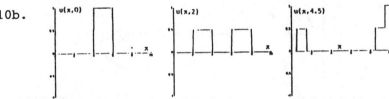

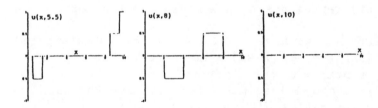

10c.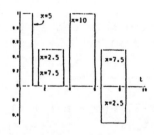

10e. Again visualize the original displacement $f(x)$ as an odd function on $-L < x < 0$ and then periodically for all x. The resulting motion is the result of waves moving left and right from this extension. Note that the original 'corners' persist for all time.

13. Using the chain rule we obtain $u_x = u_\xi \xi_x + u_\eta \eta_x =$
$u_\xi + u_\eta$ since $\xi_x = \eta_x = 1$. Differentiating a second time
gives $u_{xx} = u_{\xi\xi} + 2u_{\xi\eta} + u_{\eta\eta}$. In a similar way we obtain
$u_t = u_\xi \xi_t + u_\eta \eta_t = -au_\xi + au_\eta$, since $\xi_t = -a$, $\eta_t = a$. Thus
$u_{tt} = a^2(u_{\xi\xi} - 2u_{\xi\eta} + u_{\eta\eta})$. Substituting for u_{xx} and u_{tt}
in the wave equation, we obtain $u_{\xi\eta} = 0$. Integrating
both sides of $u_{\xi\eta} = 0$ with respect to η yields
$u_\xi(\xi,\eta) = \gamma(\xi)$ where γ is an arbitrary function of ξ.
Integrating both sides of $u_\xi(\xi,\eta) = \gamma(\xi)$ with respect to ξ
yields $u(\xi,\eta) = \int\gamma(\xi)d\xi + \psi(\eta) = \phi(\xi) + \psi(\eta)$ where $\psi(\eta)$
is an arbitrary function of η and $\int\gamma(\xi)d\xi$ is some
function of ξ denoted by $\phi(\xi)$. Thus
$u(x,t) = u(\xi(x,t),\eta(x,t)) = \phi(x - at) + \psi(x + at)$.

14. The graph of $y = \sin(x-at)$ for the various values of t is
indicated in the figure. Note that the graph of $y = \sin x$
is displaced to the right by the distance "at" for each
value of t.

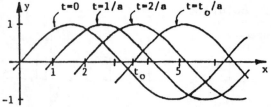

Similarly, the graph of $y = \phi(x + at)$ would be displaced
to the left by a distance "at" for each t. Thus
$\phi(x + at)$ represents a wave moving to the left.

16a. From Prob. 13 we have $u(x,t) = \phi(x-at) + \psi(x+at)$ and
thus $u_t(x,t) = -a\phi'(x-at) + a\psi'(x+at)$. Hence
$u(x,0) = \phi(x) + \psi(x) = f(x)$ and
$u_t(x,0) = -a\phi'(x) + a\psi'(x) = 0$. Dividing the last
equation by a yields the desired result.

16b. Using the hint and the first equation obtained in part(a)
leads to $\phi(x) + \psi(x) = 2\phi(x) + c = f(x)$ so
$\phi(x) = (1/2)f(x) - c/2$ and $\psi(x)$
$= (1/2)f(x) + c/2$. Hence
$u(x,t) = \phi(x - at) + \psi(x + at)$
$= (1/2)[f(x - at) - c] + (1/2)[f(x + at) + c]$
$= (1/2)[f(x - at) + f(x + at)]$.

16c. Substituting $x + at$ for x in $f(x)$ yields
$$f(x + at) = \begin{cases} 2 & -1 < x + at < 1 \\ 0 & \text{otherwise} \end{cases}$$
Subtracting "at" from both sides of the inequality then

yields

$$f(x + at) = \begin{cases} 2 & -1-at < x < 1+at \\ 0 & \text{otherwise} \end{cases}.$$

17a. As in Prob. 16a, we have $u(x,0) = \phi(x) + \psi(x) = 0$ and $u_t(x,0) = -a\phi'(x) + a\psi'(x) = g(x)$.

17b. From part (a) we have $\psi(x) = -\phi(x)$ which yields $-2a\phi(x) = g(x)$ from the second equation in part (a). Integration then yields $\phi(x) - \phi(x_0) = \dfrac{-1}{2a}\displaystyle\int_{x_0}^{x} g(\xi)d\xi$ and hence $\psi(x) = (1/2a)\displaystyle\int_{x_0}^{x} g(\xi)d\xi - \phi(x_0)$.

17c. $u(x,t) = \phi(x-at) + \psi(x+at)$

$\quad = -(1/2a)\displaystyle\int_{x_0}^{x-at} g(\xi)d\xi + \phi(x_0) + (1/2a)\displaystyle\int_{x_0}^{x+at} g(\xi)d\xi - \phi(x_0)$

$\quad = (1/2a)[\displaystyle\int_{x_0}^{x+at} g(\xi)d\xi - \int_{x_0}^{x-at} g(\xi)d\xi]$

$\quad = (1/2a)[\displaystyle\int_{x_0}^{x+at} g(\xi)d\xi + \int_{x-at}^{x_0} g(\xi)d\xi]$

$\quad = (1/2a\displaystyle\int_{x-at}^{x+at} g(\xi)d\xi.$

21. Substituting $u(r,\theta,t) = R(r)\Theta(\theta)T(t)$ into the P.D.E. yields $R''\Theta T + R'\Theta T/r + R\Theta''T/r^2 = R\Theta T''/a^2$ or equivalently $R''/R + R'/rR + \Theta''/\Theta r^2 = T''/a^2 T$. In order for this equation to be valid for $0 < r < r_0$, $0 \le \theta \le 2\pi$, $t > 0$, it is necessary that both sides of the equation be equal to the same constant $-\sigma$. Otherwise, by keeping r and θ fixed and varying t, one side would remain unchanged while the other side varied. Thus we arrive at the two equations $T'' + \sigma a^2 T = 0$ and $r^2 R''/R + rR'/R + \sigma r^2 = -\Theta''/\Theta$, where we have multiplied both sides by r^2. By an argument similar to the one above we conclude that both sides of the last equation must be equal to the same constant δ. This leads to the two equations $r^2 R'' + rR' + (\sigma r^2 - \delta)R = 0$ and $\Theta'' + \delta\Theta = 0$. Since the circular membrane is continuous, we must have $\Theta(2\pi) = \Theta(0)$, which requires $\delta = \mu^2$, μ a non-negative integer. The condition $\Theta(2\pi) = \Theta(0)$ is also known as the periodicity condition. Since we also desire solutions which vary periodically in time, it is clear that the separation constant σ should be positive, $\sigma = \lambda^2$. Thus we arrive at the three equations $r^2 R'' + rR' + (\lambda^2 r^2 - \mu^2)R = 0$, $\Theta'' + \mu^2\Theta = 0$, and $T'' + \lambda^2 a^2 T = 0$.

23a. Substituting $u(x,t) = X(x)T(t)$ into the PDE and
     separating variables we obtain $\dfrac{X''}{X} = \dfrac{T''}{\alpha^2 T} + \gamma^2 = \sigma$. The
     separation constant $\sigma = -\lambda^2$ using the same arguements as
     seen earlier. Thus $X'' + \lambda^2 X = 0$, $X(0) = X(L) = 0$ and
     $T'' + (\gamma^2 + \lambda^2)\alpha^2 T = 0$, $T'(0) = 0$. The BVP has the
     solution $\lambda = n\pi/L$ and $X(x) = \sin(n\pi x/L)$, while the IVP has
     the solution $\cos\beta t$, where $\beta = \sqrt{\gamma^2 + \lambda^2}\,\alpha = \sqrt{\gamma^2 + n^2\pi^2/L^2}\,\alpha$.
     Thus $u(x,t)$ has the form shown. Satisfying the remaining
     IC yields the shown values for c_n.

23b. Use $\sin A \cos B = [\sin(A+B) + \sin(A-B)]/2$ with $A = n\pi x/L$ and
     $B = (n\pi/L)\sqrt{1 + \dfrac{\gamma^2 L^2}{n^2\pi^2}}\,\alpha t = (n\pi/L)\alpha_n t$, where α_n is the
     speed of the wave propagation.

23c. From part (b), α_n is independent of n when $\gamma = 0$.

Section 10.8, Page 645

1a. Assuming that $u(x,y) = X(x)Y(y)$ leads to the two O.D.E.
     $X'' - \sigma X = 0$, $Y'' + \sigma Y = 0$. The B.C. $u(0,y) = 0$,
     $u(a,y) = 0$ imply that $X(0) = 0$ and $X(a) = 0$. Thus
     nontrivial solutions to $X'' - \sigma X = 0$ which satisfy these
     boundary conditions are possible only if $\sigma = -(n\pi/a)^2$,
     $n = 1,2\ldots$; the corresponding solutions for $X(x)$ are
     proportional to $\sin(n\pi x/a)$. The B.C. $u(x,0) = 0$ implies
     that $Y(0) = 0$. Solving $Y'' - (n\pi/a)^2 Y = 0$ subject to this
     condition we find that $Y(y)$ must be proportional to
     $\sinh(n\pi y/a)$. The fundamental solutions are then
     $u_n(x,y) = \sin(n\pi x/a)\sinh(n\pi y/a)$, $n = 1,2,\ldots$, which
     satisfy Laplace's equation and the homogeneous B.C. We
     assume that $u(x,y) = \displaystyle\sum_{n=1}^{\infty} c_n \sin(n\pi x/a)\sinh(n\pi y/a)$, where
     the coefficients c_n are determined from the B.C.
     $u(x,b) = \displaystyle\sum_{n=1}^{\infty} c_n \sin(n\pi x/a)\sinh(n\pi b/a) = g(x)$. Recognizing
     this as a sine series it follows that
     $c_n \sinh(n\pi b/a) = (2/a)\displaystyle\int_0^a g(x)\sin(n\pi x/a)dx$, $n = 1,2,\ldots$.

1b. Substituting for $g(x)$ in the equation for c_n we have

$$c_n \sinh(n\pi b/a) = (2/a)\left\{\int_0^{a/2} x\sin(n\pi x/a)dx + \int_{a/2}^a (a-x)\sin(n\pi x/a)dx\right\}$$

$$= [4a\sin(n\pi/2)]/n^2\pi^2, \quad n = 1,2,\ldots, \text{ so}$$

$c_n = (4a/\pi^2)[\sin(n\pi/2)]/[n^2\sinh(n\pi b/a)]$. Substituting these values for c_n in the above series yields the desired solution.

1c.

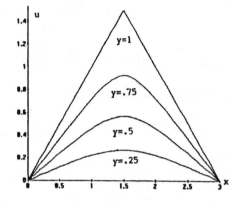

1d.

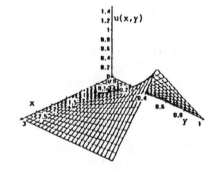

2.   In solving the D.E. $Y'' - \lambda^2 Y = 0$, one normally writes $Y(y) = c_1\sinh\lambda y + c_2\cosh\lambda y$. However, since we have

Y(b) = 0, it is advantageous to rewrite Y as $Y(y) = d_1\sinh\lambda(b-y) + d_2\cosh\lambda(b-y)$, where d_1, d_2 are also arbitrary constants and can be related to c_1, c_2 using the appropriate hyperbolic trigonometric identities. The important thing, however, is that the second form also satisfies the D.E. and thus $Y(y) = d_1\sinh\lambda(b-y)$ satisfies the D.E. and the homogeneous B.C. $Y(b) = 0$. The rest of the problem follows the pattern of Prob. 1.

3a. Let $v(x,y)$ satisfy Laplace's equation with the B.C. of Eq.(4) and let $w(x,y)$ satisfy Laplace's equation with the B.C. of Prob. 2. Then using superposition will give

$u(x,y) = v(x,y) + w(x,y)$ as the solution of the given problem.

4. Following the pattern of Prob. 3, one could consider adding the solutions of four problems, each with only one non-homogeneous B.C. It is also possible to consider adding the solutions of only two problems, each with only two non-homogeneous B.C., as long as they involve the same variable. For instance, one such problem would be $u_{xx} + u_{yy} = 0$, $u(x,0) = 0$, $u(x,b) = 0$, $u(0,y) = k(y)$, $u(a,y) = f(y)$, which has the fundamental solutions $u_n(x,y) = [c_n\sinh(n\pi x/b) + d_n\cosh(n\pi x/b)]\sin(n\pi y/b)$.

Assuming $u(x,y) = \displaystyle\sum_{n=1}^{\infty} u_n(x,y)$ and using the B.C.

$u(0,y) = k(y)$ we obtain $d_n = (2/b)\displaystyle\int_0^b k(y)\sin(n\pi y/b)dy$. Using

the B.C. $u(a,y) = f(y)$ we obtain

$c_n\sinh(n\pi a/b) + d_n\cosh(n\pi a/b) = (2/b)\displaystyle\int_0^b f(y)\sin(n\pi y/b)dy$, which

can be solved for c_n, since d_n is already known. The second problem, in this approach, would be $u_{xx} + u_{yy} = 0$, $u(x,0) = h(x)$, $u(x,b) = g(x)$, $u(0,y) = 0$ and $u(a,y) = 0$. This has the fundamental solutions $u_n(x,y) = [a_n\sinh(n\pi y/a) + b_n\cosh(n\pi y/a)]\sin(n\pi x/a$, so that

$u(x,y) = \displaystyle\sum_{n=1}^{\infty} u_n(x,y)$. Thus $u(x,0) = h(x)$ gives

$b_n = (2/a)\displaystyle\int_0^a h(x)\sin(n\pi x/a)dx$ and $u(x,b) = g(x)$ gives

$a_n\sinh(n\pi b/a) + b_n\cosh(n\pi b/a) = (2/a)\displaystyle\int_0^a g(x)\sin(n\pi x/a)dx$,

which can be solved for a_n since b_n is known.

5. Using Eq.(20) and following the same arguments as presented in the text, we find that $R(r) = k_1 r^n + k_2 r^{-n}$ and $\Theta(\theta) = c_1\sin n\theta + c_2\cos n\theta$, for n a positive integer, and $u_0(r,\theta) = 1$ for $n = 0$. Since we require that $u(r,\theta)$ be bounded as $r \to \infty$, we conclude that $k_1 = 0$. The fundamental solutions are therefore $u_n(r,\theta) = r^{-n}\cos n\theta$, $v_n(r,\theta) = r^{-n}\sin n\theta$, $n = 1,2,\ldots$ together with $u_0(r,\theta) = 1$. Assuming that u can be expressed as a linear combination of the fundamental solutions we have

$u(r,\theta) = c_0/2 + \displaystyle\sum_{n=1}^{\infty} r^{-n}(c_n\cos n\theta + k_n\sin n\theta)$. The B.C.

requires that

$$u(a,\theta) = c_0/2 + \sum_{n=1}^{\infty} a^{-n}(c_n\cos n\theta + k_n\sin n\theta) = f(\theta) \text{ for}$$

$0 \leq \theta < 2\pi$. This is precisely the Fourier series
representation for $f(\theta)$ of period 2π and thus

$$a^{-n}c_n = (1/\pi)\int_0^{2\pi} f(\theta)\cos n\theta d\theta, \quad n = 0,1,2,\ldots \text{ and}$$

$$a^{-n}k_n = (1/\pi)\int_0^{2\pi} f(\theta)\sin n\theta d\theta, \quad n = 1,2\ldots \text{ .}$$

7. Again we let $u(r,\theta) = R(r)\Theta(\theta)$ and thus we have
$r^2R'' + rR' - \sigma R = 0$ and $\Theta'' + \sigma\Theta = 0$, with $R(0)$ bounded and
the B.C. $\Theta(0) = \Theta(\alpha) = 0$. For $\sigma \leq 0$ we find that
$\Theta(0) \equiv 0$, so we let $\sigma = \lambda^2$ (λ^2 real) and thus
$\Theta(\theta) = c_1\cos\lambda\theta + c_2\sin\lambda\theta$. The B.C. $\Theta(0) = 0 \Rightarrow c_1 = 0$ and
the B.C. $\Theta(\alpha) = 0 \Rightarrow \lambda = n\pi/\alpha$, $n = 1,2,\ldots$.
Substituting these values into Eq.(31) we obtain
$R(r) = k_1 r^{n\pi/\alpha} + k_2 r^{-n\pi/\alpha}$. However $k_2 = 0$ since $R(0)$ must
be bounded, and thus the fundamental solutions are
$u_n(r,\theta) = r^{n\pi/\alpha}\sin(n\pi\theta/\alpha)$. The desired solution may now
be formed using previously discussed procedures.

8a. Separating variables, we obtain
$X'' + \lambda^2 X = 0$, $X(0) = 0$, $X(a) = 0$ and $Y'' - \lambda^2 Y = 0$, $Y(y)$ bounded
as $y \to \infty$. Thus $X(x) = \sin(n\pi x/a)$, and $\lambda^2 = (n\pi/a)^2$.
However, since neither $\sinh y$ nor $\cosh y$ are bounded as $y \to \infty$,
we must write the solution to $Y'' - (n\pi/a)^2 Y = 0$ as
$Y(y) = c_1\exp[n\pi y/a] + c_2\exp[-n\pi y/a]$. Thus we must choose
$c_1 = 0$ so that $u(x,y) = X(x)Y(y) \to 0$ as $y \to \infty$. The
fundamental solutions are then $u_n(x,t) = e^{-n\pi y/a}\sin(n\pi x/a)$.

$$u(x,y) = \sum_{n=1}^{\infty} c_n u_n(x,y) \text{ then gives}$$

$$u(x,0) = \sum_{n=1}^{\infty} c_n\sin(n\pi x/a) = f(x) \text{ so that } c_n = \frac{2}{a}\int_0^a f(x)\sin\frac{n\pi x}{a}dx.$$

8b. $c_n = \frac{2}{a}\int_0^a x(a-x)\sin\frac{n\pi x}{a}dx = \frac{4a^2}{n^3\pi^3}(1-\cos n\pi)$

8c. Using just the first term and letting $a = 5$, we have
$u(x,y) = \frac{200}{\pi^3}e^{-\pi y/5}\sin\frac{\pi x}{5}$, which, for a fixed y, has a maximum

at $x = 5/2$ and thus we need to find y such that
$u(5/2,y) = \frac{200}{\pi^3}e^{-\pi y/5} = .1$. Taking the logarithm of both

sides and solving for y yields $y_0 = 6.6315$. With an equation
solver, more terms can be used. However, to four decimal
places, three terms yield the same result as above.

13a. Assuming that $u(x,y) = X(x)Y(y)$ and substituting into
Eq.(1) leads to the two O.D.E. $X'' - \sigma X = 0$, $Y'' + \sigma Y = 0$.
The B.C. $u(x,0) = 0$, $u_y(x,b) = 0$ imply that $Y(0) = 0$ and
$Y'(b) = 0$. For nontrivial solutions to exist for
$Y'' + \sigma Y = 0$ with these B.C. we find that σ must take the
values $(2n-1)^2\pi^2/4b^2$, $n = 1,2,...$; the corresponding
solutions for $Y(y)$ are proportional to $\sin[(2n-1)\pi y/2b]$.
Solutions to $X'' - [(2n-1)^2\pi^2/4b^2]X = 0$ are of the form
$X(x) = A\sinh[(2n-1)\pi x/2b] + B\cosh[(2n-1)\pi x/2b]$. However,
the boundary condition $u(0,y) = 0$ implies that
$X(0) = B = 0$. It follows that the fundamental solutions
are $u_n(x,y) = c_n\sinh[(2n-1)\pi x/2b]\sin[(2n-1)\pi y/2b]$,
$n = 1,2,...$. To satisfy the remaining B.C. at $x = a$ we
assume that we can represent the solution $u(x,y)$ in the

form $u(x,y) = \displaystyle\sum_{n=1}^{\infty} c_n\sinh[(2n-1)\pi x/2b]\sin[(2n-1)\pi y/2b]$.

The coefficients c_n are determined by the B.C.

$u(a,y) = \displaystyle\sum_{n=1}^{\infty} c_n\sinh[(2n-1)\pi a/2b]\sin[(2n-1)\pi y/2b] = f(y)$.

By properly extending f as a periodic function of period
$4b$ as in Prob. 39, Sect. 10.4, we find that the
coefficients c_n are given by

$c_n\sinh[(2n-1)\pi a/2b] = (2/b)\displaystyle\int_0^b f(y)\sin[(2n-1)\pi y/2b]dy$,

$n = 1,2,...$.

16a. If $u(x,z) = X(x)Z(z)$ we obtain $X'' + \lambda^2 X = 0$, $X'(0) = 0$,
$X'(a) = 0$ and $Z'' - \lambda^2 Z = 0$, $Z'(0) = 0$. The first B.V.P.
yields $\lambda^2 = n^2\pi^2/a^2$ and $X(x) = \cos(n\pi x/a)$, $n = 0,1,2 \cdots$, as
in Ex.(2) of Sect. 10.6. Solving the second B.V.P we find
$Z(z) = \cosh(n\pi z/a)$, so that

$u(x,z) = c_0/2 + \displaystyle\sum_{n=1}^{\infty} c_n\cosh(n\pi z/a)\cos(n\pi x/a)$. Setting $z = b$

we get $u(x,b) = c_0/2 + \displaystyle\sum_{n=1}^{\infty} c_n\cosh(n\pi b/a)\cos(n\pi x/a) = b + \alpha x$,

where $c_0 = (2/a)\displaystyle\int_0^a (b + \alpha x)dx = 2b + \alpha a$ and

$c_n\cosh(n\pi b/a) = (2/a)\displaystyle\int_0^a (b + \alpha x)\cos(n\pi x/a)dx$

$= \begin{cases} 0 & n \text{ even} \\ -4\alpha a/n^2\pi^2 & n \text{ odd} \end{cases}$. Solving for c_n and

substituting into $u(x,z)$ yields the desired solution.

CHAPTER 11

Section 11.1, Page 662

2. Since the B.C. at $x = 1$ is nonhomogeneous, the B.V.P. is nonhomogeneous.

4. The D.E. may be written $y'' + (\lambda - x^2)y = 0$ and is thus homogeneous, as are both B.C.

5. Since the D.E. contains the nonhomogeneous term 1, the B.V.P. is nonhomogeneous.

9. If $\lambda = 0$, then $y(x) = c_1 x + c_2$ and thus $y(0) = c_2$, $y(1) = c_1 + c_2$, $y'(0) = c_1$ and $y'(1) = c_1$. Hence the B.C. yield the two equations $c_2 - c_1 = 0$ and $c_1 + c_2 + c_1 = 0$ which give $c_1 = c_2 = 0$ and thus $\lambda = 0$ is not an eigenvalue.

   If $\lambda > 0$, the general solution of the D.E. is $y = c_1 \sin\sqrt{\lambda}\, x + c_2 \cos\sqrt{\lambda}\, x$. The B.C. require that $-\sqrt{\lambda}\, c_1 + c_2 = 0$, and $(\sin\sqrt{\lambda} + \sqrt{\lambda}\cos\sqrt{\lambda})c_1 + (\cos\sqrt{\lambda} - \sqrt{\lambda}\sin\sqrt{\lambda})c_2 = 0$. Taking the determinant of the coefficients of c_1 and c_2 we see that in order to have nontrivial solutions λ must satisfy $(\lambda - 1)\sin\sqrt{\lambda} - 2\sqrt{\lambda}\cos\sqrt{\lambda} = 0$. In this case $c_2 = \sqrt{\lambda}\, c_1$ and thus $\phi_n = \sin\sqrt{\lambda_n}\, x + \sqrt{\lambda_n}\cos\sqrt{\lambda_n}\, x$. If $\lambda \neq 1$, the eigenvalue equation is equivalent to $\tan\sqrt{\lambda} = 2\sqrt{\lambda}/(\lambda - 1)$ and thus by graphing $f(\sqrt{\lambda}) = \tan\sqrt{\lambda}$ and $g(\sqrt{\lambda}) = 2\sqrt{\lambda}/(\lambda - 1)$ we can estimate the eigenvalues. Since $g(\sqrt{\lambda})$ has a vertical asymptote at $\lambda = 1$ and $f(\sqrt{\lambda})$ has a vertical asymptote at $\sqrt{\lambda} = \pi/2$, we see that $1 < \sqrt{\lambda_1} < \pi/2$. By interating numerically, we find $\sqrt{\lambda_1} \approx 1.30655$ and thus $\lambda_1 \approx 1.7071$. The second eigenvalue will lie to the right of π, the second zero of $\tan\sqrt{\lambda}$. Again iterating numerically, we find $\sqrt{\lambda_2} \approx 3.6732$ and thus $\lambda_2 \approx 13.4924$. For large values of n, we see from the graph that $\sqrt{\lambda_n} \approx (n-1)\pi$, which are the zeros of $\tan\sqrt{\lambda}$. Thus $\lambda_n \approx (n-1)^2\pi^2$ for large n.

   For $\lambda < 0$, the discussion follows the pattern of Ex. 1 yielding $y(x) = c_1 \sinh\sqrt{\mu}\, x + c_2 \cosh\sqrt{\mu}\, x$. The B.C. then yield

$-\sqrt{\mu}\, c_1 + c_2 = 0$ and
$(\sinh\sqrt{\mu} + \sqrt{\mu}\cosh\sqrt{\mu})c_1 + (\cosh\sqrt{\mu} + \sqrt{\mu}\sinh\sqrt{\mu})c_2 = 0$, which
have a non-zero solution if and only if
$(\mu+1)\sinh\sqrt{\mu} + 2\sqrt{\mu}\cosh\sqrt{\mu} = 0$. By plotting $y = \tanh\sqrt{\mu}$ and
$y = -2\sqrt{\mu}/(\mu+1)$ we see that they intersect only at $\mu = 0$, and
thus there are no negative eigenvalues.

10. If $\lambda = 0$, the general solution of the D.E. is
$y = c_1 + c_2 x$. The B.C. $y(0) + y'(0) = 0$ requires
$c_1 + c_2 = 0$ and the B.C. $y(1) = 0$ requires $c_1 + c_2 = 0$
and thus $\lambda = 0$ is an eigenvalue with corresponding
eigenfunction $\phi_0(x) = 1-x$.

If $\lambda < 0$, set $-\lambda = \mu^2$ to obtain $y = c_1\cos\mu x + c_2\sin\mu x$.
In this case the B.C. require $c_1 + \mu c_2 = 0$ and
$c_1\cos\mu + c_2\sin\mu = 0$ which yields nontrivial solutions for c_1
and c_2 (i.e., $c_1 = -\mu c_2$) if and only if $\tan\mu = \mu$. By
plotting on the same graph $f(\mu) = \mu$ and $g(\mu) = \tan\mu$, we see
that they intersect at $\mu_0 = 0$ ($\mu = 0 \Rightarrow \lambda = 0$, which has
already been discussed), $\mu_1 \approx 4.49341$ (which is just to the
left of the vertical asymptote of $\tan\mu$ at $\mu = 3\pi/2$,
$\mu_2 \approx 7.72525$ (which is just to the left of the vertical
asymptote of $\tan\mu$ at $\mu = 5\pi/2$) and for larger n values
$\mu_n \approx (2n+1)\pi/2$. Since $\lambda_n = -\mu_n^2$, we have $\lambda_1 \approx -20.1907$,
$\lambda_2 = -59.6795$, $\lambda_n \approx -(2n+1)^2\pi^2/4$ and $\phi_n = \sin\mu_n x - \mu_n\cos\mu_n x$.

If $\lambda > 0$, the general solution of the D.E. is
$y(x) = c_1\cosh\sqrt{\lambda}\, x + c_2\sinh\sqrt{\lambda}\, x$. The B.C. respectively
require that $c_1 + \sqrt{\lambda}\, c_2 = 0$ and $c_1\cosh\sqrt{\lambda} + c_2\sinh\sqrt{\lambda} = 0$ and
thus λ must satisfy $\tanh\sqrt{\lambda} = \sqrt{\lambda}$ in order to have
nontrivial solutions. The only solution of this equation is
$\lambda = 0$ and thus there are no positive eigenvalues.

11a. From Eq.(i) the coefficient of y' is μQ and from Eq.(ii) the
coefficient of y' is $(\mu P)'$. Thus $(\mu P)' = \mu Q$, which gives
Eq.(iii).

11b. Eq.(iii) is both linear and separable. Using the latter
approach we have $d\mu/\mu = [(Q-P')/P]dx$ and thus
$\ln\mu = \int_{x_0}^{x} [Q(s)/P(s)]ds - [\ln P(x) - \ln P(x_0)]$. Taking the
exponential of both sides yields Eq.(iv). The choice of

x_0 simply alters the constant of integration, which is immaterial here.

13. Since $P(x) = x^2$ and $Q(x) = x$, we find that
$\mu(x) = (1/x^2)\exp[\int_{x_0}^x (s/s^2)ds] = k/x$, where k is an
arbitrary constant which may be set equal to 1. It follows that Bessel's equation takes the form
$(xy')' + (x^2-v^2/x)y = 0$.

18a. Assuming $y = s(x)u$, we have $y' = s'u + su'$ and
$y'' = s''u + 2s'u' + su''$ and thus the D.E. becomes
$su'' + (2s'+4s)u' + [s'' + 4s' + (4+9\lambda)s]u = 0$. Setting
$2s' + 4s = 0$ we find $s(x) = e^{-2x}$ and the D.E. becomes
$u'' + 9\lambda u = 0$. The B.C. $y(0) = 0$ yields $s(0)u(0) = 0$, or
$u(0) = 0$ since $s(0) \neq 0$. The B.C. at L
is $y'(L) = s'(L)u(L) + s(L)u'(L) = e^{-2L}(-2u(L) + u'(L)) = 0$
and thus $u'(L) - 2u(L) = 0$. Thus the B.V.P. satisfied by
$u(x)$ is $u'' + 9\lambda u = 0$, $u(0) = 0$, $u'(L) - 2u(L) = 0$.

If $\lambda < 0$, the general solution of the D.E. $u'' + 9\lambda u = 0$
is $u = c_1\sinh 3\mu x + c_2\cosh 3\mu x$ where $-\lambda = \mu^2$. The B.C.
require that $c_2 = 0$, $c_1(3\mu\cosh 3\mu L - 2\sinh 3\mu L) = 0$. In
order to have nontrivial solutions μ must satisfy the
equation $3\mu/2 = \tanh 3\mu L$. A graphical analysis reveals
that for $L \leq 1/2$ this equation has no solutions for $\mu \neq 0$
so there are no negative eigenvalues for $L \leq 1/2$. If
$L > 1/2$ there is one solution and hence one negative
eigenvalue with eigenfuction $\phi_{-1}(x) = e^{-2x}\sinh 3\mu x$.

If $\lambda = 0$, the general solution of the D.E. $u'' + 9\lambda u = 0$
is $u = c_1 + c_2x$. The B.C. require that $c_1 = 0$,
$c_2(1-2L) = 0$ so nontrivial solutions are possible only if
$L = 1/2$. In this case the eigenfuction is $\phi_0(x) = xe^{-2x}$.

If $\lambda > 0$, the general solution of the D.E. $u'' + 9\lambda u = 0$
is $u = c_1\sin 3\sqrt{\lambda}\, x + c_2\cos\sqrt{\lambda}\, x$. The B.C. require that
$c_2 = 0$, $c_1(3\sqrt{\lambda}\cos 3\sqrt{\lambda}\, L - 2\sin 3\sqrt{\lambda}\, L) = 0$. In order to
have nontrivial solutions λ must satisfy the equation
$\sqrt{\lambda} = (2/3)\tan 3\sqrt{\lambda}\, L$. A graphical analysis reveals that
there is an infinite number of solutions to this
eigenvalue equation. Thus the eigenfunctions are
$\phi_n(x) = e^{-2x}\sin 3\sqrt{\lambda_n}\, x$ where the eigenvalues λ_n satisfy
$\sqrt{\lambda_n} = (2/3)\tan 3\sqrt{\lambda_n}\, L$.

20. This is an Euler equation whose characteristic equation
    has roots $r_1 = \lambda$ and $r_2 = 1$. If $\lambda = 1$ the general
    solution of the D.E. is $y = c_1 x + c_2 x \ln x$ and the B.C.
    require that $c_1 = c_2 = 0$ and thus $\lambda = 1$ is not an
    eigenvalue. If $\lambda \neq 1$, $y = c_1 x + c_2 x^\lambda$ is the general
    solution and the B.C. require that $c_1 + c_2 = 0$ and
    $2c_1 + c_2 2^\lambda - (c_1 + \lambda c_2 2^{\lambda-1}) = 0$. Thus nontrivial solutions
    exist if and only if $\lambda = 2(1-2^{-\lambda})$. The graphs of
    $f(\lambda) = \lambda$ and $g(\lambda) = 2(1-2^{-\lambda})$ intersect only at $\lambda = 1$
    (which has already been discussed) and $\lambda = 0$. Thus the
    only eigenvalue is $\lambda = 0$ with corresponding eigenfunction
    $\phi(x) = x - 1$ (since $c_1 = -c_2$).

22a. For positive λ, the general solution of the D.E. is
    $y = c_1 \sin\sqrt{\lambda}\, x + c_2 \cos\sqrt{\lambda}\, x$. The B.C. require that
    $\sqrt{\lambda}\, c_1 + \alpha c_2 = 0$, $c_1 \sin\sqrt{\lambda} + c_2 \cos\sqrt{\lambda} = 0$. Nontrivial
    solutions exist if and only if $\sqrt{\lambda}\cos\sqrt{\lambda} - \alpha\sin\sqrt{\lambda} = 0$.
    If $\alpha = 0$ this equation is satisfied by the sequence
    $\lambda_n = [(2n-1)\pi/2]^2$, $n = 1,2,\ldots$. If $\alpha \neq 0$, λ must
    satisfy the equation $\sqrt{\lambda}/\alpha = \tan\sqrt{\lambda}$. A plot of the
    graphs of $f(\sqrt{\lambda}) = \sqrt{\lambda}/\alpha$ and $g(\sqrt{\lambda}) = \tan\sqrt{\lambda}$ reveals
    that there is an infinite sequence of postive eigenvalues
    for $\alpha < 0$ and $\alpha > 0$.

22b. By procedures shown previously, the cases $\lambda < 0$ and
    $\lambda = 0$, when $\alpha < 1$, lead to only the trivial solution and
    thus by part (a) all real eigenvalues are positive. For
    $0 < \alpha < 1$, the graphs of $f(\sqrt{\lambda})$ and $g(\sqrt{\lambda})$, see part (a),
    intersect once on $0 < \sqrt{\lambda} < \pi/2$. As α approaches 1 from
    below, the slope of $f(\sqrt{\lambda})$ decreases and thus the
    intersection point approaches zero.

22c. If $\lambda = 0$, then $y(x) = c_1 + c_2 x$ and the B.C. yield
    $\alpha c_1 + c_2 = 0$ and $c_1 + c_2 = 0$, which have a non-zero
    solution if and only if $\alpha = 1$.

22d. Let $-\lambda = \mu^2$, then $y(x) = c_1 \cosh\mu x + c_2 \sinh\mu x$ and thus the
    B.C. yield $\alpha c_1 + \mu c_2 = 0$ and $(\cosh\mu)c_1 + (\sinh\mu)c_2 = 0$, which
    have non-zero solutions if and only if $\tanh\mu = \mu/\alpha$. For $\alpha>1$,
    the straight line $y = \mu/\alpha$ intersects the curve $y = \tanh\mu$ in
    one point, which increases as α increases. Thus $\lambda = -\mu^2$
    decreases as α increases.

23. Using the D.E.for ϕ_m and following the hint yields:
$\int_0^L \phi_m'' \phi_n dx + \lambda_m \int_0^L \phi_m \phi_n dx = 0$. Integrating the first term by parts
yields: $\phi_m' \phi_n \big|_0^L - \int_0^L \phi_m' \phi_n' dx = -\lambda_m \int_0^L \phi_m \phi_n dx$. Upon utilizing he
B.C. the first term on the left vanishes and thus
$\int_0^L \phi_n' \phi_m' dx = \lambda_m \int_0^L \phi_m \phi_n dx$. Similarly, the D.E. for ϕ_n yields
$\int_0^L \phi_m' \phi_n' dx = \lambda_n \int_0^L \phi_n \phi_m dx$ and thus $(\lambda_n - \lambda_m) \int_0^L \phi_m \phi_n dx = 0$. If
$\lambda_n \neq \lambda_m$ the desired result follows.

24b. The general solution of the D.E. is
$y = c_1 \sin\mu x + c_2 \cos\mu x + c_3 \sinh\mu x + c_4 \cosh\mu x$ where $\lambda = \mu^4$.
The B.C. require that $c_2 + c_4 = 0$, $-c_2 + c_4 = 0$,
$c_1 \sin\mu L + c_2 \cos\mu L + c_3 \sinh\mu L + c_4 \cosh\mu L = 0$, and
$c_1 \cos\mu L - c_2 \sin\mu L + c_3 \cosh\mu L + c_4 \sinh\mu L = 0$. The first two
equations yield $c_2 = c_4 = 0$, and the last two have nontrivial
solutions if and only if $\sin\mu L\cosh\mu L - \cos\mu L\sinh\mu L = 0$. In
this case the third equation yields $c_3 = -c_1 \sin\mu L/\sinh\mu L$ and
thus the desired eigenfunctions are obtained. The quantity
μL can be approximated by finding the intersection of
$f(x) = \tan x$ and $g(x) = \tanh x$, where $x = \mu L$. The first
intersection is at $x \approx 3.9266$, which gives $\lambda_1 \approx 237.72/L^4$ and
the second intersection is at $x \approx 7.0686$, which gives
$\lambda_2 \approx 2,496.5/L^4$.

Section 11.2, Page 675

1. We have $y(x) = c_1 \cos\sqrt{\lambda}\, x + c_2 \sin\sqrt{\lambda}\, x$ and thus $y(0) = 0$
yields $c_1 = 0$ and $y'(1) = 0$ yields $\sqrt{\lambda}\, c_2 \cos\sqrt{\lambda} = 0$. $\lambda = 0$
gives $y(x) = 0$, so is not an eigenvalue. Otherwise
$\lambda = (2n-1)^2 \pi^2/4$ and the eigenfuctions are $\sin[(2n-1)\pi x/2]$,
$n = 1,2,\ldots$. Thus, by Eq.(20), we must choose k_n so that
$\int_0^1 \{k_n \sin[(2n-1)\pi x/2]\}^2 dx = 1$, since the weight function
$r(x) = 1$ (by comparing the D.E. to Eq.(1)). Evaluating the
integral yields $k_n^2/2 = 1$ and thus $k_n = \sqrt{2}$ and the desired
normalized eigenfuctions are obtained.

3. Note here that $\phi_0(x) = 1$, the eigenfunction for $\lambda = 0$,
satisifes Eq.(20) and hence it is already normalized.

5. From Prob. 17 of Sect. 11.1 we have $e^x \sin n\pi x$, $n = 1,2,\ldots$
as the eigenfunctions and thus k_n must be chosen so that

$\int_0^1 r(x)k_n^2 e^{2x}\sin^2 n\pi x\,dx = 1$. To determine $r(x)$, we must write
the D.E. in the form of Eq.(1). That is, we multiply the
D.E. by $r(x)$ to obtain $ry'' - 2ry' + ry + \lambda ry = 0$. Now
choose r so that $(ry')' = ry'' - 2ry'$, which yields $r' = -2r$ or
$r(x) = e^{-2x}$. Thus the above integral becomes
$\int_0^1 k_n^2\sin^2 n\pi x\,dx = 1$ and $k_n = \sqrt{2}$. Hence $\phi_n(x) = \sqrt{2}\,e^x\sin n\pi x$
are the normalized eigenfunctions.

7. Using Eq.(34) with $r(x) = 1$, we find that the
    coefficients of the series (32) are determined by
    $a_n = (f,\phi_n) = \sqrt{2}\int_0^1 x\sin[(2n-1)\pi x/2]\,dx$
    $\qquad = (4\sqrt{2}/(2n-1)^2\pi^2)\sin(2n-1)\pi/2$. Thus Eq.(32) yields
    $f(x) = \dfrac{4\sqrt{2}}{\pi^2}\displaystyle\sum_{n=1}^{\infty}\dfrac{(-1)^{n-1}}{(2n-1)^2}\sqrt{2}\sin[(2n-1)\pi x/2],\ 0 \le x \le 1,$
    which agrees with the expansion using the approach
    developed in Prob. 39 of Sect. 10.4.

10. In this case $\phi_n(x) = (\sqrt{2}/\alpha_n)\cos\sqrt{\lambda_n}\,x$, where
    $\alpha_n = (1 + \sin^2\sqrt{\lambda_n})^{1/2}$. Thus Eq.(34) yields
    $a_n = (\sqrt{2}/\alpha_n)\int_0^1\cos\sqrt{\lambda_n}\,x\,dx = \sqrt{2}\sin\sqrt{\lambda_n}/\alpha_n\sqrt{\lambda_n}$.

14. In this case $L[y] = y'' + y' + 2y$ is not of the form shown
    in Eq.(3) and thus the B.V.P. is not self adjoint.

17. In this case $L[y] = [(1+x^2)y']' + y$ and thus the D.E. has
    the form shown in Eq.(3). However, the B.C. are not
    separated and thus we must determine by integration
    whether Eq.(8) is satisfied. Therefore, for u and v
    satisfying the B.C., integration by parts yields the
    following:
    $(L[u],v) = \int_0^1\{[(1+x^2)u']'+u\}v\,dx$
    $\qquad = vu'(1+x^2)\Big|_0^1 - \int_0^1\{(1+x^2)v'u'\}dx + \int_0^1 uv\,dx$
    $\qquad = vu'(1+x^2)\Big|_0^1 - uv'(1+x^2)\Big|_0^1 + \int_0^1\{[(1+x^2)v']'+v\}u\,dx$
    $\qquad = (u,L[v])$
    since the integrated terms add to zero with the given
    B.C. Thus the B.V.P. is self-adjoint.

21a. Substituting ϕ for y in the D.E., multiplying both sides
     by ϕ, and integrating form 0 to 1 yields

$\lambda \int_0^1 r\phi^2 dx = \int_0^1 \{-[p(x)\phi']'\phi + q(x)\phi^2\}dx$. Integrating the first term on the right side once by parts, we obtain
$\lambda \int_0^1 r\phi^2 dx = -p(1)\phi'(1)\phi(1) + p(0)\phi'(0)\phi(0) + \int_0^1 (p\phi'^2+q\phi^2)dx$.
If $\alpha_2 \neq 0$, $\beta_2 \neq 0$, then $\phi'(1) = -\beta_1\phi(1)/\beta_2$ and
$\phi'(0) = -\alpha_1\phi(0)/\alpha_2$ and the result follows. If $\alpha_2 = 0$,
then $\phi(0) = 0$ and the boundary term at 0 will be missing.
A similar result is obtained if $\beta_2 = 0$.

21b. From the text, $p(x) > 0$ and $r(x) > 0$ for $0 \leq x \leq 1$ [following
Eq.(4)]. If $q(x) \geq 0$ and if β_1/β_2 and $-\alpha_1/\alpha_2$ are non-negative,
then all terms in the final equation of part (a) are
non-negative and thus λ must be non-negative. Note that λ
could be zero if $q(x) = 0$, $\beta_1 = 0$, $\alpha_1 = 0$ and $\phi(x) = 1$.

21c. If either $q(x) \neq 0$ or $\alpha_1 \neq 0$ or $\beta_1 \neq 0$, there is at least one
positive term on the right and thus λ must be positive.

23a. Using $\phi(x) = U(x) + iV(x)$ in Eq.(4) we have
$L[\phi] = L[U(x) + iV(x)] = \lambda r(x)[U(x)+iV(x)]$. Using the
linearity of L and the fact that λ and $r(x)$ are real we
have $L[U(x)] + iL[V(x)] = \lambda r(x)U(x) + i\lambda r(x)V(x)$.
Equating the real and imaginary parts shows that both U
and V satisfy Eq.(1). The B.C. Eq.(2) are also satisfied
by both U and V, using the same arguments, and thus both
U and V are eigenfunctions.

23b. By Theorem 11.2.3 each eigenvalue λ has only one linearly
independent eigenfuction. By part (a) we have U and V
being eigenfunctions corresponding to λ and thus U and V
must be linearly dependent.

23c. From part (b) we have $V(x) = cU(x)$ and thus
$\phi(x) = U(x) + icuU(x) = (1+ic)U(x)$.

24. This is an Euler equation, so for $y = x^r$ we have
$r^2 - (\lambda+1)r + \lambda = 0$ or $(r-1)(r-\lambda) = 0$. If $\lambda = 1$, the general
solution to the D.E. is $y = c_1 x + c_2 x\ln x$. The B.C. require
that $c_1 = 0$, $2c_1 + 2(\ln2)c_2 = 0$ so $c_1 = c_2 = 0$ and $\lambda = 1$ is
not an eigenvalue. If $\lambda \neq 1$, the general solution to the
D.E. is $y = c_1 x + c_2 x^\lambda$. The B.C. require that $c_1 + c_2 = 0$ and
$2c_1 + 2^\lambda c_2 = 0$. Nontrivial solutions exist if and only if
$2^\lambda - 2 = 0$. If λ is real, this equation has no solution
(other than $\lambda = 1$) and again $y = 0$ is the only solution to
the boundary value problem. Suppose that $\lambda = a + bi$ with
$b \neq 0$. Then $2^\lambda = 2^{a+bi} = 2^a 2^{bi} = 2^a\exp(ib\ln2)$, which upon

substitution into $2^\lambda = 2$ yields the equation
$\exp(ib\ln2) = 2^{1-a}$. Since $e^{ix} = \cos x + i\sin x$, it follows that
$\cos(b\ln2) = 2^{1-a}$ and $\sin(b\ln2) = 0$, which yield $a = 1$ and
$b(\ln2) = 2n\pi$ or $b = 2n\pi/\ln2$, $n = \pm1, \pm2,\ldots$. Thus the only
eigenvalues of the problem are
$\lambda_n = 1 + i(2n\pi/\ln2)$, $n = \pm1, \pm2,\ldots$.

25b. For $\lambda \le 0$, there are no eigenfuctions. For $\lambda > 0$ the
general solution of the D.E. is
$y = c_1 + c_2x + c_3\sin\sqrt{\lambda}\,x + c_4\cos\sqrt{\lambda}\,x$. The B.C. require
that
$c_1+c_4 = 0$, $c_4 = 0$, $c_1 + c_2L + c_3\sin\sqrt{\lambda}\,L + c_4\cos\sqrt{\lambda}\,L = 0$,
and
$c_2 + \sqrt{\lambda}\,c_3\cos\sqrt{\lambda}\,L - \sqrt{\lambda}\,c_4\sin\sqrt{\lambda}\,L = 0$. Thus $c_1 = c_4 = 0$
and for nontrivial solutions to exist λ must satisfy the
equation $\sqrt{\lambda}\,L\cos\sqrt{\lambda}\,L - \sin\sqrt{\lambda}\,L = 0$. In this case
$c_2 = (-\sqrt{\lambda}\cos\sqrt{\lambda}\,L)(c_3)$ and it follows that the
eigenfunction ϕ_1 is given by
$\phi_1(x) = \sin(\sqrt{\lambda_1}\,x) - \sqrt{\lambda_1}\,x\cos\sqrt{\lambda_1}\,L$ where λ_1 is the
smallest positive solution of the equation
$\sqrt{\lambda}\,L = \tan\sqrt{\lambda}\,L$. A graphical or numerical estimate of λ_1
reveals that $\lambda_1 \approx (4.4934)^2/L^2$.

25c. Assuming $\lambda \ne 0$ the eigenvalue equation is
$2(1-\cos x) = x\sin x$, where $x = \sqrt{\lambda}\,L$. Graphing
$f(x) = 2(1-\cos x)$ and $g(x) = x\sin x$ we see that there is an
intersection for $6 < x < 7$. Since both $f(x)$ and $g(x)$ are
zero for $x = 2\pi$, this then is the precise root and thus
$\lambda_1 = (2\pi)^2/L^2$. In addition, it appears there might be an
intersection for $0 < x < 1$. Using a Taylor series
representation for $f(x)$ and $g(x)$ about $x = 0$, however,
shows there is no intersection for $0 < x < 1$. Of course
$x = 0$ is also an intersection, which yields $\lambda = 0$, which
gives the trivial solution and hence $\lambda = 0$ is not an
eigenvalue.

27a. Setting $c(x,t) = X(x)T(t)$ in Eq.(i) we obtain
$$\frac{T'}{DT} = \frac{X'' - (v/D)X'}{X} = -\lambda \text{ and thus } X'' - (v/D)X' + \lambda X = 0,$$
$X(0) = 0$, $X'(L) = 0$, and $T' + \lambda DT = 0$. To put the first
equation in the Sturm-Liouville form multiply by $p(x)$ and
choose p so that $(pX')' = pX'' - p(v/D)X'$. This gives
$p' = -p(v/D)$ or $p(x) = \exp(-vx/D)$.

27b. The characteristic equation of the first O.D.E. in part (a) is $r^2 - (v/D)r + \lambda = 0$, so that $r_{1,2} = (v/2D) \pm i\sqrt{\lambda - v^2/4D^2}$ and hence $X(x) = e^{vx/2D}(c_1\cos\mu x + c_2\sin\mu x)$. Now $X(0) = 0 \Rightarrow c_1 = 0$ and $X'(L) = 0$ requires that $(v/2D)\sin\mu L + \mu\cos\mu L = 0$. Thus μ satisfies $\tan\mu L = -(2D/v)\mu$.

27c. Plot on the same graph $y = \tan Lx$ and $y = -(2D/v)x$. For large n the straight line will intersect the tangent curve very close the its vertical asymptotes, which are $Lx = (2n-1)\pi/2$, or $x = (2n-1)\pi/2L$.

27d. Since $r(x) = \exp(-vx/D)$ we have from Eq.(vi)
$$\int_0^L r(x)X_n^2(x)dx = \int_0^L \sin^2(\mu_n x)dx = L/2 - (1/4\mu_n)\sin 2\mu_n L$$
$$= L/2 + (v/4D\mu_n^2)\sin^2\mu_n L,$$
using the double angle formula for sine and Eq.(vii).

27e. From part (a) we have $T' + \lambda_n DT = 0$, so that $T(t) = e^{-\lambda_n Dt}$, where λ_n is given in part (b). Thus $c(x,t) = \sum_{n=1}^{\infty} a_n T_n(t)X_n(x)$

and the a_n are chosen to satisfy $c(x,0) = \sum_{n=1}^{\infty} a_n X_n(x) = f(x)$.

Multiplying both sides of this last equation by $e^{-vx/2D}\sin\mu_n x$, integrating from 0 to L, using orthogonality, and the results of part (d) gives the desired result.

Section 11.3, Page 689

1. We must first find the eigenvalues and normalized eigenfunctions of the associated homogeneous problem $y'' + \lambda y = 0$, $y(0) = 0$, $y(1) = 0$. This problem has the solutions $\phi_n(x) = k_n\sin n\pi x$, for $\lambda_n = n^2\pi^2$, $n = 1,2,\ldots$. Choosing k_n so that $\int_0^1 \phi_n^2 dx = 1$ we find $k_n = \sqrt{2}$. Hence the solution of the original nonhomogeneous problem is given by $y = \sum_{n=1}^{\infty} b_n\phi_n(x)$, where the coefficients b_n are found from Eq.(12), $b_n = c_n/(\lambda_n - 2)$ where c_n is given by $c_n = \sqrt{2}\int_0^1 x\sin n\pi x dx$ (Eq.9). [Note that the original problem can be written as $-y'' = 2y + x$ and therefore comparison with Eq.(1) yields $r(x) = 1$ and $f(x) = x$].

Integrating the expression for c_n by parts yields

$$c_n = \sqrt{2}\,(-1)^{n+1}/n\pi \text{ and thus } y = \sum_{n=1}^{\infty} \frac{\sqrt{2}\,(-1)^{n+1}}{(n^2\pi^2-2)n\pi}\sqrt{2}\,\sin n\pi x.$$

2. From Prob. 1 of Sect. 11.2 we have
    $\phi_n = \sqrt{2}\sin[(2n-1)\pi x/2]$ for $\lambda_n = (2n-1)^2\pi^2/4$ and from
    Prob. 7 of that section we have
    $c_n = 4\sqrt{2}\,(-1)^{n+1}/(2n-1)^2\pi^2$. Substituting these values

    into $b_n = c_n/(\lambda_n-2)$ and $y = \sum_{n=1}^{\infty} b_n\phi_n$ yields the desired

    result.

3. Referring to Prob. 3 of Sect. 11.2 we have

    $y = b_0 + \sum_{n=1}^{\infty} b_n(\sqrt{2}\cos n\pi x)$, where $b_n = c_n/(\lambda_n-2)$ for

    $n = 0,1,2,\ldots$. The rest of the calculations follow
    those of Prob. 1.

5. Note that the associated eigenvalue problem is the same
    as for Prob. 1 and that $|1-2x| = 1-2x$ for $0 \leq x \leq 1/2$
    while $|1-2x| = 2x-1$ for $1/2 \leq x \leq 1$.

8. Writing the D.E. in the form Eq.(1), we have $-y'' = \mu y + f(x)$,
    so $r(x) = 1$. The associated eigenvalue problem is
    $y'' + \lambda y = 0$, $y'(0) = 0$, $y'(1) = 0$ which has the eigenvalues
    $\lambda_n = n^2\pi^2$, $n = 0,1,2\ldots$ and the normalized eigenfunctions
    $\phi_0 = 1$, $\phi_n(x) = \sqrt{2}\cos n\pi x$, $n = 1,2\ldots$, as found in Prob. 3,

    Sect. 11.2. Now, we assume $y(x) = b_0 + \sum_{n=1}^{\infty} b_n\phi_n(x) = \sum_{n=0}^{\infty} b_n\phi_n(x)$,

    Eq.(4), and thus $b_n = \dfrac{c_n}{\lambda_n-\mu}$ [Eq.(12)], where

    $c_n = \int_0^1 f(x)\phi_n(x)dx$, $n = 0,1,2\ldots$, Eq.(9). Thus

    $$y(x) = \sum_{n=0}^{\infty} \frac{c_n}{\lambda_n-\mu}\phi_n(x) = -c_0/\mu + \sqrt{2}\sum_{n=1}^{\infty} \frac{c_n\cos n\pi x}{\lambda_n-\mu},$$ where we

    have assumed $\mu \neq \lambda_n$, $n = 0,1,2\ldots$.

10. Since $\mu = \pi^2$ is an eigenvalue of the corresponding
    homogeneous equation, Theorem 11.3.1 tells us that a
    solution will exist only if $-(a+x)$ is orthogonal to the

corresponding eigenfunction $\sqrt{2}\sin n\pi x$. Thus we require $\int_0^1 (a+x)\sin n\pi x\, dx = 0$, which yields $a = -1/2$. With $a = -1/2$, we find that the particular solution is $Y = (x-1/2)/\pi^2$ and $y_c = c\sin n\pi x + d\cos n\pi x$ by methods of Chapt. 3. Setting $y = y_c + Y$ and choosing d to satisfy the B.C. we obtain the desired family of solutions.

11. Note that in this case $\mu = 4\pi^2$ and $\phi_2 = \sqrt{2}\sin 2\pi x$ are the eigenvalue and eigenfunction respectively of the corresponding homogeneous equation. However, there is no value of a for which $-\sqrt{2}\int_0^1 (a+x)\sin 2\pi x\, dx = 0$, and thus there is no solution.

12. In this case a solution will exist only if $-a$ is orthogonal to $\sqrt{2}\cos n\pi x$., that is if $\int_0^1 a\cos n\pi x\, dx = 0$. Since this condition is valid for all a, a family of solutions exists.

14. Since $\displaystyle\sum_{n=1}^{\infty} c_n\phi_n(x)$ converges to zero we have $\displaystyle\sum_{n=1}^{\infty} c_n\phi_n(x) = 0$. Multiplying and integrating as suggested yields

$$\int_0^1 \left[\sum_{n=1}^{\infty} c_n\phi_n(x)\right] r(x)\phi_m(x)\, dx = 0 \text{ or}$$

$$\sum_{n=1}^{\infty} c_n\int_0^1 r(x)\phi_n(x)\phi_m(x)\, dx = 0.$$ The integral that multiplies c_n is just δ_{nm} [Eq.(22) of Sect. 11.2]. Thus the infinite sum becomes c_m, which then yields $c_m = 0$.

18. A twice differentiable function v satisfying the boundary conditions can be found by assuming that $v = ax + b$. Thus $v(0) = b = 1$ and $v(1) = a + 1$ while $v'(1) = a$. Hence $2a + 1 = -2$ or $a = -3/2$ and $v(x) = 1 - 3x/2$. Assuming $y = u + v$ we have $(u+v)'' + 2(u+v) = u'' + 2u + 2(1-3x/2) = 2 - 4x$ or $u'' + 2u = -x$, $u(0) = 0$, $u(1) + u'(1) = 0$ which is the same as Ex. 1 of the text.

19. From Eq.(30) we assume $u(x,t) = \displaystyle\sum_{n=1}^{\infty} b_n(t)\phi_n(x)$, where the ϕ_n are the eigenfunctions of the related eigenvalue

problem $y'' + \lambda y = 0$, $y(0) = 0$, $y'(1) = 0$ and the $b_n(t)$ are given by Eq.(42). From Prob. 2, we have $\phi_n = \sqrt{2} \sin[(2n-1)\pi x/2]$ and $\lambda_n = (2n-1)^2\pi^2/4$. To evaluate Eq.(42) we need to calculate $B_n = \int_0^1 \sin(\pi x/2)\sqrt{2} \sin[(2n-1)\pi x/2 dx$ [Eq.(41) with $r(x) = 1$ and $f(x) = \sin(\pi x/2)$], which is zero except for $n = 1$ in which case $B_1 = \sqrt{2}/2$, and $\gamma_n = \int_0^1 (-x)\sqrt{2} \sin[(2n-1)\pi x/2]dx$ [Eq.(35) with $F(x,t) = -x$]. This integral is the negative of the c_n in Prob. 2 and thus $\gamma_n = -4\sqrt{2}(-1)^{n+1}/(2n-1)^2\pi^2 = -c_n$, $n = 1,2,\ldots$. Setting $\gamma_n = -c_n$ in Eq.(42) we then have

$$b_1 = \frac{\sqrt{2}}{2}e^{-\pi^2 t/4} - c_1\int_0^t e^{-\pi^2(t-s)/4}ds$$

$$= \frac{\sqrt{2}}{2}e^{-\pi^2 t/4} - \frac{4c_1}{\pi^2}e^{-\pi^2(t-s)/4}\Big|_0^t$$

$$= \frac{\sqrt{2}}{2}e^{-\pi^2 t/4} - \frac{4c_1}{\pi^2} + \frac{4c_1}{\pi^2}e^{-\pi^2 t/4} \text{ and}$$

similarly $b_n = -c_n\int_0^t e^{-\lambda_n(t-s)}ds = -(c_n/\lambda_n)e^{-\lambda_n(t-s)}\Big|_0^t$

$$= -(c_n/\lambda_n)(1-e^{-\lambda_n t}), \text{ where } \lambda_n = (2n-1)^2\pi^2/4,$$

$n = 2,3,\ldots$. Substituting these values for b_n along with $\phi_n = \sqrt{2}\sin[(2n-1)\pi x/2]$ into the series for $u(x,t)$ yields the solution to the given problem.

22. In this case $B_n = 0$ for all n and γ_n is given by

$$\gamma_n = \int_0^1 e^{-t}(1-x)\sqrt{2} \sin[(2n-1)\pi x/2]dx$$

$$= e^{-t}\int_0^1 (1-x)\sqrt{2} \sin[(2n-1)\pi x/2]dx. \text{ This last integral}$$

can be written as the sum of two integrals, each of which has been evaluated in either Prob. 6 or 7 of Sect. 11.2. Letting c_n denote the value obtained, we then have

$$\gamma_n = c_n e^{-t} \text{ and thus } b_n = c_n\int_0^t e^{-\lambda_n(t-s)}e^{-s}ds =$$

$$c_n e^{-\lambda_n t}\int_0^t e^{(\lambda_n-1)s}ds = [c_n/(\lambda_n-1)](e^{-t}-e^{-\lambda_n t}), \text{ where}$$

$\lambda_n = (2n-1)^2\pi^2/4$. Substituting these values into Eq.(30) yields the desired solution.

24. Using the approach of Prob. 23 we find that $v(x)$ satisfies
$v'' = 2$, $v(0) = 1$, $v(1) = 0$. Thus $v(x) = x^2 + c_1 x + c_2$ and
the B.C. yield $v(0) = c_2 = 1$ and $v(1) = 1 + c_1 + 1 = 0$ or
$c_1 = -2$. Hence $v(x) = x^2 - 2x + 1$ and $w(x,t) = u(x,t) - v(x)$
where, from Prob. 23, we have $w_t = w_{xx}$, $w(0,t) = 0$,
$w(1,t) = 0$ and $w(x,0) = x^2 - 2x + 2 - v(x) = 1$. This last
problem can be solved by methods of this section or by
methods of Chapt. 10. Using the approach of this section we
have $w(x,t) = \sum\limits_{n=1}^{\infty} b_n(t)\phi_n(x)$ where $\phi_n(x) = \sqrt{2}\, \sin n\pi x$

[which are the normalized eigenfunctions of the
associated eigenvalue problem $y'' + \lambda y = 0$, $y(0) = 0$,
$y(1) = 0$] and the b_n are given by Eq.(42). Since the
P.D.E. for $w(x,t)$ is homogeneous Eq.(42) reduces to
$b_n = B_n e^{-\lambda_n t}$ ($\lambda_n = n^2\pi^2$ from the above eigenvalue problem),
where
$B_n = \int_0^1 1\cdot\sqrt{2}\, \sin n\pi x\, dx = \sqrt{2}\,[1-(-1)^n]/n\pi$. Thus
$$u(x,t) = x^2 - 2x + 1 + \sum_{n=1}^{\infty} \frac{\sqrt{2}\,[1-(-1)^n]}{n\pi}\, e^{-n^2\pi^2 t}\sqrt{2}\,\sin n\pi x,$$
which simplifies to the desired solution.

28a. Since $y_c = c_1 + c_2 x$, we assume that
$Y(x) = u_1(x) + x u_2(x)$. Then $Y' = u_2$ since we require
$u_1' + x u_2' = 0$. Differentiating again yields $Y'' = u_2'$ and
thus $u_2' = -f(x)$ by substitution into the D.E. Hence
$u_2(x) = -\int_0^x f(s)ds$, $u_1' = xf(x)$, and $u_1(x) = \int_0^x sf(s)ds$.
Therefore $Y = \int_0^x sf(s)ds - x\int_0^x f(s)ds = -\int_0^x (x-s)f(s)ds$ and
$\phi(x) = c_1 + c_2 x - \int_0^x (x-s)f(s)ds$.

28b. From part (a) we have $y(0) = c_1 = 0$. Thus
$y(x) = c_2 x - \int_0^x (x-s)f(s)ds$ and hence
$y(1) = c_2 - \int_0^1 (1-s)f(s)ds = 0$, which yields the desired
value of c_2.

28c. From parts (a) and (b) we have
$$\phi(x) = x\int_0^1 (1-s)f(s)ds - \int_0^x (x-s)f(s)ds$$
$$= \int_0^x x(1-s)f(s)ds + \int_x^1 x(1-s)f(s)ds - \int_0^x (x-s)f(s)ds$$

$$= \int_0^x (x-xs-x+s)f(s)ds + \int_x^1 x(1-s)f(s)ds$$

$$= \int_0^x s(1-x)f(s)ds + \int_x^1 x(1-s)f(s)ds.$$

28d. We have $\phi(x) = \int_0^x s(1-x)ds + \int_x^1 x(1-s)ds$

$$= \int_0^x G(x,s)f(s)ds + \int_x^1 G(x,s)f(s)ds$$

$$= \int_0^1 G(x,s)f(s)ds.$$

30b. In this case $y_1(x) = \sin x$ and $y_2(x) = \sin(1-x)$ [assume $y_2(x) = c_1\cos x + c_2\sin x$, let $x = 1$, solve for c_2 in terms of c_1 using $y(1) = 0$ and then let $c_1 = \sin 1$]. Using these functions for y_1 and y_2 we find $W(y_1,y_2) = -\sin 1$ and thus $G(x,s) = -\sin s \, \sin(1-x)/(-\sin 1)$, since $p(x) = 1$, for $0 \le s \le x$. Interchanging the x and s verifies $G(x,s)$ for $x \le s \le 1$.

30c. Since $W(y_1,y_2)(x) = y_1(x)y_2'(x) - y_2(x)y_1'(x)$ we find that

$$[p(x)W(y_1,y_2)(x)]' = p'(x)[y_1(x)y_2'(x) - y_2(x)y_1'(x)]$$
$$+ p(x)[y_1'(x)y_2'(x) + y_1(x)y_2''(x) - y_2'(x)y_1'(x) - y_2(x)y_1''(x)]$$
$$= y_1[py_2']' - y_2[py_1']' = y_1[q(x)y_2] - y_2[q(x)y_1] = 0.$$

30d. Let $c = p(x)W(y_1,y_2)(x)$. If $0 \le s \le x$, then $G(x,s) = -y_1(s)y_2(x)/c$. Since the first argument in $G(s,x)$ is less than the second argument, the bottom expression of formula (iv) must be used to determine $G(s,x)$. Thus, $G(s,x) = -y_1(s)y_2(x)/c$. A similar argument holds if $x \le s \le 1$.

30e. We have $\phi(x) = \int_0^1 G(x,s)f(s)ds$

$$= -\int_0^x \frac{y_1(s)y_2(x)f(s)}{c}ds - \int_x^1 \frac{y_1(x)y_2(s)f(s)}{c}ds$$

(where $c = p(x)W(y_1,y_2)$ and thus, by Leibnitz's rule,

$$c\phi'(x) = -y_1(x)y_2(x)f(x) - \int_0^x y_1(s)y_2'(x)f(s)ds + y_1(x)y_2(x)f(x)$$
$$-\int_x^1 y_1'(x)y_2(s)f(s)ds. \quad \text{From this we obtain}$$

$$-c(p\phi')' = (py_2')'\int_0^x y_1(s)f(s)ds + py_2'y_1 f(x)$$

$$+ (py_1')'\int_x^1 y_2(s)f(s)ds - py_1'y_2 f(x). \quad \text{Dividing by}$$

c and adding $q(x)\phi(x)$ we get

$$(-p\phi')'+q\phi = \frac{(py_2')'}{c}\int_0^x y_1(s)f(s)ds - \frac{qy_2}{c}\int_0^x y_1(s)f(s)ds$$

$$+ \frac{(py_1')'}{c}\int_x^1 y_2(s)f(s)ds - \frac{qy_1}{c}\int_x^1 y_2(s)f(s)ds + f(x)$$

$$= \frac{(py_2')'-qy_2}{c}\int_0^x y_1(s)f(s)ds + \frac{(py_1')'-qy_1}{c}\int_x^1 y_2(s)f(s)ds + f(x)$$

$= f(x)$, since y_1 and y_2 satisfy $L[y] = 0$. Using $\phi(x)$ and $\phi'(x)$ as found above, the B.C. are both satisfied since $y_1(x)$ satisfies one B.C. and $y_2(x)$ satisfies the other B.C.

33. In general $y(x) = c_1\cos x + c_2\sin x$. For $y'(0) = 0$ we must choose $c_2 = 0$ and thus $y_1(x) = \cos x$. For $y(1) = 0$ we have $c_1\cos 1 + c_2\sin 1 = 0$, which yields $c_2 = -c_1(\cos 1)/\sin 1$ and thus $y_2(x) = c_1\cos x - c_1(\cos 1)\sin x/\sin 1$

$$= c_1(\sin 1\cos x - \cos 1\sin x)/\sin 1$$

$$= \sin(1-x) \text{ [by setting } c_1 = \sin 1].$$

Furthermore, $W(y_1,y_2) = -\cos 1$ and thus

$$G(x,s) = \begin{cases} \dfrac{\cos s\,\sin(1-x)}{\cos 1} & 0 \le s \le x \\[3mm] \dfrac{\cos x\,\sin(1-s)}{\cos 1} & x \le s \le 1 \end{cases},$$

and hence

$$\phi(x) = \int_0^x [\cos s\,\sin(1-x)f(s)/\cos 1]ds$$

$$+ \int_x^1 [\cos x\,\sin(1-s)f(s)/\cos 1]ds$$

is the solution of the given B.V.P.

Section 11.4, Page 700

2a. The D.E. is the same as Eq.(7) and thus, from Eq.(9), the general solution of the D.E. is
$y = c_1 J_0(\sqrt{\lambda}\,x) + c_2 Y_0(\sqrt{\lambda}\,x)$. The B.C. at $x = 0$ requires that $c_2 = 0$, and the B.C. at $x = 1$ requires
$c_1\sqrt{\lambda}\,J_0'(\sqrt{\lambda}) = 0$. For $\lambda = 0$ we have $\phi_0(x) = J_0(0) = 1$

and if λ_n is the nth positive root of $J_0'(\sqrt{\lambda}) = 0$ then
$\phi_n(x) = J_0(\sqrt{\lambda_n}\,x)$. Note that for $\lambda = 0$ the D.E. becomes
$(xy')' = 0$, which has the general solution $y = c_1 \ln x + c_2$.
To satisfy the bounded conditions at $x = 0$ we must choose
$c_1 = 0$, thus obtaining the same solution as above.

2b. For $n \neq 0$, set $y = J_0(\sqrt{\lambda_n}\,x)$ in the D.E. and integrate
from 0 to 1 to obtain $-\int_0^1 (xJ_0')'dx = \lambda_n \int_0^1 xJ_0(\sqrt{\lambda_n}\,x)dx$.
Integrating the left side of this equation yields
$\int_0^1 (xJ_0')'dx = xJ_0'(\sqrt{\lambda_n}\,x)\Big|_0^1 = J_0'(\sqrt{\lambda_n}) - 0 = 0$ since the λ_n
are eigenvalues from part (a). Thus $\int_0^1 xJ_0(\sqrt{\lambda_n}\,x)dx = 0$,
which is the desired result for $m = 0$, $n \neq 0$. For other
n and m, we let $L[y] = -(xy')'$. Then
$L[J_0(\sqrt{\lambda_n}\,x)] = \lambda_n xJ_0(\sqrt{\lambda_n}\,x)$ and
$L[J_0(\sqrt{\lambda_m}\,x)] = \lambda_m xJ_0(\sqrt{\lambda_m}\,x)$. Multiply the first equation
by $J_0(\sqrt{\lambda_m}\,x)$, the second by $J_0(\sqrt{\lambda_n}\,x)$, subtract the
second from the first, and integrate from 0 to 1 to
obtain
$\int_0^1 \{J_0(\sqrt{\lambda_m}\,x)L[J_0(\sqrt{\lambda_n}\,x)] - J_0(\sqrt{\lambda_n}\,x)L[J_0(\sqrt{\lambda_m}\,x)]\}dx =$
$(\lambda_n - \lambda_m)\int_0^1 xJ_0(\sqrt{\lambda_n}\,x)J_0(\sqrt{\lambda_m}\,x)dx$. Again the left side
is zero after each term is integrated by parts once, as
was done above. If $\lambda_n \neq \lambda_m$, the result follows with
$\phi_n(x) = J_0(\sqrt{\lambda_n}\,x)$.

2c. Since $\lambda = 0$ is an eigenvalue we assume that
$y = b_0 + \sum_{n=1}^{\infty} b_n J_0(\sqrt{\lambda_n}\,x)$. Since
$-[xJ_0'(\sqrt{\lambda_n}\,x)]' = \lambda_n xJ_0(\sqrt{\lambda_n}\,x)$, $n = 0,1,\ldots$, we find that
$-(xy')' = x\sum_{n=1}^{\infty} \lambda_n b_n J_0(\sqrt{\lambda_n}\,x)$ [note that $\lambda_0 = 0$ and b_0 are
missing on the right]. Now assume
$f(x)/x = c_0 + \sum_{n=1}^{\infty} c_n J_0(\sqrt{\lambda_n}\,x)$. Multiplying both sides by
$xJ_0(\sqrt{\lambda_m}\,x)$, integrating from 0 to 1 and using the
orthogonality relations of part (b), we find

$$c_n = \int_0^1 f(x)J_0(\sqrt{\lambda_n}\,x)dx / \int_0^1 xJ_0^2(\sqrt{\lambda_n}\,x)dx, \quad n = 0,1,2,\ldots \; .$$

[Note that $c_0 = 2\int_0^1 f(x)dx$ since the denominator can be integrated.] Substituting the series for y and $f(x)/x$ into the D.E., using the above result for $-(xy')'$, and simplifying we find that

$$(\mu b_0 + c_0) + \sum_{n=1}^{\infty} [c_n - b_n(\lambda_n - \mu)]J_0(\sqrt{\lambda_n}\,x) = 0. \quad \text{Thus}$$

$b_0 = -c_0/\mu$ and $b_n = c_n/(\lambda_n-\mu)$, $n = 1,2,\ldots$, where $\sqrt{\lambda_n}$ are obtained from $J_0'(\sqrt{\lambda_n}) = 0$.

4a. Let $L[y] = -[(1-x^2)y']'$. Then $L[\phi_n] = \lambda_n\phi_n$ and $L[\phi_m] = \lambda_m\phi_m$. Multiply the first equation by ϕ_m, the second by ϕ_n, subtract the second from the first, and integrate from 0 to 1 to obtain
$\int_0^1 (\phi_m L[\phi_n] - \phi_n L[\phi_m])dx = (\lambda_n - \lambda_m)\int_0^1 \phi_n\phi_m dx$. The integral on the left side can be shown to be 0 by integrating each term once by parts. Since $\lambda_n \neq \lambda_m$ if $m \neq n$, the result follows. Note that the result may also be written as
$\int_0^1 P_{2m-1}(x)P_{2n-1}(x)dx = 0$, $m \neq n$.

4b. First let $f(x) = \sum_{n=1}^{\infty} c_n\phi_n(x)$, multiply both sides by $\phi_m(x)$, and integrate term by term from $x = 0$ to $x = 1$. The orthogonality condition yields

$$c_n = \int_0^1 f(x)\phi_n(x)dx / \int_0^1 \phi_n^2(x)dx, \quad n = 1,2,\ldots \text{ where it is}$$

understood that $\phi_n(x) = P_{2n-1}(x)$. Now assume
$y = \sum_{n=1}^{\infty} b_n\phi_n(x)$. As in Prob. 2 and in the text

$-[(1-x^2)y']' = \sum_{n=1}^{\infty} \lambda_n b_n\phi_n$ since the ϕ_n are eigenfunctions.

Thus, substitution of the series for y and f into the D.E.
and simplification yields $\sum_{n=1}^{\infty} [b_n(\lambda_n-\mu) - c_n]\phi_n(x) = 0$. Hence
$b_n = c_n/(\lambda_n - \mu)$, $n = 1,2,\ldots$ and the desired solution is obtained [after setting $\phi_n(x) = P_{2n-1}(x)$].

Section 11.5, Page 705

1a. Since $u(x,0) = 0$ we have $Y(0) = 0$. However, since the
     other two boundaries are given by $y = 2x$ and $y = 2(x-2)$
     we cannot separate x and y dependence and thus neither X
     nor Y satisfy homogeneous B.C. at both end points.

1b. The line $y = 2x$ is transformed into $\xi = 0$ and $y = 2(x-2)$ is
     transformed into $\xi = 2$. The lines $y = 0$ and $y = 2$ are
     transformed into $\eta = 0$ and $\eta = 2$ respectively, so the
     parallelogram is transformed into a square of side 2. From
     the given equations, we have $x = \xi + \eta/2$ and $y = \eta$. Thus
     $u_\xi = u_x x_\xi + u_y y_\xi = u_x$ and
     $u_\eta = u_x x_\eta + u_y y_\eta = (1/2)u_x + u_y$. Likewise
     $u_{\xi\xi} = u_{xx} - x_\xi + u_{xy}y_\xi = u_{xx}$
     $u_{\xi\eta} = u_{xx}x_\eta + u_{xy}y_\eta = (1/2)u_{xx} + u_{xy}$ and
     $u_{\eta\eta} = (1/2)u_{xx}x_\eta + (1/2)u_{xy}y_\eta + u_{yx}x_\eta + u_{yy}y_\eta$
         $= (1/4)u_{xx} + u_{xy} + u_{yy}$. Therefore,
     $(5/4)u_{\xi\xi} - u_{\xi\eta} + u_{\eta\eta} = u_{xx} + u_{yy} = 0$. The B.C. become
     $U(\xi,0) = 0$, $U(\xi,2) = f(\xi+1)$ (since $x = \xi + \eta/2$), $U(0,\eta) = 0$,
     and $U(2,\eta) = 0$.

1c. Substituting $u(\xi,\eta) = U(\xi)V(\eta)$ into the equation of part (b)
     yields $5/4U''V - U'V' + UV'' = 0$ or upon dividing by UV
     $$\frac{5}{4}\frac{U''}{U} + \frac{V''}{V} = \frac{U'V'}{UV},$$ which is not separable.

2.   This problem is very similar to the example worked in the
     text. The fundamental solutions satisfying the
     P.D.E.(3), the B.C. $u(1,t) = 0$, $t \geq 0$ and the finiteness
     condition are given by Eqs.(15) and (16). Thus assume
     $u(r,t)$ is of the form given by Eq.(17). The I.C. require
     that $u(r,0) = \sum\limits_{n=1}^{\infty} c_n J_0(\lambda_n r) = 0$, so $c_n = 0$, and
     $u_t(r,0) = \sum\limits_{n=1}^{\infty}\lambda_n a k_n J_0(\lambda_n r) = g(r)$. From Eq.(26) of
     Sect. 11.4 we obtain
     $\lambda_n k_n a = \int_0^1 rg(r)J_0(\lambda_n r)dr / \int_0^1 rJ_0^2(\lambda_n r)dr$, $n = 1,2,\ldots$.

4.   This problem is the same as Prob. 21 of Sect. 10.7. The
     periodicity condition requires that μ of that problem be
     an integer and thus substituting $\mu^2 = n^2$ into the
     previous results yields the given equations.

5a. Substituting $u(r,\theta,z) = R(r)\Theta(\theta)Z(z)$ into Laplace's
     equation yields $R''\Theta Z + R'\Theta Z/r + R\Theta''Z/r^2 + R\Theta Z'' = 0$ or
     equivalently $R''/R + R'/rR + \Theta''/r^2\Theta = -Z''/Z = \sigma$. In order
     to satisfy arbitrary B.C. it can be shown that σ must be
     negative, so assume $\sigma = -\lambda^2$, and thus $Z'' - \lambda^2 Z = 0$ and,
     after some algebra, it follows that
     $r^2R''/R + rR'/R + \lambda^2r^2 = -\Theta''/\Theta = \alpha$. The periodicity
     condition $\Theta(0) = \Theta(2\pi)$ requires that $\sqrt{\alpha}$ be an integer n
     so $\alpha = n^2$. Thus $r^2R'' + rR' + (\lambda^2r^2 - n^2)R = 0$,
     $\Theta'' + n^2\Theta = 0$, and $Z'' - \lambda^2 Z = 0$.

5b. If $u(r,\theta,z)$ is independent of θ, then the $\Theta''/r^2\Theta$ term
     does not appear in the second equation of part (a) and
     thus $R''/R + R'/rR = -Z''/Z = -\lambda^2$, from which the desired
     result follows.

6.   Assuming that $u(r,z) = R(r)Z(z)$ it follows from Prob. 5
     that $R = c_1 J_0(\lambda r) + c_2 Y_0(\lambda r)$, from Eq.(13), and
     $Z = k_1 e^{-\lambda z} + k_2 e^{\lambda z}$. Since $u(r,z)$ is bounded as $r \to 0$ and
     approaches zero as $z \to \infty$ we require that $c_2 = 0$, $k_2 = 0$.
     The B.C. $u(1,z) = 0$ requires that $J_0(\lambda) = 0$ leading to an
     infinite set of discrete positive eigenvalues
     $\lambda_1, \lambda_2, \ldots \lambda_n \ldots$. The fundamental solutions of the
     problem are then $u_n(r,z) = J_0(\lambda_n r)e^{-\lambda_n z}$, $n = 1,2,\ldots$.

     Thus assume $u(r,z) = \displaystyle\sum_{n=1}^{\infty} c_n J_0(\lambda_n r)e^{-\lambda_n z}$. The B.C.

     $u(r,0) = f(r)$, $0 \le r \le 1$ requires that

     $u(r,0) = \displaystyle\sum_{n=1}^{\infty} c_n J_0(\lambda_n r) = f(r)$ so

     $c_n = \displaystyle\int_0^1 rf(r)J_0(\lambda_n r)dr \Big/ \int_0^1 rJ_0^2(\lambda_n r)dr$, $n = 1,2,\ldots$.

7b. Again, Θ periodic of period 2π implies $\lambda^2 = n^2$. Thus the
     solutions to the D.E. are $R(r) = c_1 J_n(kr) + c_2 Y_n(kr)$ (note
     that λ and k here are the reverse of Prob. 3 of Sect. 11.4)
     and $\Theta(\theta) = d_1 \cos n\theta + d_2 \sin n\theta$, $n = 0,1,2\ldots$. For the solution
     to remain bounded, $c_2 = 0$ and thus

     $v(r,\theta) = (1/2)c_0 J_0(kr) + \displaystyle\sum_{m=1}^{\infty} J_m(kr)(b_m \sin m\theta + c_m \cos m\theta)$.

     Hence $v(c,\theta)$ is then a Fourier Series of period 2π and
     the coefficients are found as in Sect. 10.2, Eqs.(13),
     (14) and Prob. 27.

9a. Substituting $u(\rho,\theta,\phi) = P(\rho)\Theta(\theta)\Phi(\phi)$ into Laplace's equation leads to
$\rho^2 P''/P + 2\rho P'/P = -(\csc^2\phi)\Theta''/\Theta - \Phi''/\Phi - (\cot\phi)\Phi'/\Phi = \sigma$.
In order to satisfy arbitrary B.C. it can be shown that σ must be positive, so assume $\sigma = \mu^2$.
Thus $\rho^2 P'' + 2\rho P' - \mu^2 P = 0$. Then we have
$(\sin^2\phi)\Phi''/\Phi + (\sin\phi\cos\phi)\Phi'/\Phi + \mu^2\sin^2\phi = -\Theta''/\Theta = \alpha$.
The periodicity condition $\Theta(0) = \Theta(2\pi)$ requires that $\sqrt{\alpha}$ be an integer λ so $\alpha = \lambda^2$. Hence $\Theta'' + \lambda^2\Theta = 0$ and
$(\sin^2\phi)\Phi'' + (\sin\phi\cos\phi)\Phi' + (\mu^2\sin^2\phi - \lambda^2)\Phi = 0$.

10. Since u is independent of θ, only the first and third of the Eqs. in Prob. 9a hold. The general solution to the Euler equation is

$P = c_1\rho^{r_1} + c_2\rho^{r_2}$ where $r_1 = (-1+\sqrt{1+4\mu^2})/2 > 0$ and

$r_2 = (-1-\sqrt{1+4\mu^2})/2 < 0$. Since we want u to be bounded

as $\rho \to 0$, we set $c_2 = 0$. As found in Prob. 22 of
Sect. 5.3, the solutions of Legendre's equation, Prob.9c, are either singular at 1, at -1, or at both unless
$\mu^2 = n(n+1)$, where n is an integer. In this case, one of the two linearly independent solutions is a polynomial denoted by P_n (Probs. 23 and 24 of Sect. 5.3). Since

$r_1 = (-1 + \sqrt{1+4n(n+1)})/2 = n$, the fundamental solutions

of this problem satisfying the finiteness condition are
$u_n(\rho,\phi) = \rho^n P_n(s) = \rho^n P_n(\cos\phi)$, $n = 1,2,\ldots$. It can be

shown that an arbitrary piecewise continuous function on
$[-1,1]$ can be expressed as a linear combination of
Legendre polynomials. Hence we assume that

$u(\rho,\phi) = \sum_{n=1}^{\infty} c_n\rho^n P_n(\cos\phi)$. The B.C. $u(1,\phi) = f(\phi)$ requires

that $u(1,\phi) = \sum_{n=1}^{\infty} c_n P_n(\cos\phi) = f(\phi)$, $0 \le \phi \le \pi$. From

Prob. 28 of Sect. 5.3 we know that $P_n(x)$ are orthogonal.
However here we have $P_n(\cos\phi)$ and thus we must rewrite
the equation in Prob.9b to find $-[(\sin\phi)\Phi']' = \mu^2(\sin\phi)\Phi$.
Thus $P_n(\cos\phi)$ and $P_m(\cos\phi)$ are orthogonal with weight
function $\sin\phi$. Thus we must multiply the series
expansion for $f(\phi)$ by $\sin\phi P_m(\cos\phi)$ and integrate from 0
to π to obtain
$c_m = \int_0^\pi f(\phi)\sin\phi P_m(\cos\phi)d\phi / \int_0^\pi \sin\phi P_m^2(\cos\phi)d\phi$. To obtain
the answer as given in the text let $s = \cos\phi$.

Section 11.6, Page 715

2a. $b_m = \sqrt{2} \int_0^1 x\sin m\pi x\, dx = \sqrt{2}(-1)^{m+1}/m\pi$ and thus

$$S_n = \frac{2}{\pi} \sum_{m=1}^{n} \frac{(-1)^{m+1}\sin m\pi x}{m}.$$

2b. We have $R_n = \int_0^1 [x - S_n(x)]^2 dx$, where $S_n(x)$ is given in part (a). Using appropriate computer software we find $R_1 = .1307$, $R_2 = .0800$, $R_5 = .0367$, $R_{10} = .0193$, $R_{15} = .0131$ and $R_{19} = .0104$.

2c. As in part (b), we find $R_{20} = .0099$, and thus $n = 20$ will insure a mean square error less than .01.

4a. Writing $S_n(x)$ as a quotient we have $S_n(x) = \dfrac{n\sqrt{x}}{e^{nx^2/2}}$. Now use L'Hopital's Rule with respect to n to obtain

$$\lim_{n\to\infty} S_n(x) = \lim_{n\to\infty} \frac{\sqrt{x}}{\dfrac{x^2}{2}e^{nx^2/2}} = 0 \text{ for } x \neq 0. \text{ For } x = 0,$$

$S_n(0) = 0$ and thus $\lim_{n\to\infty} S_n(x) = 0$ for all x in $[0,1]$.

$R_n = \int_0^1 [0-S_n(x)]^2 dx = n^2 \int_0^1 x e^{-nx^2} dx$

$= -\dfrac{n}{2} e^{-nx^2} \Big|_0^1 = \dfrac{n}{2} - \dfrac{ne^{-n}}{2}$. Since $ne^{-n} \to 0$ as $n \to \infty$, we have that $R_n \to \infty$ as $n \to \infty$.

5. Expanding the integrand of Eq. (6), we get

$$R_n = \int_0^1 r(x)[f(x) - S_n(x)]^2 dx$$

$$= \int_0^1 r(x)f^2(x)dx - 2\sum_{i=1}^{n} c_i \int_0^1 r(x)f(x)\phi_i(x)dx$$

$$+ \sum_{i=1}^{n}\sum_{j=1}^{n} c_i c_j \int_0^1 r(x)\phi_i(x)\phi_j(x)dx,$$

where the last term is obtained by calculating $S_n^2(x)$. Using Eqs.(1) and (9) this becomes

$$R_n = \int_0^1 r(x)f^2(x)dx - 2\sum_{i=1}^{n} c_i a_i + \sum_{i=1}^{n} c_i^2$$

$$= \int_0^1 r(x)f^2(x)dx - \sum_{i=1}^{n} a_i^2 + \sum_{i=1}^{n}(c_i - a_i)^2, \text{ by completing}$$

the square. Since all terms involve a real quantity squared (and $r(x) > 0$) we may conclude R_n is minimized by choosing $c_i = a_i$. This can also be shown by solving $\partial R_n / \partial c_i = 0$, which gives $2(c_i - a_i) = 0$.

7b. From part (a) we have $f_0(x) = 1$ and thus $f_1(x) = c_1 + c_2 x$ must satisfy $(f_0, f_1) = \int_0^1 (c_1 + c_2 x) dx = 0$ and
$(f_1, f_1) = \int_0^1 (c_1 + c_2 x)^2 dx = 1$. Evaluating the integrals yields $c_1 + c_2/2 = 0$ and $c_1^2 + c_1 c_2 + c_2^2/3 = 1$, which have the solution $c_1 = \sqrt{3}$, $c_2 = -2\sqrt{3}$ and thus $f_1(x) = \sqrt{3}(1-2x)$. Notice that $\sqrt{3}(2x - 1)$ also satisfies the given conditions.

7c. $f_2(x) = c_1 + c_2 x + c_3 x^2$ must satisfy $(f_0, f_2) = 0$, $(f_1, f_2) = 0$ and $(f_2, f_2) = 1$.

7d. For $g_2(x) = c_1 + c_2 x + c_3 x^2$ we have $(g_0, g_2) = 0$ and $(g_1, g_2) = 0$, which yield the same ratio of coefficients as found in part (c). Thus $g_2(x) = cf_2(x)$, where c may now be found from $g_2(1) = 1$.

8. This problem follows the pattern of Prob. 7c except now the limits on the orthogonality integral are from -1 to 1. That is $(P_i, P_j) = \int_{-1}^1 P_i(x) P_j(x) dx = 0$, $i \neq j$. For $i = 0$ we have $P_0(x) = 1$ and for $i = 0$ and $j = 1$ we have $P_1(x) = c_1 + c_2 x$ and hence
$(P_0, P_1) = \int_{-1}^1 (c_1 + c_2 x) dx = (c_1 x + c_2 x^2/2) \Big|_{-1}^1 = 2c_1 = 0$ and thus $P_1(1) = 1$ yields $P_1(x) = x$. The others follow in a similar fashion.

9a. This part has essentially been worked in Prob. 5 by setting $c_i = a_i$.

9b. Eq.(6) shows that $R_n \geq 0$ since $r(x) \geq 0$ and thus from part (a) $\int_0^1 r(x) f^2(x) dx - \sum_{i=1}^n a_i^2 \geq 0$. The result follows.

9c. Since f is square integrable [a condition of Thm.11.6.1], $\int_0^1 r(x)f^2(x)dx = M < \infty$ and therefore the monotone increasing sequence of partial sums $T_n = \sum_{i=1}^{n} a_i^2$ is bounded above. Thus $\lim_{n\to\infty} T_n$ exists, which proves the convergence of the given sum.

9d. This result follows from part (a) and part (c).

9e. By definition if $\sum_{i=1}^{\infty} a_i\phi_i(x)$ converges to f(x) in the mean, then $R_n \to 0$ as $n \to \infty$. Hence $\int_0^1 r(x)f^2(x)dx = \sum_{i=1}^{\infty} a_i^2$.

Conversely, if $\int_0^1 r(x)f^2(x)dx = \sum_{i=1}^{\infty} a_i^2$, $\lim_{n\to\infty} R_n = 0$ and $\sum_{i=1}^{\infty} a_i\phi_i(x)$ converges to f(x) in the mean.

10. Bessel's inequality implies that $\sum_{i=1}^{\infty} a_i^2$ converges and thus the nth term $a_n \to 0$ as $n \to \infty$.

12. If the series were the eigenfunction series for a square integrable function, the series $\sum_{i=1}^{\infty} a_i^2$ would have to converge. But $a_0 = 1$, $a_1 = 1/\sqrt{2}$, ..., $a_n = 1/\sqrt{n}$, ..., and $\sum_{n=1}^{\infty} a_n^2 = \sum_{n=1}^{\infty} 1/n$ is the well-known harmonic series which does not converge.

Notes

Notes

Notes